ENVIRONMENTAL ENGINEERING, SIXTH EDITION

ENVIRONMENTAL ENGINEERING, SIXTH EDITION

Water, Wastewater, Soil and
Groundwater Treatment
and Remediation

EDITED BY NELSON L. NEMEROW, FRANKLIN J. AGARDY,
PATRICK SULLIVAN, AND JOSEPH A. SALVATO

WILEY

John Wiley & Sons, Inc.

Library of Congress Cataloging-in-Publication Data:

Environmental engineering. Water, wastewater, soil, and groundwater treatment
and remediation / edited by Franklin J. Agardy and Patrick Sullivan.—6th ed.
    p. cm.
  Selected and revised from earlier work: Environmental engineering /
[edited by] Joseph A. Salvato, Nelson L. Nemerow, Franklin J. Agardy. 5th ed. 2005.
  Includes bibliographical references and index.
  ISBN 978-0-470-08303-1 (cloth)
  1. Water—Purification. 2. Water treatment plants. 3. Sewage—Purification.
4. Sewage disposal plants. 5. Pollution. 6. Soil remediation. I. Agardy,
Franklin J. II. Sullivan, Patrick J., Ph.D. III.
Title: Water, wastewater, soil, and groundwater treatment and remediation.
  TD430.E58 2009
  628.1—dc22

                                                  2008032160

Printed in the United States of America

SKY10063210_122123

Doctors Agardy and Sullivan would like to dedicate this sixth edition of *Environmental Engineering* to Nelson L. Nemerow who passed away in December of 2006. Dr. Nemerow was born on April 16, 1923 and spent most of his productive years as an educator and prolific author. He spent many years teaching at Syracuse University, the University of Miami, North Carolina State, Florida International, and Florida Atlantic University. He authored some 25 books dedicated to advancing the art of waste disposal and utilization. His passion was waste minimization and the title of one of his most recent publications, *Zero Pollution for Industry*, summed up more than fifty years of teaching and consulting. A devoted husband and father, he divided his time between residences in Florida and Southern California. Nelson served in the United States Merchant Marine during World War II. His commitment to excellence was second to none.

CONTENTS

CHAPTER 3 WASTEWATER TREATMENT AND DISPOSAL   **283**
John R. Kiefer

PREFACE

As the global population grows and many developing countries modernize, the importance of water supply and water treatment becomes a much greater factor in the welfare of nations. In similar fashion, the need to address both domestic and industrial wastes generated by these nations moves higher on the scale of importance. Clearly, in today's world the competition for water resources coupled with the unfortunate commingling of wastewater discharges with freshwater supplies creates additional pressure on treatment systems.

This volume attempts to address issues of water supply including the demand for fresh water, the treatment technologies available to treat water, and the treatment and disposal of community-generated wastewaters. The focus is the practicality and appropriateness of treatment—in sufficient detail so that the practicing public health official, water treatment engineer and plant operator, as well as those in the domestic and industrial waste treatment professions, can address their problems in a practical manner. The emphasis is on basic principles and practicality.

FRANKLIN J. AGARDY
PATRICK SULLIVAN
NELSON L. NEMEROW

CONTRIBUTORS

T. DAVID CHINN Senior Vice President, HDR Engineering, Austin, Texas, Tim.Chinn@hdrinc.com

JOHN R. KIEFER Consulting Environmental Engineer, Greenbrae, California, john.kiefer@sbcglobal.net

CHAPTER 1

WATER SUPPLY

T. DAVID CHINN
Professional Engineer, Senior Vice President, HDR Engineering, Austin, Texas

INTRODUCTION

A primary requisite for good health is an adequate supply of water that is of satisfactory sanitary quality. It is also important that the water be attractive and palatable to induce its use; otherwise, consumers may decide to use water of doubtful quality from a nearby unprotected stream, well, or spring. Where a municipal water supply passes near a property, the owner of the property should be urged to connect to it because such supplies are usually under competent supervision.

When a municipal water supply is not available, the burden of developing a safe water supply rests with the owner of the property. Frequently, private supplies are so developed and operated that full protection against dangerous or objectionable pollution is not afforded. Failure to provide satisfactory water supplies in most instances must be charged either to negligence or ignorance because it generally costs no more to provide a satisfactory installation that will meet good health department standards.

The following definitions are given in the National Drinking Water Regulations as amended through July, 2002:

Public water system means either a community or noncommunity system for the provision to the public of water for human consumption through pipes or other constructed conveyances, if such system has at least 15 service connections, or regularly serves an average of at least 25 individuals daily at least 60 days out of the year. Such a term includes (1) any collection, treatment, storage, and distribution facilities under the control of the operator of such system and used primarily in connection with such system, and

(2) any collection or pretreatment storage facilities not under such control which are used primarily in connection with such system.

A *community water system* has at least 15 service connections used by year-round residents, or regularly serves at least 25 year-round residents. These water systems generally serve cities and towns. They may also serve special residential communities, such as mobile home parks and universities, which have their own drinking water supply.

A *noncommunity water system* is a public water system that is not a community water system, and can be either a "transient noncommunity water system" (TWS) or a "non-transient noncommunity water system" (NTNCWS). TWSs typically serve travelers and other transients at locations such as highway rest stops, restaurants, and public parks. The system serves at least 25 people a day for at least 60 days a year, but not the same 25 people. On the other hand, NTNCWSs serve the same 25 persons for at least 6 months per year, but not year round. Some common examples of NTNCWSs are schools and factories (or other workplaces) that have their own supply of drinking water and serve 25 of the same people each day.

In 2007 there were approximately 156,000 public water systems in the United States serving water to a population of nearly 286 million Americans. There were approximately 52,110 community water systems, of which 11,449 were surface water supplies and 40,661 were groundwater supplies. There were 103,559 noncommunity water systems, of which 2557 were surface water supplies and 101,002 were groundwater supplies. Of the community water systems, 43,188 are small systems that serve populations less than 3300; 4822 are medium systems and serve populations between 3300 and 10,000; and 4100 are large systems serving populations over 10,000. In terms of numbers, the small and very small community and noncommunity water systems represent the greatest challenge to regulators and consultants—both contributing to over 88 percent of the regulatory violations in 2007.[1]

In addition to public water systems, the U.S. Geological Survey estimated that 43.5 million people were served by their own individual water supply systems in 2000. These domestic systems are—for the most part—unregulated by either state or county health departments.[2]

A survey made between 1975 and 1977 showed that 13 to 18 million people in communities of 10,000 and under used individual wells with high contamination rates.[3] The effectiveness of state and local well construction standards and health department programs has a direct bearing on the extent and number of contaminated home well-water supplies in specific areas.

A safe and adequate water supply for 2.4 billion people,[4] about one-third of the world's population, is still a dream. The availability of any reasonably clean water in the less-developed areas of the world just to wash and bathe would go a long way toward the reduction of such scourges as scabies and other skin diseases, yaws and trachoma, and high infant mortality. The lack of safe water makes high incidences of shigellosis, amebiasis, schistosomiasis,* leptospirosis, infectious hepatitis, giardiasis, typhoid, and paratyphoid fever commonplace.[5] Ten million persons suffer from dracunculiasis or guinea worm disease in Africa and parts of Asia.[6] The World Health Organization (WHO) estimates that some 3.4 million people die each year from water-borne diseases caused by microbially contaminated water supplies or due to a lack of access to sanitation facilities. Tragically, over one half of these deaths are children under the age of five years old.[7] Three-fourths of all illnesses in the developing world are associated with inadequate water and sanitation.[8] It is believed that the provision of safe water supplies, accompanied by a program of proper excreta disposal and birth control, could vastly improve the living conditions of millions of people in developing countries of the world.[9] In 1982, an estimated 46 percent of the population of Latin America and the Caribbean had access to piped water supply and 22 percent had access to acceptable types of sewage disposal.[10]

The diseases associated with the consumption of contaminated water are discussed in Chapter 1 of *Environmental Engineering, Sixth Edition: Prevention and Response to Water-, Food-, Soil,- and Air-Borne Disease and Illness* and summarized in Table 1.4 of that volume.

Groundwater Pollution Hazard

Table 1.1 shows a classification of sources and causes of groundwater pollution. The 20 million residential cesspool and septic tank soil absorption systems alone discharge about 400 billion gallons of sewage per day into the ground, which in some instances may contribute to groundwater pollution. This is in addition to sewage from restaurants, hotels, motels, resorts, office buildings, factories, and other establishments not on public sewers.[11] The contribution from industrial and other sources shown in Table 1.1 is unknown. It is being inventoried by the EPA, and is estimated at 900 billion gal/year,[12] the EPA, with state participation, is also developing a groundwater protection strategy. Included in the strategy is the classification of all groundwater and protection of existing and potential drinking water sources and "ecologically vital" waters.

Groundwater pollution problems have been found in many states. Primarily, the main cause is organic chemicals, such as trichloroethylene, 1,1,1-trichloroethane, benzene, perchlorate, gasoline (and gasoline additives such as MTBE), pesticides and soil fumigants, disease-causing organisms, and nitrates. Other sources are industrial and municipal landfills; ponds, pits, and lagoons; waste oils and highway deicing compounds; leaking underground storage tanks and pipelines; accidental spills; illegal dumping; and abandoned oil and gas wells. With 146 million people in the United States dependent on groundwater

---

*Two hundred million cases of schistosomiasis worldwide were estimated in 2004, spread mostly through water contact (Centers for Disease Control and Prevention).

TABLE 1.1　Classification of Sources and Causes of Groundwater Pollution Used in Determining Level and Kind of Regulatory Control

Wastes		Nonwastes	
Category I[a]	Category II[b]	Category III[c]	Category IV[d]
Land application of wastewater: spray irrigation, infiltration–percolation basins, overland flow	Surface impoundments: waste-holding ponds, lagoons, and pits	Buried product storage tanks and pipelines	Saltwater intrusion: seawater encroachment, upward coning of saline groundwater
Subsurface soil absorption systems: septic systems	Landfills and other excavations: landfills for industrial wastes, sanitary landfills for municipal solid wastes, municipal landfills	Stockpiles: highway deicing stockpiles, ore stockpiles	River infiltration
Waste disposal wells and brine injection wells	Water and wastewater treatment plant sludges, other excavations (e.g., mass burial of livestock)	Application of highway deicing salts	Improperly constructed or abandoned wells
Drainage wells and sumps	—	Product storage ponds	Farming practices (e.g., dryland farming)
Recharge wells	Animal feedlots	Agricultural activities: fertilizers and pesticides, irrigation return flows	
	Leaky sanitary sewer lines Acid mine drainage Mine spoil pipes and tailings	Accidental spills	

[a]Systems, facilities, or activities designed to discharge waste or wastewaters (residuals) to the land and groundwaters.

[b]Systems, facilities, or activities that may discharge wastes or wastewaters to the land and groundwaters.

[c]Systems, facilities, or activities that may discharge or cause a discharge of contaminants that are not wastes to the land and groundwaters.

[d]Causes of groundwater pollution that are not discharges.

Source: The Report to Congress, Waste Disposal Practices and Their Effects on Ground Water, Executive Summary, U.S. Environmental Protection Agency, Washington, DC, January 1977, p. 39.

sources for drinking water,* these resources must be protected from physical, chemical, radiological, and microbiological contamination.

Whereas surface water travels at velocities of feet per second, groundwater moves at velocities that range from less than a fraction of a foot per day to several feet per day. Groundwater organic and inorganic chemical contamination may persist for decades or longer and, because of the generally slow rate of movement of groundwater, may go undetected for many years. Factors that influence the movement of groundwater include the type of geological formation and its permeability, the rainfall and the infiltration, and the hydraulic gradient. The slow uniform rate of flow, usually in an elongated plume, provides little opportunity for mixing and dilution, and the usual absence of air in groundwater to decompose or break down the contaminants add to the long-lasting problem usually created. By contrast, dilution, microbial activity, surface tension and attraction to soil particles, and soil adsorptive characteristics might exist that could modify, immobilize, or attenuate the pollutant travel. More attention *must* be given to the *prevention* of ground-water pollution and to wellhead protection.

TRAVEL OF POLLUTION THROUGH THE GROUND

Identification of the source of well pollution and tracing the migration of the incriminating contaminant are usually not simple operations. The identification of a contaminant plume and its extent can be truly complex. Comprehensive hydrogeological studies and proper placement and construction of an adequate number of monitoring wells are necessary.

Geophysical methods to identify and investigate the extent and characteristics of groundwater pollution include geomagnetics, electromagnetics, electrical resistivity, ground-probing radar, and photoionization meters.[13] *Geomagnetics* uses an instrument producing a magnetic field to identify and locate buried metals and subsurface materials that are not in their natural or undisturbed state. *Electromagnetics* equipment measures the difference in conductivity between buried materials such as the boundaries of contaminated plumes or landfills saturated with leachate and uncontaminated materials. *Electrical resistivity* measures the resistance a material offers to the passage of an electric current between electric probes, which can be interpreted to identify or determine rock, clay and other materials, porosity, and groundwater limits. *Ground-probing radar* uses radar energy to penetrate and measure reflection from the water table and subsurface materials. The reflection from the materials varies with depth and the nature of the material, such as sandy soils versus saturated clays. *Photoionization meters* are used to detect the presence of specific volatile organic compounds such as gasoline, and methane in a landfill, through the use of shallow boreholes. Other detection methods are remote imagery and aerial photography, including infrared.

---

*Ninety-eight percent of the rural population in the United States and 32 percent of the population served by municipal water systems use groundwater (U.S. Geological Survey, 2000).

Sampling for contaminants must be carefully designed and performed. Errors can be introduced: Sampling from an unrepresentative water level in a well, contamination of sampling equipment, and incorrect analysis procedure are some potential sources of error. The characteristics of a pollutant, the subsurface formation, the hydraulic conductivity of the aquifer affected, groundwater slope, rainfall variations, and the presence of geological fractures, faults, and channels make determination of pollution travel and its sampling difficult. Geophysical techniques can help, and great care must be used in determining the number, spacing, location, and depths of sampling wells and screen entry levels. As a rule, monitoring wells and borings will be required to confirm and sample subsurface contamination.

Since the character of soil and rock, quantity of rain, depth of groundwater, rate of groundwater flow, amount and type of pollution, absorption, adsorption, biological degradation, chemical changes, and other factors usually beyond control are variable, one cannot say with certainty through what thickness or distance sewage or other pollutants must pass to be purified. Microbiological pollution travels a short distance through sandy loam or clay, but it will travel indefinite distances through coarse sand and gravel, fissured rock, dried-out cracked clay, or solution channels in limestone. Acidic conditions and lack of organics and certain elements such as iron, manganese, aluminum, and calcium in soil increase the potential of pollution travel. Chemical pollution can travel great distances.

The Public Health Service (PHS) conducted experiments at Fort Caswell, North Carolina, in a sandy soil with groundwater moving slowly through it. The sewage organisms (coliform bacteria) traveled 232 feet, and chemical pollution as indicated by uranin dye traveled 450 feet.[14] The chemical pollution moved in the direction of the groundwater flow largely in the upper portion of the groundwater and persisted for 2-1/2 years. The pollution band did not fan out but became narrower as it moved away from the pollution source. It should be noted that in these tests there was a small draft on the experimental wells and that the soil was a sand of 0.14 mm effective size and 1.8 uniformity coefficient. It should also be noted that, whereas petroleum products tend to float on the surface, halogenated solvents gradually migrate downward.

Studies of pollution travel were made by the University of California using twenty-three 6-inch observation wells and a 12-inch gravel-packed recharge well. Diluted primary sewage was pumped through the 12-inch recharge well into a confined aquifer having an average thickness of 4.4 feet approximately 95 feet below ground surface. The aquifer was described as pea gravel and sand having a permeability of 1900 gal/ft^2/day. Its average effective size was 0.56 mm and uniformity coefficient was 6.9. The medium effective size of the aquifer material from 18 wells was 0.36 mm. The maximum distance of pollution travel was 100 feet in the direction of groundwater flow and 63 feet in other directions. It was found that the travel of pollution was affected not by the groundwater velocity but by the organic mat that built up and filtered out organisms, thereby preventing them from entering the aquifer. The extent of the pollution then regressed as the organisms died away and as pollution was filtered out.[15]

Butler, Orlob, and McGauhey[16] made a study of the literature and reported the results of field studies to obtain more information about the underground travel of harmful bacteria and toxic chemicals. The work of other investigators indicated that pollution from dry-pit privies did not extend more than 1 to 5 feet in dry or slightly moist fine soils. However, when pollution was introduced into the underground water, test organisms (*Balantidium coli*) traveled to wells up to 232 feet away.[17] Chemical pollution was observed to travel 300 to 450 feet, although chromate was reported to have traveled 1,000 feet in 3 years, and other chemical pollution 3 to 5 miles. Leachings from a garbage dump in groundwater reached wells 1,476 feet away, and a 15-year-old dump continued to pollute wells 2,000 feet away. Studies in the Dutch East Indies (Indonesia) report the survival of coliform organisms in soil 2 years after contamination and their extension to a depth of 9 to 13 feet, in decreasing numbers, but increasing again as groundwater was approached. The studies of Butler et al. tend to confirm previous reports and have led the authors to conclude "that the removal of bacteria from liquid percolating through a given depth of soil is inversely proportional to the particle size of the soil."[18]

Knowledge concerning viruses in groundwater is limited, but better methodology for the detection of viruses is improving this situation. Keswick and Gerba[19] reviewed the literature and found 9 instances in which viruses were isolated from drinking water wells and 15 instances in which viruses were isolated from beneath land treatment sites. Sand and gravel did not prevent the travel of viruses long distances in groundwater. However, fine loamy sand over coarse sand and gravel effectively removed viruses. Soil composition, including the presence of clay, is very important in virus removal, as it is in bacteria removal. The movement of viruses through soil and in groundwater requires further study. Helminth eggs and protozoa cysts do not travel great distances through most soils because of their greater size but can travel considerable distances through macropores and crevices. However, nitrate travel in groundwater may be a major inorganic chemical hazard. In addition, organic chemicals are increasingly being found in groundwater. See (1) "Removal of Gasoline, Fuel Oil, and Other Organics in an Aquifer"; (2) "Prevention and Removal of Organic Chemicals"; and (3) "Synthetic Organic Chemicals Removal" in Chapter 2.

When pumping from a deep well, the direction of groundwater flow around the well within the radius of influence, not necessarily circular, will be toward the well. Since the level of the water in the well will probably be 25 to 150 feet, more or less, below the ground surface, the drawdown cone created by pumping may exert an attractive influence on groundwater, perhaps as far as 100 to 2,000 feet or more away from the well, because of the hydraulic gradient, regardless of the elevation of the top of the well. The radius of the drawdown cone or circle of influence may be 100 to 300 feet or more for fine sand, 600 to 1,000 feet for coarse sand, and 1,000 to 2,000 feet for gravel. See Figure 1.1. In other words, distances and elevations of sewage disposal systems and other sources of pollution must be considered relative to the hydraulic gradient and elevation of the water level in the well, while it is being pumped. It must also be recognized that pollution can travel in three dimensions in all or part of the aquifer's vertical thickness, dependent on the contaminant viscosity and density, the formation transmissivity,

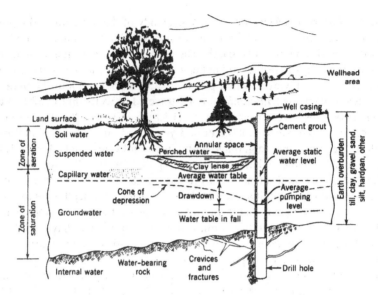

FIGURE 1.1 A geologic section showing groundwater terms. (*Source: Rural Water Supply*, New York State Department of Health, Albany, 1966.)

and the groundwater flow. Liquids lighter than water, such as gasoline, tend to collect above the groundwater table. Liquids heavier or more dense tend to pass through the groundwater and accumulate above an impermeable layer.

A World Health Organization (WHO) report reminds us that, in nature, atmospheric oxygen breaks down accessible organic matter and that topsoil (loam) contains organisms that can effectively oxidize organic matter.[20] However, these benefits are lost if wastes are discharged directly into the groundwater by way of sink holes, pits, or wells or if a subsurface absorption system is water-logged.

From the investigations made, it is apparent that the safe distance between a well and a sewage or industrial waste disposal system is dependent on many variables, including chemical, physical, and biological processes.* These four factors should be considered in arriving at a satisfactory answer:

1. The amount of sand, clay, organic (humus) matter, and loam in the soil, the soil structure and texture, the effective size and uniformity coefficient, groundwater level, and unsaturated soil depth largely determine the ability of the soil to remove microbiological pollution deposited in the soil.

2. The volume, strength, type, and dispersion of the polluting material, rainfall intensity and infiltration, and distance, elevation, and time for pollution to travel with relation to the groundwater level and flow and soil penetrated are important. Also important is the volume of water pumped and well drawdown.

---

*A summary of the distances of travel of underground pollution is also given in Task Group Report, "Underground Waste Disposal and Control," *J. Am. Water Works Assoc.*, **49**, (October 1957): 1334–1341.

3. The well construction, tightness of the pump line casing connection, depth of well and well casing, geological formations penetrated, and sealing of the annular space have a very major bearing on whether a well might be polluted by sewage, chemical spills or wastes, and surface water.

4. The well recharge (wellhead) area, geology, and land use possibly permit groundwater pollution. Local land-use and watershed control is essential to protect and prevent pollution of well-water supplies.

Considerable professional judgment is needed to select a proper location for a well. The limiting distances given in Table 1.2 for private dwellings should

TABLE 1.2 Minimum Separation Distances (feet) from On-Site Wastewater Sources

Sources	To Well or Suction Line[a]	To Stream, Lake, or Water Course	To Property Line or Dwelling
House sewer (water-tight joints)	25 if cast iron pipe or equal, 50 otherwise	25	—
Septic tank	50	50	10
Effluent line to distribution box	50	50	10
Distribution box	100	100	20
Absorption field	100[b]	100	20
Seepage pit or cesspool	150[b] (more in coarse gravel)	100	20
Dry well (roof and footing)	50	25	20
Fill or built-up system	100	100	20
Evapotranspiration–absorption system	100	50	20
Sanitary privy pit	100	50	20
Privy, water-tight vault	50	50	10
Septic privy or aqua privy	50	50	10

[a]Water service and sewer lines may be in the same trench if cast-iron sewer with water-tight joints is laid at all points 12 in. below water service pipe; or sewer may be on dropped shelf at one side at least 12 in. below water service pipe, provided that sewer pipe is laid below frost with tight and root-proof joints and is not subject to settling, superimposed loads, or vibration. Water service lines under pressure shall not pass closer than 10 ft of a septic tank, absorption tile field, leaching pit, privy, or any other part of a sewage disposal system.

[b]Sewage disposal systems located of necessity upgrade or in the general path of drainage to a well should be spaced 200 ft or more away and not in the direct line of drainage. Wells require a minimum 20 ft of casing extended and sealed into an impervious stratum. If subsoil is coarse sand or gravel, do not use seepage pit; use absorption field with 12 in. medium sand on bottom of trench. Also require oversize drill hole and grouted well to a safe depth. See Table 1.15.

be used as a guide. Experience has shown them to be reasonable and effective in most instances *when coupled with a sanitary survey of the drainage area and proper interpretation of available hydrologic and geologic data and good well construction, location, and protection.*[21] See Figure 1.1 for groundwater terms. Well location and construction for public and private water systems should follow regulatory standards. See "Source and Protection of Water Supply" later in this chapter.

Disease Transmission

Water, to act as a vehicle for the spread of a specific disease, must be contaminated with the associated disease organism or hazardous chemical. Disease organisms can survive for days to years, depending on their form (cyst, ova) and environment (moisture, competitors, temperature, soil, and acidity) and the treatment given the wastewater. All sewage-contaminated waters must be presumed to be potentially dangerous. Other impurities, such as inorganic and organic chemicals and heavy concentrations of decaying organic matter, may also find their way into a water supply, making the water hazardous, unattractive, or otherwise unsuitable for domestic use unless adequately treated. The inorganic and organic chemicals causing illness include mercury, lead, chromium, nitrates, asbestos, polychlorinated biphenyl (PCB), polybrominated biphenyl (PBB), mirex, Kepone vinyl chloride, trichloroethylene, benzene, and others.

Communicable and noninfectious diseases that may be spread by water are discussed in Table 1.4 in Chapter 1 of *Environmental Engineering, Sixth Edition: Prevention and Response to Water-, Food-, Soil,- and Air-Borne Disease and Illness.*

WATER QUANTITY AND QUALITY

Water Cycle and Geology

The movement of water can be best illustrated by the hydrologic, or water, cycle shown in Figure 1.2. Using the clouds and atmospheric vapors as a starting point, moisture condenses out under the proper conditions to form rain, snow, sleet, hail, frost, fog, or dew. Part of the precipitation is evaporated while falling; some of it reaches vegetation foliage, the ground, and other surfaces. Moisture intercepted by surfaces is evaporated back into the atmosphere. Part of the water reaching the ground surface runs off to streams, lakes, swamps, or oceans whence it evaporates; part infiltrates the ground and percolates down to replenish the groundwater storage, which also supplies lakes, streams, and oceans by underground flow. Groundwater in the soil helps to nourish vegetation through the root system. It travels up the plant and comes out as transpiration from the leaf structure and then evaporates into the atmosphere. In its cyclical movement, part of the water is temporarily retained by the earth, plants, and animals to sustain life. The average annual precipitation in the United States is about 30 inches, of which 72 percent evaporates from water and land surfaces and transpires from

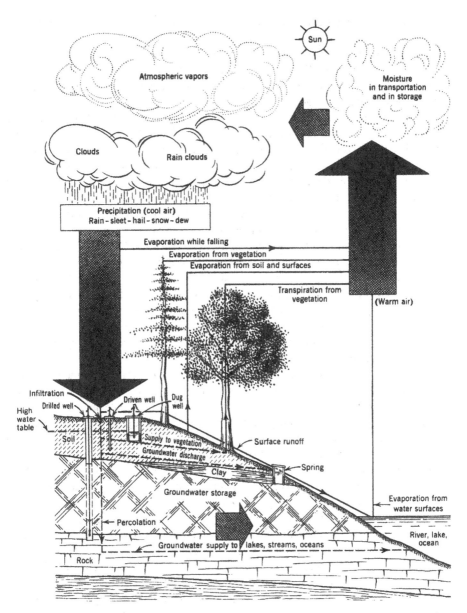

FIGURE 1.2 Figure hydrologic or (water) cycle. The oceans hold 317,000,000 mi³ of water. Ninety-seven percent of the Earth's water is salt water; 3 percent of the Earth's fresh water is groundwater, snow and ice, fresh water on land, and atmospheric water vapor; 85 percent of the fresh water is in polar ice caps and glaciers. Total precipitation equals total evaporation plus transpiration. Precipitation on land equals 24,000 mi³/year. Evaporation from the oceans equals 80,000 mi³/year. Evaporation from lakes, streams, and soil and transpiration from vegetation equal 15,000 mi³.

plants and 28 percent contributes to the groundwater recharge and stream flow.[22] See also "Septic Tank Evapotranspiration System," in Chapter 3.

The volume of fresh water in the hydrosphere has been estimated to be $8,400,000 \text{ mi}^3$ with $5,845,000 \text{ mi}^3$ in ice sheets and glaciers, $2,526,000 \text{ mi}^3$ in groundwater, $21,830 \text{ mi}^3$ in lakes and reservoirs, $3,095 \text{ mi}^3$ in vapors in the atmosphere, and 509 mi^3 in river water.[23]

When speaking of water, we are concerned primarily with surface water and groundwater, although rainwater and saline water are also considered. In falling through the atmosphere, rain picks up dust particles, plant seeds, bacteria, dissolved gases, ionizing radiation, and chemical substances such as sulfur, nitrogen, oxygen, carbon dioxide, and ammonia. Hence, rainwater is not pure water as one might think. It is, however, very soft. Water in streams, lakes, reservoirs, and swamps is known as *surface water*. Water reaching the ground and flowing over the surface carries anything it can move or dissolve. This may include waste matter, bacteria, silt, soil, vegetation, and microscopic plants and animals and other naturally occurring organic matter. The water accumulates in streams or lakes. Sewage, industrial wastes, and surface and groundwater will cumulate, contribute to the flow, and be acted on by natural agencies. On the one hand, water reaching lakes or reservoirs permit bacteria, suspended matter, and other impurities to settle out. On the other hand, microscopic as well as macroscopic plant and animal life grow and die, thereby removing and contributing impurities in the cycle of life.

Part of the water reaching and flowing over the ground infiltrates and percolates down to form and recharge the groundwater, also called underground water. In percolating through the ground, water will dissolve materials to an extent dependent on the type and composition of the strata through which the water has passed and the quality (acidity) and quantity of water. Groundwater will therefore usually contain more dissolved minerals than surface water. The strata penetrated may be unconsolidated, such as sand, clay, and gravel, or consolidated, such as sandstone, granite, and limestone. A brief explanation of the classification and characteristics of formations is given next.

Igneous rocks are those formed by the cooling and hardening of molten rock masses. The rocks are crystalline and contain quartz, feldspar, mica, hornblende, pyroxene, and olivene. Igneous rocks are not usually good sources of water, although basalts are exceptions. Small quantities of water are available in fractures and faults. Examples are granite, dioxite, gabbro, basalt, and syenite.

Sedimentary formations are those resulting from the deposition, accumulation, and subsequent consolidation of materials weathered and eroded from older rocks by water, ice, or wind and the remains of plants, animals, or material precipitated out of solution. Sand and gravel, clay, silt, chalk, limestone, fossils, gypsum, salt, peat, shale, conglomerates, loess, and sandstone are examples of sedimentary formations. Deposits of sand and gravel generally yield large quantities of water. Sandstones, shales, and certain limestones may yield abundant groundwater, although results may be erratic, depending on bedding planes and joints, density, porosity, and permeability of the rock.

Metamorphic rocks are produced by the alteration of igneous and sedimentary rocks, generally by means of heat and pressure. Gneisses and schists, quartzites, slates, marble, serpentines, and soapstones are metamorphic rocks. A small quantity of water is available in joints, crevices, and cleavage planes.

Karst areas are formed by the movement of underground water through carbonate rock fractures and channels, such as in limestone and gypsum, forming caves, underground channels, and sink holes. Because karst geology can be so porous, groundwater movement can be quite rapid (several feet per day). Therefore, well water from such sources is easily contaminated from nearby and distant pollution sources.

Glacial drift is unconsolidated sediment that has been moved by glacier ice and deposited on land or in the ocean.

Porosity is a measure of the amount of water that can be held by a rock or soil in its pores or voids, expressed as a percentage of the total volume. The volume of water that will *drain* freely out of a saturated rock or soil by gravity, expressed as a percentage of the total volume of the mass, is the *effective porosity* or *specific yield*. The volume of water retained is the *specific retention*. This is due to water held in the interstices or pores of the rock or soil by molecular attraction (cohesion) and by surface tension (adhesion). For example, plastic clay has a porosity of 45 to 55 percent but a specific yield of practically zero. In contrast, a uniform coarse sand and gravel mixture has a porosity of 30 to 40 percent with nearly all of the water capable of being drained out.

The *permeability* of a rock or soil, expressed as the standard coefficient of permeability or *hydraulic conductivity*, is the rate of flow of water at 60°F (16°C), in gallons per day, through a vertical cross-section of 1 ft^2, under a head of 1 foot, per foot of water travel. There is no direct relationship between permeability, porosity, and specific yield.

Transmissivity is the hydraulic conductivity times the saturated thickness of the aquifer.

Groundwater Flow

The flow through an underground formation can be approximated using Darcy's law,[24] expressed as $Q = KIA$, where

> Q = quantity of flow per unit of time, gpd
> K = hydraulic conductivity (water-conducting capacity) of the formation, gpd/ft^2 (see Table 1.3)
> I = hydraulic gradient, ft/ft (may equal slope of groundwater surface)
> A = cross-sectional area through which flow occurs, ft^2, at right angle to flow direction

For example, a sand aquifer within the floodplain of a river is about 30 feet thick and about a mile wide. The aquifer is covered by a confining unit of glacial till, the bottom of which is about 45 feet below the land surface. The difference in

TABLE 1.3 Porosity, Specific Yield, and Hydraulic Conductivity of Some Materials

Material	Porosity (vol %)	Specific Yield (%)	Hydraulic Conductivity or Permeability Coefficient,[a] K (gpd/ft^2)
Soils	55[b] 50–60[e]	40[b]	10^{-5}–10 (glacial till)
Clay	50[b] 45[d] 45–55[c]	2[b] 3[d] 1–10[e]	10^{-2}–10^2 (silt, loess) 10^{-6}–10^{-2} (clay)
Sand	25[b] 35[d] 30–40[c]	22[b] 25[d] 10–30[c]	1–10^2 (silty sand) 10–10^4
Gravel	20[b] 25[d] 30–40[c]	19[b] 22[d] 15–30[c]	10^3–10^5
Limestone	20[b] 5[d] 1–10[c]	18[b] 2[d] 0.5–5[e]	10^{-3}–10^5 (fractured to cavernous, carbonate rocks)
Sandstone	11[b] 15[d] 10–20[c]	6[b] 8[d] 5–15[e]	10^{-4}–10 (fractured to semiconsolidated)
Shale	5[d] 1–10[c]	2[d] 0.5–5[c]	10^{-7}–10^{-3} (unfractured to fractured)
Granite	0.1[b] 0.1[d] 1[c]	0.09[b] 0.5[d]	10^{-7}–10^2 (unfractured to fractured, igneous and metamorphic)
Basalt	11[b]	8[d]	10^{-7}–10^5 (unfractured, fractured, to lava)

[a] *Protection of Public Water Supplies from Ground-Water Contamination*, Seminar Publication, EPA/625/4-85/016, Center for Environmental Research Information, Cincinnati, OH, September 1985, p. 11.
[b] R. C. Heath, *Basic Ground-Water Hydrology*, U.S. Geological Survey Paper 2220, U.S. Government Printing Office, Washington, DC, 1983.
[c] H. Ries and T. L. Watson, *Engineering Geology*, Wiley, New York, 1931.
[d] R. K. Linsley and J. B. Franzini, *Water Resources Engineering*, McGraw-Hill, New York, 1964.
[e] F. G. Driscoll, *Groundwater and Wells*, 2nd ed., Johnson Division, St. Paul, MN, 1986, p. 67.
Source: D. K. Todd, *Ground Water Hydrology*, 2nd ed., Wiley, New York, 1980.

water level between two wells a mile apart is 10 feet. The hydraulic-conductivity of the sand is 500 gpd/ft^2. Find Q:

$$Q = KIA$$

$$= 500 \, \text{gpd/ft}^2 \times (10 \, \text{ft}/5280 \, \text{ft}) \times 5{,}280 \, \text{ft} \times 30 \, \text{ft}$$

$$= 150{,}000 \, \text{gpd}$$

Also,

$$v = \frac{KI}{7.48n}$$

where

v = groundwater velocity, ft/day
n = effective porosity as a decimal

Find v:

$$v = \frac{500 \, \text{gpd} \times 10 \, \text{ft}/5280 \, \text{ft}}{7.48 \, \text{g/ft}^3 \times 0.2}$$

$$= 0.63 \, \text{ft/day}$$

Another example is given using Figure 1.3 and Darcy's law, expressed as

$$v = Ks$$

where

v = velocity of flow through an aquifer
K = coefficient of permeability (hydraulic conductivity)
s = hydraulic gradient

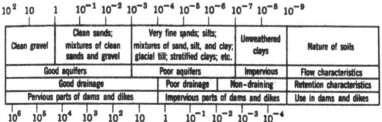

FIGURE 1.3 Magnitude of coefficient of permeability for different classes of soils. (*Source*: G. M. Fair, J. C. Geyer, and D. A. Okun, *Water and Wastewater Engineering*, Wiley, New York, 1966, pp. 9–13.)

Also,

$$Q = va$$

where

Q = discharge
a = cross-sectional area of aquifer

Example: (1) Estimate the velocity of flow (ft/day) and the discharge (gpd) through an aquifer of very coarse sand 1,000 feet wide and 50 feet thick when the slope of the groundwater table is 20 ft/m.

(2) Find the standard coefficient of permeability and the coefficient of transmissibility on the assumption that the water temperature is 60°F (16°C).

1. From Figure 1.3, choose a coefficient of permeability K = 1.0 cm/sec = 2835 ft/day. Because s = 20/5280, v = 2835 × 20/5280 = 11 ft/day and Q = 11 × 1000 × 50 × 7.5 × 10^{-6} = 4.1 mgd.
2. The standard coefficient of permeability is 2,835 × 7.5 = 2.13 × 10^4, and the coefficient of transmissibility becomes 2.13 × 10^4 × 50 = 1.06 × 10^6.

The characteristics of some materials are given in Table 1.3.

Groundwater Classification

The EPA has proposed the following groundwater classification system:

Class I: *Special Ground Water* are those which are highly vulnerable to contamination because of the hydrological characteristics of the areas in which they occur *and* which are also characterized by either of the following two factors:

a. Irreplaceable, in that no reasonable alternative source of drinking water is available to substantial populations; or

b. Ecologically vital, in that the aquifer provides the base flow for a particularly sensitive ecological system that, if polluted, would destroy a unique habitat.

Class II: *Current and Potential Sources of Drinking Water and Waters Having Other Beneficial Uses* are all other groundwaters which are currently used or are potentially available for drinking water or other beneficial use.

Class III: *Ground Waters Not Considered Potential Sources of Drinking Water and of Limited Beneficial Use* are ground waters which are heavily saline, with Total Dissolved Solids (TDS) levels over 10,000 mg/1, or are otherwise contaminated beyond levels that allow cleanup using methods employed in public water system treatment. These ground waters also must not migrate to Class I or Class II ground waters or have a discharge to surface water that could cause degradation.[25]

This classification system has been debated at great length. Some states have adopted stricter standards and eliminated class III, whereas others have added classifications.

Water Quality

The cleanest available sources of groundwater and surface water should be protected, used, and maintained for potable water supply purposes. Numerous parameters are used to determine the suitability of water and the health significance of contaminants that may be found in untreated and treated water. Watershed and wellhead protection regulations should be a primary consideration.

Microbiological, physical, chemical, and microscopic examinations are discussed and interpreted in this chapter under those respective headings. Water quality can be best assured by maintaining water clarity, chlorine residual in the distribution system, confirmatory absence of indicator organisms, and low bacterial population in the distributed water.[26]

Table 1.4 shows the standards for drinking water coming out of a tap served by a public water system. These are based on the National Primary Drinking Water Standards developed under the Safe Drinking Water Act of 1974 as amended in 1986 and 1996. The maximum contaminant level goals (MCLGs) in Table 1.4 are nonenforceable health goals that are to be set at levels at which no known or anticipated adverse health effects occur and that allow an adequate margin of safety. Maximum contaminant levels (MCLs) are enforceable and must be set as close to MCLGs as is feasible, based on the use of best technology, treatment techniques, analytical capabilities, costs, and other means. The EPA has based the MCLs on the potential health effects from the ingestion of a contaminant on the assumption that the effects observed (of a high dose) in animals may occur (at a low dose) in humans. This assumption has engendered considerable debate.

Secondary regulations, shown in Table 1.5, have also been adopted, but these are designed to deal with taste, odor, and appearance of drinking water and are not mandatory unless adopted by a state. Although not mandatory, these parameters have an important indirect health significance. Water that is not palatable is not likely to be used for drinking, even though reported to be safe, in both developed and underdeveloped areas of the world. A questionable or contaminated water source may then be inappropriately used. Water industry professionals in the United States should adhere to the USEPA primary and secondary standards without deviation or risk jeopardizing public health, either acutely (in the short term) or chronically (exposure over a long period.) It is also important to note that while each of the 50 states (and territories) must adopt and enforce USEPA's standards, they are free to either promulgate standards that are more stringent that USEPA or regulate contaminants that are of particular concern in their state. California, for example, regulates perchlorate even though there is no federal mandate to do so.

Tables 1.6 to 1.10 give World Health Organization (WHO) water-quality guidelines. It is not intended that the individual values in Tables 1.6 to 1.10 be

TABLE 1.4 Summary of National Primary Drinking Water Regulations, March 2008 (USEPA)

Name of Contaminant	Maximum Contaminant Level (MCL) (mg/1 unless noted)	Health Effects of Contaminant
	Inorganic Chemicals	
Antimony	0.006	Decreased longevity, blood effects
Arsenic	0.010	Cancer risk/cardiovascular and dermal effects
Asbestos (fiber length >10 μm)	7 million fibers per liter (MFL)	Lung tumors/cancer risk
Barium	2	Circulatory/gastrointestinal effects
Beryllium	0.004	Bone/lung effects/cancer risk
Cadmium	0.005	Liver/kidney/bone/circulatory effects
Chromium (total)	0.1	Liver/kidney/circulatory effects
Copper	Treatment technique (action limit 1.3)	Gastrointestinal/liver/kidney effects
Cyanide	0.2	Thyroid/neurologic effects
Fluoride	4	Skeletal effects
Lead	Treatment technique (action limit 0.015)	Cancer risk/kidney/nervous system effects Highly toxic to infants
Mercury (inorganic)	0.002	Kidney damage
Nickel	0.1	Nervous system/liver/heart effects/dermatitis
Nitrate (as N)	10	Methemoglobinemia (blue baby syndrome)/diuresis
Nitrite (as N)	1	Methemoglobinemia (blue baby syndrome)
Selenium	0.05	Nervous system/kidney/liver/circulatory effects
Thallium	0.002	Kidney/liver/brain/intestine effects
Radionuclides		
Combined radium-226 and radium-228	5 pCi/1	Cancer risk
Gross alpha (excluding radon and uranium)	15 pCi/1	Cancer risk

Beta particle and photon	4 mrem/year	Cancer risk
Radioactivity, uranium	0.030	Kidney effects/cancer risk

SYNTHETIC ORGANIC COMPOUNDS

Organic Chemicals

2,3,7,8-TCDD (Dioxin)	0.00000003	Cancer risk/reproductive effects
2,4,5-TP (Silvex)	0.05	Liver/kidney effects
2,4-D	0.07	Nervous system/liver/kidney effects
Acrylamide	Treatment technique	Cancer risk/nervous system effects
Alachlor	0.002	Cancer risk/liver/kidney/spleen effects
Aldicarb[a]	0.003	Nervous system effects
Aldicarb sulfone[a]	0.003	Nervous system effects
Aldicarb sulfoxide[a]	0.004	Nervous system effects
Atrazine	0.003	Cardiologic effects/cancer risk/muscular degeneration
Carbofuran	0.04	Nervous/reproductive system effects
Chlordane	0.002	Cancer risk/liver/kidney/spleen effects
Dalapon	0.2	Kidney effects
Di(2-ethylhexyl)adipate	0.4	Liver/bone effects/cancer risk
1,2-Dibromo-3-chlorpropane (DBCP)	0.0002	Cancer risk/kidney/reproductive effects
Di(2-ethylhexyl)phthalate (DEHP)	0.006	Cancer risk/liver/reproductive effects
Dinoseb	0.007	Thyroid/reproductive organ effects
Diquat	0.02	Ocular effects
Endothall	0.1	Kidney/liver/gastrointestinal effects
Endrin	0.002	Liver effects
Epichlorohydrin	Treatment technique	Cancer risk/circulatory/stomach effects
Ethylene dibromide (EDB)	0.00005	Cancer risk/liver/heart/kidney/nervous system effects
Glyphosate	0.7	Kidney/reproductive effects

(continues)

TABLE 1.4 (*continued*)

Name of Contaminant	Maximum Contaminant Level (MCL) (mg/l unless noted)	Health Effects of Contaminant
Heptachlor	0.0004	Cancer risk/liver effects
Heptachlor epoxide	0.0002	Cancer risk
Hexachlorobenzene	0.001	Cancer risk/liver/reproductive effects
Hexachlorocyclopentadiene (HEX)	0.05	Kidney/stomach effects
Lindane	0.0002	Kidney/liver/nervous/circulatory effects
Methoxychlor	0.04	Kidney/liver/nervous/developmental effects
Oxamyl (Vydate)	0.2	Nervous system effects
PAHs (benzo[*a*]pyrene)	0.0002	Cancer risk/developmental/reproductive effects
PCBs (polychlorinated biphenyls)	0.0005	Cancer risk/liver/gastrointestinal effects
Pentachlorophenol	0.001	Cancer risk/reproductive/liver/kidney effects
Picloram	0.5	Nervous system/liver effects
Simazine	0.004	Cancer risk/liver/kidney/thyroid effects
Toxaphene	0.003	Cancer risk/liver/kidney/nervous system effects

VOLATILE ORGANIC COMPOUNDS

1,1-Dichloroethylene	0.007	Kidney/liver effects/cancer risk
1,1,1-Trichloroethane	0.2	Liver/circulatory/nervous system effects
1,1,2-Trichloroethane	0.005	Kidney/liver effects/cancer risk
1,2-Dichloroethane	0.005	Cancer risk
1,2-Dichloroprapane	0.005	Cancer risk/liver/kidney/gastrointestinal effects
1,2,4-Trichlorobenzene	0.07	Kidney/liver/adrenal gland effects
Benzene	0.005	Cancer risk/nervous system effects
Carbon tetrachloride	0.005	Cancer risk/liver effects
Chlorobenzene	0.1	Nervous system/liver effects
Cis-1,2-dichloroethylene	0.07	Liver/nervous/circulatory effects

Dichloromethane	0.005	Cancer risk/liver effects
Ethylbenzene	0.7	Kidney/liver/nervous system effects
Ortho-dichlorobenzene	0.6	Kidney/liver/blood cell/nervous system effects
Para-dichlorobenzene	0.075	Cancer risk/liver/circulatory effects
Styrene	0.1	Liver/nervous system effects/cancer risk
Tetrachloroethylene (PCE)	0.005	Cancer risk/liver/kidney/nervous system effects
Toluene	1	Kidney/liver effects
Trans-1,2-dichloroethylene	0.1	Nervous system/liver/circulatory effects
Trichloroethylene (TCE)	0.005	Cancer risk/liver effects
Vinyl chloride	0.002	Cancer risk/neurologic/liver effects
Xylenes (total)	10	kidney/liver/nervous system effects

Microbiologic Contaminants

TOTAL COLIFORM RULE (TCR)

| Total coliforms; fecal coliforms; *Escherichia coli* | Less than 40 samples/month; no more than one positive for total coliforms. If 40 samples or more per month, or more than 5% positive. Maximum contaminate level goal (MCLG) = 0 for total coliform, fecal coliform, and *E. coli*. Every sample containing total coliforms must be analyzed for fecal coliforms. | Presence of fecal coliform or *E. coli* may indicate potential contamination that can cause diarrhea, cramps, nausea, headaches, or other symptoms. |

SURFACE WATER TREATMENT RULE

Turbidity	Treatment technique	None; interferes with disinfection
Giardia	Treatment technique (MCLG = 0)	Giardiasis
Enteric Viruses	Treatment technique (MCLG = 0)	Gastrointestinal and other viral infections
Legionella	Treatment technique (MCLG = 0)	Legionnaire's disease
Heterotrophic plate count (HPC)	Treatment technique (MCLG = none)	Gastrointestinal infections

INTERIM ENHANCED SURFACE WATER TREATMENT RULE (IESWTR)

| Turbidity | Treatment technique | None; interferes with disinfection |
| *Cryptosporidium* | Treatment technique (MCLG = 0) | Cryptosporidiosis |

(continues)

TABLE 1.4 (*continued*)

Name of Contaminant	Maximum Contaminant Level (MCL) (mg/1 unless noted)	Health Effects of Contaminant
FILTER BACKWASH RULE		
Cryptosporidium	Treatment technique (MCLG = 0)	Cryptosporidiosis
LONG TERM 2 ENHANCED SURFACE WATER TREATMENT RULE (LT2ESWTR)		
Cryptosporidium	Treatment technique (MCLG = 0) PWSs to monitor for cryptosporidium between 2006–2010 (staggered by system size). Additional treatment (if necessary) in place by 2012-2014.	Cryptosporidiosis
Giardia	Treatment technique (MCLG = 0)	Giardiasis
Viruses	Treatment technique (MCLG = 0)	Gastrointestinal and other viral infections
Disinfectants and Disinfection By-Products: Stage 1,2 D/DBPR		
DISINFECTANTS		
Chlorine	Maximum disinfectant residual level (MRDL)-4.0 (as Cl$_2$)	Eye/nose irritation, stomach discomfort
Chloramines	MRDL-4.0 (as Cl$_2$)	Hemolytic anemia in dialysis
Chlorine dioxide	MRDL-0.8 (as ClO$_2$)	Anemia; nervous system effects
DISINFECTION BY-PRODUCTS		
Total trihalomethanes (TTHMs)	0.080	Cancer risk/reproductive effects
Haloacetic acids (HAA5)	0.060	Cancer risk
Chlorite	1.0	Cancer risk
Bromate	0.010	Cancer risk/nervous system/liver effects
Total organic carbon (TOC)	Treatment technique	

a Aldicarb and metabolites are presently stayed, pending reproposal.

TABLE 1.5   Secondary Drinking Water Regulations, 2008 (USEPA)

Contaminant	Effect	Level
pH	Water should not be to acidic or too basic.	6.5–8.5
Aluminum	Colored water	0.05–0.2 mg/l
Chloride	Taste and corrosion of pipes	250 mg/l
Copper	Taste and staining of porcelain	1 mg/l
Foaming agents	Aesthetic	0.5 mg/l
Sulfate	Taste and laxative effects	250 mg/l
Total dissolved solids (hardness)	Taste and possible relation between low hardness and cardiovascular disease, also an indicator of corrosivity (related to lead levels in water); can damage plumbing and limit effectiveness of soaps and detergents	500 mg/l
Silver	Skin and eye discoloration	0.1 mg/l
Zinc	Taste	5 mg/l
Fluoride	Dental fluorosis (a brownish discoloration of the teeth)	2 mg/l
Color	Aesthetic	15 color units
Corrosivity	Aesthetic and health related (corrosive water can leach pipe materials, such as lead, into the drinking water)	Noncorrosive
Iron	Taste	0.3 mg/l
Manganese	Taste	0.05 mg/l
Odor	Aesthetic	3 threshold odor number

Source: U.S. Environmental Protection Agency, Fact Sheet, Office of Ground Water and Drinking Water, Washington, DC, March 2008.

used directly. Guideline values in the tables must be used and interpreted in conjunction with the information contained in the appropriate sections of Chapters 2 to 5 of *Guidelines for Drinking-Water Quality*, 2nd ed., volume 2, WHO, Geneva, 1996, 1998. Water treatment plant designers, operators and regulators worldwide should evaluate their water-quality goals and strive to produce the best water quality possible given the available technology, regardless of regulatory parameters.

TABLE 1.6 Microbiological and Biological Quality (WHO)

	Unit	Guideline Value	Remarks
		I. Microbiological Quality	
A. PIPED WATER SUPPLIES			
A.1 Treated Water Entering Distribution System			
Fecal coliforms	Number/100 ml	0	Turbidity <1 NTU; for disinfection with chlorine, pH preferably <8.0; free chlorine residual 0.2–0.5 mg/l following 30 min (minimum) contact.
Coliform organisms	Number/100 ml	0	—
A.2 Untreated Water Entering Distribution System			
Fecal coliforms	Number/100 ml	0	
Coliform organisms	Number/100 ml	0	In 98% of samples examined throughout the year, in the case of large supplies when sufficient samples are examined.
		3	In an occasional sample, but not in consecutive samples.
A.3 Water in Distribution System			
Fecal coliforms	Number/100 ml	0	
Coliform organisms	Number/100 ml	0	In 95% of samples examined throughout year, in the case of large supplies when sufficient samples are examined.
		3	In an occasional sample, but not in consecutive samples.

B. UNPIPED WATER SUPPLIES

Fecal coliforms	Number/100 ml	0	
Coliform organisms	Number/100 ml	10	Should not occur repeatedly; if occurrence is frequent and if sanitary protection cannot be improved, an alternative source must be found if possible.

C. BOTTLED DRINKING WATER

Fecal coliforms	Number/100 ml	0	Source should be free from fecal contamination.
Coliform organisms	Number/100 ml	0	—

D. EMERGENCY WATER SUPPLIES

Fecal coliforms	Number/100 ml	0	Advise public to boil water in case of failure to meet guideline values.
Coliform organisms	Number/100 ml	0	—
Enteroviruses	—	No guideline value set	—

II. Biological Quality

Protozoa (pathogenic)	—	No guideline value set	—
Helminths (pathogenic)	—	No guideline value set	—
Free-living organisms (algae, others)	—	No guideline value set	—

Source: Guidelines for Drinking-Water Quality, Vol. 1: Recommendations, World Health Organization, Geneva, 1984, Table 1. Reproduced with permission.

TABLE 1.7   Inorganic Constituents of Health Significance (WHO)

Constituent	Unit	Guideline Value
Arsenic	mg/l	0.05
Asbestos	—	No guideline value
Barium	—	No guideline value
Beryllium	—	No guideline value
Cadmium	mg/l	0.005
Chromium	mg/l	0.05
Cyanide	mg/l	0.1
Fluoride[a]	mg/l	1.5
Hardness	—	No health-related guideline value
Lead	mg/l	0.05
Mercury	mg/l	0.001
Nickel	—	No guideline value
Nitrate	mg/l (N)	10
Nitrite	—	No guideline value
Selenium	mg/l	0.01
Silver	—	No guideline value
Sodium	—	No guideline value

[a]Natural or deliberately added; local or climatic conditions may necessitate adaptation.

Source: Guidelines for Drinking-Water Quality, Vol. 1: Recommendations, World Health Organization, Geneva, 1984, Table 2. Reproduced with permission.

National secondary drinking water regulations shown in Table 1.5 are federally nonenforceable regulations that control contaminants in drinking water affecting the aesthetic qualities related to public acceptance of drinking water. These levels represent reasonable goals for drinking water quality. States may establish higher or lower levels, which may be appropriate, depending on local conditions such as unavailability of alternate source waters or other compelling factors, provided that public health and welfare are not adversely affected.

It is recommended that the parameters in these regulations be monitored at intervals no less frequent than the monitoring performed for inorganic chemical contaminants listed in the National Primary Drinking Water Regulations as applicable to community water systems. More frequent monitoring would be appropriate for specific parameters such as pH, color, and odor under certain circumstances as directed by the state.

Sampling and Quality of Laboratory Data

Raw and finished water should be continually monitored. Prior arrangements should also be made for the treatment plant to be immediately notified by upstream dischargers in case of wastewater treatment plant operational failures or accidental releases of toxic or other hazardous substances. A water treatment plant should have a well-equipped laboratory, certified operator, and qualified chemist. Disinfectant residual, turbidity, and pH should be monitored continuously where

TABLE 1.8 Organic Constituents of Health Significance (WHO)

	Unit	Guideline Value	Remarks
Aldrin and dieldrin	$\mu g/l$	0.03	
Benzene	$\mu g/l$	10^a	
Benzol[a]pyrene	$\mu g/l$	0.01^a	
Carbon tetrachloride	$\mu g/l$	3^a	Tentative guideline value[b]
Chlordane	$\mu g/l$	0.3	—
Chlorobenzenes	$\mu g/l$	No health-related guideline value	Odor threshold concentration between 0.1 and 3 $\mu g/l$
Chloroform	$\mu g/l$	30^a	Disinfection efficiency must not be compromised when controlling chloroform content
Chlorophenols	$\mu g/l$	No health-related guideline value	Odor threshold concentration 0.1 $\mu g/l$
2,4-D	$\mu g/l$	100^c	
DDT	$\mu g/l$	1	
1,2-Dichloroethane	$\mu g/l$	10^a	
1,1-Dichloroethene[d]	$\mu g/l$	0.3^a	
Heptachlor and heptachlor epoxide	$\mu g/l$	0.1	
Hexachlorobenzene	$\mu g/l$	0.01^a	
Gamma-HCH (lindane)	$\mu g/l$	3	
Methoxychlor	$\mu g/l$	30	
Pentachlorophenol	$\mu g/l$	10	
Tetrachloroethene[e]	$\mu g/l$	10^a	Tentative guideline value[b]
Trichloroethene[f]	$\mu g/l$	30^a	Tentative guideline value[b]
2,4,6-Trichlorophenol	$\mu g/l$	$10^{a,c}$	Odor threshold concentration, 0.1 $\mu g/l$
Trihalomethanes		No guideline value	See chloroform

[a] These guideline values were computed from a conservative hypothetical mathematical model that cannot be experimentally verified and values should therefore be interpreted differently. Uncertainties involved may amount to two orders of magnitude (i.e., from 0.1 to 10 times the number).

[b] When the available carcinogenicity data did not support a guideline value but the compounds were judged to be of importance in drinking water and guidance was considered essential, a tentative guideline value was set on the basis of the available health-related data.

[c] May be detectable by taste and odor at lower concentrations.

[d] Previously known as 1,1-dichloroethylene.

[e] Previously known as tetrachloroethylene.

[f] Previously known as trichloroethylene.

Source: Guidelines for Drinking-Water Quality, Vol. 1: *Recommendations*, World Health Organization, Geneva, 1984, Table 3. Reproduced with permission.

TABLE 1.9   Aesthetic Quality (WHO)

Characteristic	Unit	Guideline Value	Remarks
Aluminum	mg/l	0.2	
Chloride	mg/l	250	
Chlorobenzenes and chlorophenols	—	No guideline value	These compounds may affect taste and odor.
Color	True color units (TCU)	15	
Copper	mg/l	1.0	
Detergents	—	No guideline value	There should not be any foaming or taste and odor problems.
Hardness	mg/l (as $CaCO_3$)	500	
Hydrogen sulfide	—	Not detectable by consumers	
Iron	mg/l	0.3	
Manganese	mg/l	0.1	
Oxygen, dissolved	—	No guideline value	
pH	—	6.5–8.5	
Sodium	mg/l	200	
Solids, total dissolved	mg/l	1000	
Sulfate	mg/l	400	
Taste and odor	—	Inoffensive to most consumers	
Temperature	—	No guideline value	
Turbidity	Nephelometric turbidity units (NTU)	5	Preferably <1 for disinfection efficiency.
Zinc	mg/l	5.0	

Source: Guidelines for Drinking-Water Quality, Vol. 1: *Recommendations*, World Health Organization, Geneva, 1984, Table 4. Reproduced with permission.

TABLE 1.10   Radioactive Constituents (WHO)

	Unit[a]	Guideline Value
Gross alpha activity	Bq/l	0.1
Gross beta activity	Bq/l	1

[a]*Notes:* (a) If the levels are exceeded, more detailed radionuclide analysis may be necessary, (b) Higher levels do not necessarily imply that the water is unsuitable for human consumption [a]One bequerel (Bq) = 2.7×10^{-11} curie.

Source: Guidelines for Drinking-Water Quality, Vol. 1: *Recommendations*, World Health Organization, Geneva, 1984, Table 5. Reproduced with permission.

possible. In addition to routine testing equipment, equipment at large plants usually include a zeta meter for coagulant dosing measurements, a nephelometer for turbidity readings, a flame spectrophotometer for measuring inorganic chemicals, and a gas chromatograph with spectrophotometer instrument to measure organic chemicals in low concentrations (micrograms per liter or less). The analytical methods for MCL determination approved by the EPA for volatile chemicals include gas chromatography and gas chromatography–spectrometry techniques. The MCLG for a probable human carcinogen is proposed to be "zero," the limit of detection for regulatory purposes. The MCLGs are unenforceable health goals for public water systems that cause no known or adverse health effects and incorporate an adequate margin of safety. The MCL is an enforceable standard established in the primary drinking water regulations that takes economic factors into consideration, in addition to no unreasonable risk to health. It should be understood that failure to report the presence of certain chemicals or microorganisms does not mean they are not present if the laboratory does not examine for them. All examinations should be made in accordance with the procedures given in *Standard Methods for the Examination of Water and Wastewater*, latest edition or one approved by the EPA (see the Bibliography).

Water samples may be continuous (such as for turbidity or particle counting), grab (instantaneous), composite (an accumulation of grab samples of equal volume), or flow-weighted composite (proportional to volume of flow). Most drinking water samples are grab, although this can be misleading when sampling for organic chemicals or heavy metals. Wastewater samples are composite or flow-weighted composite. When sampling, laboratory collection procedures should be followed.

Drinking water samples should be collected at times of maximum water usage from representative locations including residences. The sampling tap should be clean, not leaking (except in the case of lead and copper monitoring), and flushed for two to three minutes before sample collection. A 1-inch air space should be left on top of the bottle for a bacteriological sample. The bottle should be completely filled for a chemical sample; there must be no air bubble at the top. A laboratory-prepared bottle should be used.

Examination of a nonrepresentative sample is a waste of the sample collector's and the laboratory's time. It will give misleading information that can lead to incorrect and costly actions, discredit the agency or organization involved, and destroy a legal action or research conclusion.

There is a tendency to collect more samples and laboratory data than are needed. The tremendous resources in money, manpower, and equipment committed to the proper preparation, collection, and shipment of the samples and to the analytical procedures involved are lost sight of or misunderstood. Actually, a few carefully selected samples of good quality can usually serve the intended purpose.

The purpose or use of the laboratory data should determine the number of samples and quality of the laboratory work. Data of high quality are needed for official reporting and to support enforcement action or support a health effects study, while data of lesser quality may be acceptable for trend, screening, or monitoring purposes. High-quality legal data must follow official sample collection, identification, shipment, and analytical procedures exactly and without deviation.

The goal of a quality assurance program is to obtain scientifically valid, defensible data of known precision and accuracy to fulfill the water and/or wastewater utility's responsibility to protect and enhance the nation's environment.[27]

The laboratory is an essential ingredient of the effectiveness of the environmental program. However, the laboratory must resist the temptation to become involved in program operation and regulation activities, since its function does not involve sanitary surveys, routine inspection, performance evaluation, program enforcement, responsibility, regulation continuity, and effectiveness. In addition, its limited resources would be misdirected and diluted to the detriment of its primary function. This does not mean that the laboratory should not be involved in training, treatment plant laboratory certification, and solving difficult water plant operational problems.

Sanitary Survey and Water Sampling

A sanitary survey is necessary to determine the reliability of a water system to continuously supply safe and adequate water to the consumer.[28] It is also necessary to properly interpret the results of water analyses and evaluate the effects of actual and potential sources of pollution on water quality. The value of the survey is dependent on the training and experience of the investigator. When available, one should seek the advice of the regulatory agency sanitary engineer or sanitarian. Watershed protection includes enactment of watershed rules and regulations and regular periodic surveillance and inspections. It, in effect, becomes epidemiological surveillance and is a study of environmental factors that may affect human health. Watershed rules and regulations are legal means to control land use that might cause pollution of the water draining off and into the watershed of the water supply source.

If the source of water is a natural or manmade lake, attention would be directed to the following, for each contributes distinctive characteristics to the water: entire drainage basin and location of sewage and other solid and liquid waste disposal or treatment systems; bathing areas; stormwater drains; sewer outfalls; swamps; cultivated areas; feed lots; sources of erosion, sediment and pesticides; and wooded areas, in reference to the pump intake. When water is obtained from a stream or creek, all land and habitation above the water supply intake should be investigated. This means inspection of the entire watershed drainage area so that actual and potential sources of pollution can be determined and properly evaluated and corrective measures instituted. All surface-water supplies must be considered of doubtful sanitary quality unless given adequate treatment, depending on the type and degree of pollution received.

Sanitary surveys have usually emphasized protection of surface-water supplies and their drainage areas. Groundwater supplies such as wells, infiltration galleries, and springs have traditionally been protected by proper construction and location (at an arbitrary "safe" distance from potential sources of pollution and not directly downgrade). The rule-of-thumb distance of 75, 100, or 200 feet, coupled with well construction precautions, has usually served this purpose in most instances, such as for on-site residential wells, in the absence of hydrogeological and engineering investigation and design. However, greater attention is being given to potential distant sources of pollution, especially chemical sources.

The 1996 amendments to the Safe Drinking Water Act require a more sophisticated approach referred to as wellhead protection of groundwater sources. The wellhead is defined as "the surface and subsurface area surrounding a water well or wellfield supplying a public water system through which contaminants are reasonably likely to move toward and reach such well or wellfield."* Determination of the aquifer limits and the drainage area tributary to a well or wellfield, an infiltration gallery, or spring, and the reasonable time of potential contaminants' travel, requires knowledge of the geological formations in the area and the groundwater movement in adjacent and distant tributary areas. In confined or artesian aquifers, this is not readily apparent. The water may originate nearby or at a considerable distance, depending on the extent to which the aquifer formation is confined, channeled, or fractured and on its depth. The U.S. Geological Survey and state geological and water resources agencies may be able to provide information on the local geology and the aquifers. Protection of the tributary wellhead area would require governmental land-use controls, watershed rules and regulations, water purveyor ownership, and public cooperation. To accomplish this, it is first necessary to geographically identify the wellhead area, including groundwater flow, and all existing and potential sources of contamination in that area. This must be supplemented by the controls mentioned, including enactment of watershed (wellhead) protection rules and regulations, and their enforcement. See "Source and Protection of Water Supply," later in this chapter.

The sanitary survey would include, in addition to the source as already noted, the potential for and effects of accidental chemical spills and domestic sewage or industrial waste discharges and leachate from abandoned and existing hazardous waste and landfill sites. Included in the survey would be inspection and investigation of the reservoir, intake, pumping station, treatment plant, and adequacy of each unit process; operation records; distribution system carrying capacity, head losses, and pressures; storage facilities; emergency source of water and plans to supply water in emergency; integrity of laboratory services; connections with other water supplies; and actual or possible cross-connections with plumbing fixtures, tanks, structures, or devices that might permit backsiphonage or backflow. Certification of operators, the integrity and competence of the person in charge of the plant, and adequacy of budgetary support are important factors. Consideration should also be given to land-use plans and the purchase of hydrogeologically sensitive areas and zoning controls.

---

*Also defined as the area between a well and the 99 percent theoretical maximum extent of the stabilized cone of depression. CFR Title 40, Subchapter D, Part 141, U.S. Government Printing Office, July 1999.

Water samples are collected as an adjunct to the sanitary survey as an aid in measuring the quality of the raw water and effectiveness of treatment given. Microbiological examinations; chemical, radiochemical, and physical analyses; and microscopic examinations may be made, depending on the sources of water, climate, geology, hydrology, waste disposal practices on the watershed, problems likely to be encountered, and purpose to be served. In any case, all samples should be properly collected, transported, and preserved as required, and tests should be made by an approved laboratory in accordance with the procedures provided in the latest edition of *Standard Methods for the Examination of Water and Wastewater*[29] or as approved by the EPA.

A sanitary technique and a glass or plastic sterile bottle supplied and prepared by the laboratory for the purpose should be used when collecting a water sample for bacteriological examination. Hands or faucet must not touch the edge of the lip of the bottle or the plug part of the stopper. The sample should be taken from a clean faucet that does not have an aerator or screen and that is not leaking or causing condensation on the outside. Flaming of the tap is optional. The water should be allowed to run for about two to three minutes to get a representative sample. To check for metals and bacteria in household plumbing, the sample must be taken as a "standing" sample without preliminary running of water. A household water softener or other treatment unit may introduce contamination. If a sample from a lake or stream is to be collected, the bottle should be dipped below the surface with a forward sweeping motion so that water coming in contact with the hands will not enter the bottle. When collecting a sample for bacteriological examination, there should be an air space in the bottle. When collecting samples of chlorinated water, the sample bottle should contain sodium thiosulfate to dechlorinate the water. It is recommended that all samples be examined promptly after collection and within 6 to 12 hours if possible. After 24 to 48 hours, examinations may not be reliable.

The chemical and physical analyses may be for industrial or sanitary purposes, and the determinations made will be either partial or complete, depending on the information desired. Water samples for inorganic chemical analyses are usually collected in 1-liter polyethylene containers, new or acid washed if previously used. Samples for lead in drinking water at a tap or from a drinking fountain should be collected in the morning before the system has been used and flushed out and also during the day when the water is being used. Samples for organic chemical analyses are usually collected in 40-ml glass vials or 1-liter glass bottles with Teflon-lined closure.[30] Special precautions are necessary to ensure collection of representative samples free of incidental contamination and without loss of volatile fractions.[31] Containers must be completely filled. A special preservative is added for certain tests, and delivery time to the laboratory is sometimes specified. Samples are also collected for selected tests to control routine operation of a water plant and to determine the treatment required and its effectiveness.

Samples for microscopic examination should be collected in clean, wide-mouth bottles having a volume of 1 or 2 liters from depths that will yield representative organisms. Some organisms are found relatively close to the surface, whereas others are found at middepth or near the bottom, depending on the food, type of organism, and clarity and temperature of the water. Microscopic examinations can determine the changing types, concentrations, and locations of microscopic organisms, control measures or treatment indicated, and time to start treatment. A proper program can prevent tastes and odors by eliminating the responsible organisms that secrete certain oils before they can cause the problem. In addition, objectionable appearances in a reservoir or lake are prevented and sedimentation and filter runs are improved. Attention should also be given to elimination of the conditions favoring the growth of the organisms. See also "Microscopic Examination" in this chapter and "Control of Microorganisms", in Chapter 2.

Sampling Frequency

The frequency with which source and distribution system water samples are collected and used for bacteriologic, chemical, radiologic, microscopic, and physical analyses is usually determined by the regulatory agency, the water quality historical record, plant operational control requirements, and special problems. Operators of public water systems and industrial and commercial water systems will want to collect more frequent but carefully selected samples and make more analyses to detect changes in raw water quality to better control treatment, plant operation, and product quality.

The number of distribution system samples is usually determined by the population served, quality of the water source, treatment, past history, and special problems. Table 1.11 shows the minimum required sampling frequency for coliform density at community water systems in the United States. If routine sampling results in a "positive" indication of coliform bacteria, repeat sampling must be performed to verify the presence of actual bacteria. Table 1.11a presents the number of repeat samples necessary to verify whether or not the system is contaminated. At noncommunity water supplies a sample is collected in each quarter during which the system provides water to the (traveling) public. The minimum sampling frequency recommended by the WHO is shown in Table 1.12. Sampling points should reflect the quality of the water in the distribution system and be at locations of greatest use.

Fecal coliforms/E. coli; Heterotrophic Bacteria (HPC)

- If any routine or repeat sample is total coliform positive, the system must also analyze that total coliform positive culture to determine if fecal coliforms or *E. coli* are present. If fecal coliforms or *E. coli* are detected, the system must notify the state before the end of the same business day, or, if detected after the close of business for the state, by the end of the next business day.
- If any repeat sample is fecal coliform or *E. coli* positive or if a fecal coliform- or *E. coli*-positive original sample is followed by a total coliform-positive repeat sample and the original total coliform-positive sample is not invalidated, it is an acute violation of the MCL for total coliforms.

TABLE 1.11   Total Coliform Sampling Requirements According to Population Served

Population Served	Minimum Number of of Routine Samples per Month[a]	Population Served	Minimum Number of Routine Samples per Month[a]
25–1000[b]	1[c]	59,001–70,000	70
1001–2500	2	70,001–83,000	80
2501–3300	3	83,001–96,000	90
3301–4100	4	96,001–130,000	100
4101–4900	5	130,001–220,000	120
4901–5800	6	220,001–320,000	150
5801–6700	7	320,001–450,000	180
6701–7600	8	450,001–600,000	210
7601–8500	9	600,001–780,000	240
8501–12900	10	780,001–970,000	270
12,901–17,200	15	970,001–1,230,000	300
17,201–21,500	20	1,230,001–1,520,000	330
21,501–25,000	25	1,520,001–1,850,000	360
25,001–33,000	30	1,850,001–2,270,000	390
33,00–41,000	40	2,270,001–3,020,000	420
41,001–50,000	50	3,020,001–3,960,000	450
50,001–59,000	60	3,960,001 or more	480

[a] In lieu of the frequency specified, a noncommunity water system (NCWS) using groundwater and serving 1000 persons or fewer may monitor at a lesser frequency specified by the state until a sanitary survey is conducted and reviewed by the state. Thereafter, NCWS using groundwater and serving 1,000 persons or fewer must monitor in each calendar quarter during which the system provides water to the public, unless the state determines that some other frequency is more appropriate and notifies the system (in writing). Five years after promulgation, NCWSs using groundwater and serving 1,000 persons or fewer must monitor at least once a year. A NCWSs using surface water or groundwater under the direct influence of surface water, regardless of the number of persons served, must monitor at the same frequency as a like-sized community water system (CWS). A NCWS using groundwater and serving more than 1000 persons during any month must monitor at the same frequency as a like-sized CWS, except that the state may reduce the monitoring frequency for any month the system serves 1,000 persons or fewer.

[b] Include public water systems that have at least 15 service connections but serve fewer than 25 persons.

[c] For CWS serving 25–1,000 persons, the state may reduce this sampling frequency if a sanitary survey conducted in the last 5 years indicates that the water system is supplied solely by a protected groundwater source and is free of sanitary defects. However, in no case may the state reduce the frequency to less than once a quarter.

Source: Fact Sheet, Drinking Water Regulations under the Safe Drinking Water Act, Office of Drinking Water, U.S. Environmental Protection Agency, Washington, DC, May 1990, p. 22.

**TABLE 1.11a Monitoring and Repeat Sample Frequency after Total
Coliform-Positive Routine Sample**

Samples per Month	Number of Repeat Samples[a]	Number of Routine Samples Next Month[b]
1	4	5
2	3	5
3	3	5
4	3	5
5 or greater	3	See Table 1.11

[a]Number of repeat samples in the same month for each total coliform-positive routine sample.
[b]Except where state has invalidated the original routine sample, substitutes an on-site evaluation of the problem or waives the requirement on a case-by-case basis.
Source: Fact Sheet, Drinking Water Regulations under the Safe Drinking Water Act, Office of Drinking Water, U.S. Environmental Protection Agency, Washington, DC, December 1990, pp. 23–25.

TABLE 1.12 Distribution System Sampling

Population Served	Minimum Number of Samples
<5000	1 per month
5000–100,000	1 per 5000 population per month
100,000	1 per 10,000 population per month

Source: Guidelines for Drinking-Water Quality, Vol. 1: *Recommendations*, World Health Organization, Geneva, 1984, p. 24.

- The state has the discretion to allow a water system, on a case-by-case basis, to forgo fecal coliform or *E. coli* testing on total coliform-positive samples if the system complies with all sections of the rules that apply when a sample is fecal coliform positive.
- State invalidation of the routine total coliform-positive sample invalidates subsequent fecal coliform- or *E. coli*-positive results on the same sample.
- Heterotrophic bacteria can interfere with total coliform analysis. Therefore, if the total coliform sample produces (1) a turbid culture in the absence of gas production using the multiple-tube fermentation (MTF) technique; (2) a turbid culture in the absence of an acid reaction using the presence–absence (PA) coliform test; or (3) confluent growth or a colony number that is "too numerous to count" using the membrane filter (MF) technique, the sample is invalid (unless total coliforms are detected, in which case the sample is valid). The system must collect another sample within 24 hours of being notified of the result from the same location as the original sample and have it analyzed for total coliforms.

Analytical Methodology

- Total coliform analyses are to be conducted using the 10-tube MTF technique, the MF Technique, the PA coliform test, or the minimal media ONPG-MUG test (Autoanalysis Colilert System). The system may also use the five-tube MTF technique (20-ml sample portions) or a single culture bottle containing the MTF medium as long as a 100-ml water sample is used in the analysis.
- A 100-ml standard sample volume must be used in analyzing for total coliforms, regardless of the analytical method used.

Invalidation of Total Coliform-Positive Samples

- All total coliform-positive samples count in compliance calculations, except for those samples invalidated by the state. Invalidated samples do not count toward the minimum monitoring frequency.
- A state may invalidate a sample only if (1) the analytical laboratory acknowledges that improper sample analysis caused the positive result; (2) the system determines that the contamination is a domestic or other nondistribution system plumbing problem; or (3) the state has substantial grounds to believe that a total coliform-positive result is due to some circumstance or condition not related to the quality of drinking water in the distribution system if (a) this judgment is explained in writing, (b) the document is signed by the supervisor of the state official who draws this conclusion, and (c) the documentation is made available to the EPA and the public.

Variances and Exemptions: None Allowed Sanitary Surveys

- Periodic sanitary surveys are required for all systems collecting fewer than five samples a month every 5 years at community water systems and every 10 years at noncommunity water systems using protected and disinfected groundwater.

Water Analyses

All analyses should be made in accordance with *Standard Methods* [32] in order to provide confidence in the analytical results. As indicated previously, the interpretation of water analyses is based primarily on the sanitary survey of the water system and an understanding of the criteria used in the regulatory development of the drinking water standards. A water supply that is coagulated and filtered would be expected to be practically clear, colorless, and free of iron, whereas the presence of some turbidity, color, and iron in an untreated surface water supply may be accepted as normal. A summary is given in this section of the constituents and concentrations considered significant in water examinations. Other compounds and elements not mentioned are also found in water. The effectiveness of unit treatment processes can be measured using the tests for total coliforms, fecal coliforms, fecal streptococci, and the standard plate count 6 months prior to and 12 months after the process is put into use.

A properly developed, protected, and chlorinated well-water supply showing an absence of coliform organisms can usually be assumed to be free of viruses, protozoa, and helminths if supported by a satisfactory sanitary survey. This is not necessarily so with a surface-water supply. Chemical examinations are needed to ensure the absence of toxic organic and inorganic chemicals.

A final point: The results of a microbiological or chemical examination reflect the quality of the water only at the time of sampling and must be interpreted in the light of the sanitary survey. However, inorganic chemical examination results from well-water supplies are not likely to change significantly from day to day or week to week when collected under the same conditions. Nevertheless, any change is an indication of probable contamination and reason for investigation to determine the cause. The chemical characteristics of well water are a reflection of the geological formations penetrated. Some bacterial and chemical analyses are shown in Table 1.13.

Heterotrophic Plate Count — The Standard Plate Count

The standard plate count is the total colonies of bacteria developing from measured portions (two 1 ml and two 0.1 ml) of the water being tested, which have been planted in petri dishes with a suitable culture medium (agar) and incubated for 48 hours at 95°F (35°C). Bottled water is incubated at 35°C for 72 hours.[33] Only organisms that grow on the media are measured. Drinking water will normally contain some nonpathogenic bacteria; it is almost never sterile.

The test is of significance when used for comparative purposes under known or controlled conditions to show changes from the norm and determine if follow-up investigation and action are indicated. It can monitor changes in the quality (organic nutrients) of the water in the distribution system and storage reservoirs; it can be used to detect the presence of *Pseudomonas, Flavobacterium*, and other secondary invaders that could pose a health risk in the hospital environment; it can call attention to limitations of the coliform test when the average of heterotrophic plate counts in a month exceeds 100 to 500 per ml; it can show the effectiveness of distribution system residual chlorine and possible filter breakthrough; it can show distribution system deterioration, main growth, and sediment accumulation; and it can be used to assess the quality of bottled water. Large total bacterial populations (greater than 1,000 per ml) may also support or suppress growth of coliform organisms. Taste, odor, or color complaints may also be associated with bacterial or other growths in mains or surface-water sources.[34] Bacterial counts may increase in water that has been standing if nutrients are present, such as in reservoirs after copper sulfate treatment and algae destruction or in dead-end mains. These are of no sanitary significance. Mesophilic fungi and actinomycetes, sometimes associated with tastes and odors, may be found in treated water.

TABLE 1.13 Some Bacterial and Chemical Analyses

Source of Sample	Dug Well	Lake	Reservoir	Deep Well	Deep Well
Time of year	—	April	October	—	—
Treatment	None	Chlorine	None	None	None
Bacteria per milliliters agar, 35°C, 24 hr	—	3	—	1	>5000
Coliform MPN per 100 ml	—	<2.2	—	<2.2	≥2400
Color, units	0	15	30	0	0
Turbidity, units	Trace	Trace	Trace	Trace	5.0
Odor					
Cold	2 vegetative	2 aromatic	1 vegetative	1 aromatic	3 disagreeable
Hot	2 vegetative	2 aromatic	1 vegetative	1 aromatic	3 disagreeable
Iron, mg/1	0.15	0.40	0.40	0.08	0.2
Fluorides, mg/1	<0.05	0.005	—	—	—
Nitrogen as ammonia, free, mg/1	0.002	0.006	0.002	0.022	0.042
Nitrogen as ammonia, albuminoid, mg/1	0.026	0.128	0.138	0.001	0.224
Nitrogen as nitrites, mg/1	0.001	0.001	0.001	0.012	0.030
Nitrogen as nitrates, mg/1	0.44	0.08	0.02	0.02	0.16
Oxygen consumed, mg/1	1.1	2.4	7.6	0.5	16.0
Chlorides, mg/1	17.0	5.4	2.2	9.8	6.6
Hardness (as $CaCO_3$), total, mg/1	132.0	34.0	84.0	168.0	148.0
Alkalinity (as $CaCO_{,3}$), mg/1	94.0	29.0	78.0	150.0	114.0
pH value	7.3	7.6	7.3	7.3	7.5

Bacterial Examinations

The bacterial examinations for drinking water quality should always include, as a minimum, tests for total organisms of the coliform group, which are *indicative* of fecal contamination or sewage pollution. They are a normal inhabitant of the intestinal tract of humans and other animals. The goal is no coliform organisms in drinking water. In the past, the coliform group was referred to as the *B. coli* group and the *coli–aerogenes* group. The count for the total coliform group of organisms may include *Escherichia coli*, which is most common in the feces of humans and other warm-blooded animals; *Klebsiella pneumoniae*,* which is

*May have been identified in the past as *Aerobacter aerogenes*.

found in feces and sputum, on fresh vegetables, and in organically rich surface water; *Enterobacter cloacae*, which is found in feces of warm-blooded animals in smaller number than *E. coli*, also in pipe joints, soil, and vegetation; *Citrobacter freundii*, which is normally found in soil and water, also in feces of humans and other warm-blooded animals; and *Enterobacter aerogenes*, which is found in human and other warm-blooded animal feces, soil, pipe joints, and vegetation.* Coliforms are also found in slimes, pump leathers, swimming pool ropes, stormwater drainage, surface waters, and elsewhere.

The tests for fecal coliforms, *E. coli*, fecal streptococci, and *Clostridium perfringens* may be helpful in interpreting the significance of surface-water tests for total coliforms and their possible hazard to the public health. Tests for *Pseudomonas* spp. may indicate the condition in water mains.

Coliform bacteria are not normally considered disease organisms. However, pathogenic (enterotoxigenic) strains of *E. coli* have caused outbreaks of "traveler's diarrhea" and gastroenteritis in institutions and in communities associated with food, raw milk, water, or fomites. The enteropathogenic strains have been associated with outbreaks in newborn nurseries. The test for *E. coli* at 95°F (35°C) is recommended as being a more specific indicator of fecal contamination in Denmark, Belgium, England, France,[35] and the United States. More extensive laboratory procedures are needed to identify *E. coli* and the enteropathogenic *E. coli*. *Escherichia coli* makes up about 95 percent of the fecal coliforms.

The coliform group of organisms includes all of the aerobic and facultative anaerobic, gram-negative, non-spore-forming, rod-shaped bacteria that ferment lactose with acid and gas formation within 24 to 48 hours at 95° to 90°F (35°–37°C). This is the presumptive test that can be confirmed and completed by carrying the test further, as outlined in *Standard Methods*.[36] Coliform species identification is useful in interpreting the significance of the total coliform test where the cause is unclear. Differentiation can confirm the presence of *E. coli*, and hence fecal contamination, or other types of coliforms as previously explained. Prior to December 31, 1990, the results in the MTF were reported as the most probable number (MPN) of coliform bacteria, a statistical number most likely to produce the test results observed, per 100 ml of sample.

A review of the coliform rule by the EPA, as required by the 1986 amendment to the Safe Drinking Water Act, led to the development of a new regulatory standard effective December 31, 1990. This new standard is based on the presence or absence of total coliform bacteria rather than bacterial density. The new standard sets the MCL for total coliforms as follows:

Monthly Number of Samples	MCL
Fewer than 40	No more than 1 positive sample
40 or more	No more than 5.0% positive

Enterobacter and *Klebsiella* are not considered pathogenic to humans, but may be associated with disease-causing organisms found in feces.

In addition, an acute violation necessitates immediate public notification via broadcast media if a routine sample tests positive for total coliforms and for fecal coliforms or *E. coli* and any repeat sample tests positive for total coliforms or a routine sample tests positive for total coliforms and negative for fecal coliforms or *E. coli* and any repeat sample is positive for fecal coliforms or *E. coli*.

If the MTF method is used, the sample size is 100 ml. Either five 20-ml portions or ten 10-ml portions can be used. If any tube has gas formation, the sample is total coliform positive.

If the membrane filter technique is used, the coliform bacteria trapped on the filter produce dark colonies with a metallic sheen within 24 hour (18–22 hours) on an Endo-type medium containing lactose when placed in a 35°C incubator. The dark colonies are presumed to be of the coliform group and the sample is reported as coliform positive. The test can be carried further for coliform differentiation by following the procedure in *Standard Methods*.[37] Suspended matter, algae, and bacteria in large amounts interfere with the membrane filter (0.45 μm) procedure. Bacterial overgrowth on the filter would indicate an excessive bacterial population that should be investigated as to cause and significance.

For many years, the MTF test and the membrane filtration (MF) test have been the approved methods for detecting the presence of coliform organisms. Another test, known as the Colilert test, was approved by the EPA in 1989 for the presence or absence of total coliform. A 100-ml sample and one 100-ml tube with a specially prepared media or a set of five 10-ml tubes* are used to which the test water is added and incubated at 95 to 99°F (35°–37°C). A sterile technique must of course be used. The results are available within 24 hours or may be extended to 48 hours. The presence of coliform is shown by a color change to yellow, the absence by no color change. The presence of *E. coli* is also shown by fluorescence of the tube when viewed under ultraviolet (UV) light. Heterotrophic bacteria levels of 5,000 to 700,000 per ml did not interfere with the Colilert test.

The *fecal coliform test* involves incubation at 112°F (44.5°C) for 24 hours and measures mostly *E. coli* in a freshly passed stool of humans or other warm-blooded animals. A loop of broth from each positive presumptive tube incubated at 95°F (35°C) in the total coliform test is transferred to EC (*E. coli*) broth and incubated at 112°F (44.5°C) in a waterbath; formation of gas within 24 hours indicates the presence of fecal coliform and hence also possibly dangerous contamination. Maintenance of 112°F (44.5 ± 0.2°C) is critical. Nonfecal organisms generally do not produce gas at 112°F (44.5°C). The test has greatest application in the study of stream pollution, raw water sources, sea waters, wastewaters, and the quality of bathing waters. An average individual contributes about 2 billion coliform per day through excrement.

The *fecal streptococci* test (enterococci) uses special agar media incubated at 95°F (35°C) for 48 hours. Dark red to pink colonies are counted as fecal streptococci. They are also normally found in the intestinal tract of warm-blooded

---

*Standard tables are used to determine the MPN when more than one tube is used.

animals, including humans. Most (about 80 percent) of the human fecal streptococci are *Streptococcus faecalis; Streptococcus bovis* is associated with cows, and *Streptococcus equinus* with horses. These organisms may be more resistant to chlorine than coliform and survive longer in some waters but usually die off quickly outside the host. If found, it would indicate recent pollution. An average individual contributes approximately 450 million *fecal streptococci* per day.

The test for *C. perfringens (Clostridium welchii)*, which is found in the intestines of humans and animals, may be of value in the examination of polluted waters and waters containing certain industrial wastes. Clostridia sporulate under unfavorable conditions and can survive indefinitely in the environment; they are more resistant than escherichia and streptococci. Therefore, their presence indicates past or possibly intermittent pollution.

In domestic sewage, the fecal coliform concentration is usually at least four times that of the fecal streptococci and may constitute 30 to 40 percent of the total coliforms. In stormwater and wastes from livestock, poultry, animal pets, and rodents, the fecal coliform concentration is usually less than 0.4 of the fecal streptococci. In streams receiving sewage, fecal coliforms may average 15 to 20 percent of the total coliforms in the stream. The presence of fecal coliform generally indicates fresh and possibly dangerous pollution. The presence of intermediate *aerogenes–cloacae* (IAC) subgroups of coliform organisms suggests past pollution or, in a municipal water supply, defects in treatment or in the distribution system.[38] A ratio of fecal coliforms to *C. perfringens* greater than 100 indicates sewage discharge.

The presence of any coliform organism in drinking water is a danger sign: It must be carefully interpreted in the light of water turbidity, chlorine residual, bacterial count, and sanitary survey, and it must be promptly eliminated. There may be some justification for permitting a low coliform density in developing areas of the world where the probability of other causes of intestinal diseases greatly exceeds those caused by water, as determined by epidemiological information. The lack of any water for washing promotes disease spread.

It must be understood and emphasized that the absence of coliform organisms or other indicators of contamination does not in and of itself ensure that the water is always safe to drink unless it is supported by a satisfactory, comprehensive sanitary survey of the drainage area, treatment unit processes, storage, and distribution system (including backflow prevention). Nor does the absence of coliforms ensure the absence of viruses, protozoa, or helminths unless the water is coagulated, flocculated, settled, gravity filtered, and chlorinated to yield a free residual chlorine of at least 0.5 mg/1, preferably for 1 hour before it is available for consumption. The WHO recommends a free residual chlorine of at least 0.5 mg/1 with a contact period of at least 30 minutes at a pH below 8.0 and a nephelometric turbidity unit (NTU) of 1 or less. A free ozone of 0.2 to 0.4 mg/1 for 4 minutes has been found to be effective to inactivate viruses in clean water (ref. 39, Vol. 2, p. 28). Chlorine dioxide and chloramine treatment may also be used. See "Disinfection," in Chapter 2.

Biological Monitoring

A seven-day biological toxicity test of raw water may be useful to measure chronic effects. Indicators may include the fathead minnow and *Ceriodaphnia*, their survival, growth rate, and reproduction. In some instances, biological monitoring will be more meaningful than environmental monitoring: It can measure the combined effect of air, water, and food pollutants on an organism or animal; this information can be more closely related to potential human health effects.

Virus Examination

The examination of water for enteroviruses has not yet been simplified to the point where the test can be made routinely for compliance monitoring, as for coliform. Viruses range in size from 0.02 to 0.1 μm. There are more than 100 different types of enteric viruses known to be infective. Fecal wastes may contain enteroviruses (echoviruses, polioviruses, and coxsackieviruses—groups A and B) as well as adenoviruses, reoviruses, rotaviruses, Norwalk viruses, and infectious hepatitis viruses (viral hepatitis A).

Enteroviruses may be more resistant to treatment and environmental factors than fecal bacteria, persist longer in the water environment, and remain viable for many months, dependent on temperature and other factors. Enteric viruses, such as protozoa (*Giardia lamblia, Entamoeba histolytica*, and *Cryptosporidium* spp.), may be present even if coliform are absent.

Normally, a large volume of water (100–500 gal) must be sampled and an effective system used to capture, concentrate, and identify viruses. Results may not be available until one or two weeks later.[39] Special analytical laboratory facilities and procedures are required. See *Standard Methods*.[40] A virus standard for drinking water has not been established. A goal of zero to not more than one plaque-forming unit (pfu) per 1,000 gal of drinking water has been suggested.

Since monitoring for enteric viruses is not feasible for routine control of water treatment plant operation, the EPA is requiring specific treatment, or the equivalent, of all surface waters and mandatory chlorination, or equivalent protection, of all groundwaters. Coagulation, flocculation, settling, and rapid sand filtration; slow sand filtration; and lime-soda softening process remove 99 percent or more of the viruses. A pH above 11 inactivates viruses.

Free chlorine is more effective than combined chlorine in inactivating viruses and is more effective at low pH. Turbidity can shield viruses and make chlorination only partially effective. Based on available information, the WHO considers treatment adequate if a turbidity of 1 NTU or less is achieved and the free residual chlorine is at least 0.5 mg/l after a contact period of at least 30 minutes at a pH below 8.0. Prudence would dictate that water obtained from a source known to receive sewage wastes should be coagulated, flocculated, settled, filtered, and disinfected to produce at least 0.4 mg/l free residual chlorine for 2 hours before delivery. Ozone is also an effective disinfectant for clean water if residuals of 0.2 to 0.4 mg/l are maintained for 4 min, but the residual does not remain in the distribution system.[41] The EPA requires 99.99 percent removal and/or inactivation of enteric viruses.

Protozoa and Helminths Examination

The complex procedure to sample, collect, prepare, and positively identify the protozoan cysts of *Giardia lamblia* is impractical for the routine control of water treatment. Because of this, the EPA requires complete treatment of surface waters unless the absence of giardia cysts can be demonstrated and assured by other acceptable means. Sampling for giardia cysts usually involves the filtration of about 500 gal of the water through a 1-μm-pore-size cartridge filter at a rate of about 1 gal/min. The filter extract and sediment collected are concentrated, slides are prepared, and the giardia cyst identified microscopically. Giardia cysts cannot be cultured. Ongerth[42] developed a procedure using a 5-μm-pore-size filter and a 10-gal sample that was reported to be efficient in recovering giardia cysts. Reservoir retention of 30 to 200 days did not reduce cyst concentration. It should be noted that whereas the giardia cyst is about 10 to 15 μm in size, the cryptosporidium oocyst is about 3 to 6 μm in size. The absence of coliform organisms does not indicate the absence of protozoa. Waterborne diseases caused by protozoa include amebic dysentery (amebiasis, *E. histolytica*), giardiasis (*G. lamblia*), cryptosporidiosis (*Cryptosporidium* spp.), meningoencephalitis (*Naegleria fowleri* and *Acanthamoeba culbertsoni*), and balantidiasis (*B. coli.*) Person-to-person contact, poor personal hygiene, and food are also common means of transmission of the diseases. Meningoencephalitis, also known as primary amebic meningoencephalitis, a rare but almost always fatal disease, is associated with swimming or bathing in warm, fresh, and brackish water. Immersion of the head (nasal passages) in the contaminated water is usually involved. The organism is commonly found in soil, fresh water, and decaying vegetation.

The helminths include roundworms, tapeworms, and flukes. The most common disease, spread by *Dracunculus medinensis* in drinking water, is dracontiasis, also known as Guinea-worm infection. Other helminths, such as *Fasciola, Schistosoma, Fasciolopsis, Echinococcus*, and *Ascaris*, are more likely to be transmitted by contaminated food and hand to mouth, particularly in areas where sanitation and personal hygiene are poor. Helminths are 50 to 60 μm in size.

Because of the resistance of the protozoa and helminths to normal chlorination and the lack of routine analytical procedures for water-treatment plant operation control, complete water treatment is required for drinking water.

Specific Pathogenic Organisms

It is not practical to routinely test for and identify specific disease organisms causing typhoid, paratyphoid, infectious hepatitis A, shigellosis, cholera, and others. (See Figure 1.2 for water treatment plant operation control.) The procedures would be too complex and time consuming for routine monitoring. However, laboratory techniques, media, and equipment are available for special studies and investigations where specific organism identification is indicated.

Physical Examinations

Odor Odor should be absent or very faint for water to be acceptable, less than 3 threshold odor number (TON). Water for food processing, beverages, and pharmaceutical manufacture should be essentially free of taste and odor. The test is very subjective, being dependent on the individual senses of smell and taste. The cause may be decaying organic matter, wastewaters including industrial wastes, dissolved gases, and chlorine in combination with certain organic compounds such as phenols. Odors are sometimes confused with tastes. The sense of smell is more sensitive than taste. Activated carbon adsorption, aeration, chemical oxidation (chlorine, chlorine dioxide, ozone, potassium permanganate), and coagulation and filtration will usually remove odors and tastes. Priority should first be given to a sanitary survey of the watershed drainage area and the removal of potential sources or causes of odors and tastes.

A technique for determining the concentration of odor compounds from a water sample to anticipate consumer complaints involves the "stripping" of odor compounds from a water sample that is adsorbed onto a carbon filter. The compounds are extracted from the filter and injected into a gas chromatograph-mass spectrometer for identification and quantification.[43]

Taste The taste of water should not be objectionable; otherwise, the consumer will resort to other sources of water that might not be of satisfactory sanitary quality. Algae, decomposing organic matter, dissolved gases, high concentrations of sulfates, chlorides, and iron, or industrial wastes may cause tastes and odors. Bone and fish oil and petroleum products such as kerosene and gasoline are particularly objectionable. Phenols in concentrations of 0.2 ppb in combination with chlorine will impart a phenolic or medicinal taste to drinking water. The taste test, like the odor test, is very subjective and may be dangerous to laboratory personnel. As in odor control, emphasis should be placed on the removal of potential causes of taste problems. See discussions of causes and methods to remove or reduce tastes and odors, later in this chapter.

Turbidity Turbidity is due to suspended material such as clay, silt, or organic and inorganic materials. Enhanced surface-water regulations in the United States require that the maximum contaminant level for turbidity not exceed 0.5 NTU in 95 percent of the samples taken every month and must never exceed 1 NTU. Additionally, the utility must maintain a minimum of 0.2 mg/l free chlorine residual at representative points within the distribution system. Turbidity measurements are made in terms of nephelometric turbidity units (NTU), Formazin turbidity units (FTU), and Jackson turbidity units (JTU). The lowest turbidity value that can be measured directly on the Jackson candle turbidimeter is 25 units. There is no direct relationship between NTU or FTU readings and JTU readings.[44] The NTU is the standard measure, requiring use of a nephelometer, which measures the amount of light scattered, usually at 90° from the light direction, by suspended particles in the water test sample. It can measure turbidities of less than 1 unit and differences of 0.02 unit. Secondary turbidity measurement standards calibrated against the Formazin standard may also be accepted by the EPA.

The public demands sparkling clear water. This implies a turbidity of less than 1 unit; a level of less than 0.1 unit, which is obtainable when water is coagulated, flocculated, settled, and filtered, is practical. Turbidity is a good measure of sedimentation, filtration, and storage efficiency, particularly if supplemented by the total microscopic and particle count. Increased chlorine residual, bacteriological sampling, and main flushing is indicated when the maximum contaminant level for turbidity is exceeded in the distribution system until the cause is determined and eliminated. Turbidity will interfere with proper disinfection of water, harbor microorganisms, and cause tastes and odors. As turbidity increases, coliform masking in the membrane filter technique is increased.

The American Water Works Association recommends an operating level of no more than 0.3 NTU in filter plant effluent and a goal of no more than 0.2 NTU.

An increase in the turbidity of well water after heavy rains may indicate the entrance of inadequately purified groundwater.

Color   Color should be less than 15 true color units* (sample is first filtered), although persons accustomed to clear water may notice a color of only 5 units. The goal is less than 3 units. Water for industrial uses should generally have a color of 5 to 10 or less. Color is caused by substances in solution, known as true color, and by substances in suspension, mostly organics causing apparent or organic color. Iron, copper, manganese, and industrial wastes may also cause color.

Water that has drained through peat bogs, swamps, forests, or decomposing organic matter may contain a brownish or reddish stain due to tannates and organic acids dissolved from leaves, bark, and plants. Excessive growths of algae or microorganisms may also cause color.

Color resulting from the presence of organics in water may also cause taste, interfere with chlorination, induce bacterial growth, make water unusable by certain industries without further treatment, foul anion exchange resins, interfere with colorimetric measurements, limit aquatic productivity by absorbing photosynthetic light, render lead in pipes soluble, hold iron and manganese in solution causing color and staining of laundry and plumbing fixtures, and interfere with chemical coagulation. Chlorination of natural waters containing organic water color (and humic acid) results in the formation of trihalomethanes, including chloroform. This is discussed later.

Color can be controlled at the source by watershed management. Involved is identifying waters from sources contributing natural organic and inorganic

---

*Cobalt platinum units.

color and excluding them, controlling beaver populations, increasing water flow gradients, using settling basins at inlets to reservoirs, and blending water.[45] Coagulation, flocculation, settling, and rapid sand filtration should reduce color-causing substances in solution to less than 5 units, with coagulation as the major factor. Slow sand filters should remove about 40 percent of the total color. True color is costly to remove. Oxidation (chlorine, ozone) or carbon adsorption also reduces color.

Temperature The water temperature should preferably be less than 60°F (16°C). Groundwaters and surface waters from mountainous areas are generally in the temperature range of 50° to 60°F (10°–16°C). Design and construction of water systems should provide for burying or covering of transmission mains to keep drinking water cool and prevent freezing in cold climates or leaks due to vehicular traffic. High water temperatures accelerate the growth of nuisance organisms, and taste and odor problems are intensified. Low temperatures somewhat decrease the disinfection efficiency.

Microscopic Examination

Microscopic and macroscopic organisms that may be found in drinking water sources include bacteria, algae, actinomycetes, protozoa, rotifers, yeasts, molds, small crustacea, worms, and mites. Most algae contain chlorophyll and require sunlight for their growth. The small worms are usually insect larvae. Larvae, crustacea, worms, molds or fungi, large numbers of algae, or filamentous growths in the drinking water would make the water aesthetically unacceptable and affect taste and odor. Immediate investigation to eliminate the cause would be indicated.

The term *plankton* includes algae and small animals such as cyclops and daphnia. Plankton are microscopic plants and animals suspended and floating in fresh and salt water and are a major source of food for fish. Algae include diatoms, cyanophyceae or blue-green algae (bacteria), and chlorophyceae or green algae; they are also referred to as phytoplankton. Protozoan and other small animals are referred to as zooplankton. They feed on algae and bacteria. The microbial flora in bottom sediments are called the benthos.[46] Phototrophic microorganisms are plankton primarily responsible for the production of organic matter via photosynthesis.

Algal growths increase the organic load in water, excrete oils that produce tastes and odors, clog sand filters, clog intake screens, produce slimes, interfere with recreational use of water, may cause fish kills when in "bloom" and in large surface "mats" by preventing replenishment of oxygen in the water, become attached to reservoir walls, form slimes in open reservoirs and recirculating systems, and contribute to corrosion in open steel tanks[47] and disintegration of concrete. Algae increase oxygen, and heavy concentrations reduce hardness and salts. In the absence of carbon dioxide, algae break down bicarbonates to carbonates, thereby raising the water pH to 9 or higher. Algae also contribute organics, which on chlorination add to trihalomethane formation.

Microscopic examination involves collection of water samples from specified locations and depths. The sample is preserved by the addition of formaldehyde if not taken immediately to the laboratory. At the laboratory, the plankton in the sample is concentrated by means of a centrifuge or a Sedgwick–Rafter sand filter. A 1-ml sample of the concentrate is then placed in a Sedgwick-Rafter counting cell for enumeration using a compound microscope fitted with a Whipple ocular micrometer. The Lackey Drop Microtransect Counting Method is also used, particularly with samples containing dense plankton populations.[48] Enumeration methods include total cell count, clump count, and areal standard unit count.

Examinations of surface-water sources, water mains, and well-water supplies, which are sources of difficulty, should be made weekly to observe trends and determine the need for treatment or other controls and their effectiveness before the organisms reach nuisance proportions. The "areal standard unit" represents an area 20 microns (μm) square or 400 μm^2. One micrometer equals 0.001 mm. Microorganisms are reported as the number of areal standard units per milliliter. Protozoa, rotifers, and other animal life are individually counted. Material that cannot be identified is reported as areal standard units of amorphous matter (detritus). The apparatus, procedure, and calculation of results and conversion to "Cubic Standard Units" is explained in *Standard Methods*.[49]

When more than 300 areal standard units, or organisms, per milliliter is reported, treatment with $CuSO_4$ is indicated to prevent possible trouble with tastes and odors or short filter runs. When more than 500 areal standard units or cells per milliliter is reported, complaints can be expected and the need for immediate action is indicated. A thousand units or more of amorphous matter indicates probable heavy growth of organisms that have died and disintegrated or organic debris from decaying algae, leaves, and similar materials.

The presence of asterionella, tabellaria, synedra, beggiatoa, crenothrix, *Sphaerotilis natans*, mallomonas, anabaena, aphanizomenon, volvox, ceratium, dinobryon, synura, uroglenopsis, and others, some even in small concentrations, may cause tastes and odors that are aggravated where marginal chlorine treatment is used. Free residual chlorination will usually reduce the tastes and odors. More than 25 areal standard units per milliliter of synura, dinobryon, or uroglena, or 300 to 700 units of asterionella, dictyosphaerium, aphanizomenon, volvox, or ceratium in chlorinated water will usually cause taste and odor complaints. The appearance of even 1 areal standard unit of a microorganism may be an indication to start immediate copper sulfate treatment if past experience indicates that trouble can be expected.

The blue-green algae, anabaena, microcystis (polycystis), nodularia, gloeotrichia, coelosphaerium, *Nostoc rivulare*, and aphanizomenon in large concentrations have been responsible for killing fish and causing illness in horses, sheep, dogs, ducks, chickens, mice, and cattle.[50] Illness in humans from these causes has been suspected, but confirmatory evidence is limited.[51] Gorham[52] estimated that the oral minimum lethal dose of decomposing toxic microcystis bloom for a 150-lb man is 1 to 2 quarts of thick, paintlike suspension and concluded that toxic waterblooms of blue-green algae in public water supplies are not a significant health hazard. Red tides caused by the dinoflagellates *Gonyaulax monilata* and *Gymnodinium brevis* have been correlated with mass mortality of fish.[53] Coagulation, flocculation, sedimentation, and filtration do not remove algal toxins, nor does the usual activated carbon treatment.

Investigation of conditions contributing to or favoring the growth of plankton in a reservoir and their control should reduce dependence on copper sulfate treatment. See "Control of Microorganisms", in Chapter 2.

Chemical Examinations*

The significance of selected chemical elements and compounds in drinking water is discussed next. An intake of 2 liters of water per day per person is assumed in determining health effects. The MCL is the National Drinking Water Regulation maximum contaminant level. The maximum contaminant level goal (MCLG) is a desirable one and is nonmandatory unless specifically made so by a state. The WHO level represents a guideline value "of a constituent that ensures an aesthetically pleasing water and does not result in any significant risk to the health of the consumer."[54] A value in excess of the guideline value does not in itself imply that the water is unsuitable for consumption. A comprehensive discussion of health-related inorganic and organic constituents can be found in *Guidelines for Drinking-Water Quality*, Vol. 2, WHO, Geneva, 1984.[55] Gas chromatographic mass spectrometry is considered the best method for identifying and quantifying specific organic compounds in an unknown sample. The removal of organic and inorganic chemicals from drinking water is reviewed later in this chapter.

Albuminoid Ammonia   Albuminoid ammonia represents "complex" organic matter and thus would be present in relatively high concentrations in water-supporting algae growth, receiving forest drainage, or containing other organic matter. Concentrations of albuminoid ammonia higher than about 0.15 mg/l, therefore, should be appraised in the light of origin of the water and the results of microscopic examination. In general, the following concentrations serve as a guide: low—less than 0.06 mg/l; moderate—0.06 to 0.15 mg/l; high—0.15 mg/l or greater. When organic nitrogen and ammonia nitrogen forms are found together, they are measured as Kjeldahl nitrogen.

Alkalinity   The alkalinity of water passing through distribution systems with iron pipe should be in the range of 30 to 100 mg/l, as $CaCO_3$, to prevent serious corrosion; up to 500 mg/l is acceptable, although this factor must be appraised from the standpoint of pH, hardness, carbon dioxide, and dissolved-oxygen content. Corrosion of iron pipe is prevented by the maintenance of calcium carbonate stability. Undersaturation will result in corrosive action in iron water mains and

---

*Results are reported as milligrams per liter (mg/l), which for all practical purposes can be taken to be the same as parts per million (ppm), except when the concentrations of substances in solution approach or exceed 7000 mg/l, when a density correction should be made.

cause red water. Oversaturation will result in carbonate deposition in piping and water heaters and on utensils. See "Corrosion Cause and Control", in Chapter 2. Potassium carbonate, potassium bicarbonate, sodium carbonate, sodium bicarbonate, phosphates, and hydroxides cause alkalinity in natural water. Calcium carbonate, calcium bicarbonate, magnesium carbonate, and magnesium bicarbonate cause hardness, as well as alkalinity. Sufficient alkalinity is needed in water to react with added alum to form a floc in water coagulation. Insufficient alkalinity will cause alum to remain in solution. Bathing or washing in water of excessive alkalinity can change the pH of the lacrimal fluid around the eye, causing eye irritation.

Aluminum The EPA-recommended goal is less than 0.05 mg/1; the WHO guideline is 0.2 mg/1.[56] Aluminum is not found naturally in the elemental form, although it is one of the most abundant metals on the earth's surface. It is found in all soils, plants, and animal tissues. Aluminum-containing wastes concentrate in and can harm shellfish and bottom life.[57] Alum as aluminum sulfate is commonly used as a coagulant in water treatment; excessive aluminum may pass through the filter with improper pH control. Precipitation may take place in the distribution system or on standing when the water contains more than 0.5 mg/1. Its presence in filter plant effluent is used as a measure of filtration efficiency. Although ingested aluminum does not appear to be harmful, aluminum compounds have been associated with neurological disorders in persons on kidney dialysis machines. Aluminum in the presence of iron may cause water discoloration. There may be an association between aluminum and Alzheimer's disease, but this has not been confirmed.[58]

Arsenic The MCL for arsenic in drinking water was lowered from 0.05 mg/1 to 0.01 mg/1 by the EPA in January 2001. The WHO guideline is also 0.01 mg/1. (The Occupational Safety and Health Administration (OSHA) standard is 10 μg/m^3 for occupational exposure to inorganic arsenic in air over an 8-hour day; 2 μg/m^3 for 24 hour exposure to ambient air.[59]) A probable lethal oral dose is 5 to 20 mg/kg, depending on the compound and individual sensitivity. Sources of arsenic are natural rock formations (phosphate rock), industrial wastes, arsenic pesticides, fertilizers, detergent "presoaks," and possibly other detergents. It is also found in foods, including shellfish and tobacco, and in the air in some locations.

There is ample evidence that defines a relationship between certain cancers (e.g., skin, bladder, kidney, lung, liver) and high levels of arsenic in drinking water (i.e., above 0.2 mg/1). There is significant debate, however, if these cancers are seen at lower levels of arsenic. Arsenic occurs naturally as arsenic, +3 (arsenite) and arsenic, +5 (arsenate). Arsenites are more toxic than arsenates. Arsenic may be converted to dimethylarsine by anaerobic organisms and accumulate in fish, similar to methylmercury.[60] After many years of scientific research and debate, the USEPA concluded that a concentration of 10 μg/1 (0.01 mg/1) is protective of public health. Promulgated in 2001, the lowered MCL required over 3,000 public water systems to install removal systems (or blend or abandon the high arsenic wells) by the Rule deadline of February 2006. For treatment, see "Removal of Inorganic Chemicals", in Chapter 2.

Asbestos   Most asbestos-related diseases (mesotheliomas) are associated with the breathing of air containing asbestos fibers as long as 20 years earlier. Sources of exposure include working or living in the immediate vicinity of crocidolite mines, asbestos insulation and textile factories, and shipyards. Asbestos in drinking water may come from certain naturally occurring silicate materials in contact with water or from eroded asbestos cement pipe. A study (1935–1973) on the incidence of gastrointestinal cancer and use of drinking water distributed through asbestos cement (A/C) pipe reached the preliminary conclusion that "no association was noted between these asbestos risk sources and gastrointestinal tumor incidence."[61] A subsequent study concluded, "The lack of coherent evidence for cancer risk from the use of A/C pipe is reassuring."[62] An EPA study shows no statistical association between deaths due to certain types of cancer and the use of A/C pipe. British researchers reported that the cancer risk was "sensibly zero" or exceedingly low[63]: "Available studies on humans and animals do not provide evidence to support the view that ingestion of drinking water containing asbestos causes organ-specific cancers." Nevertheless, exposure to the asbestos fibers in drinking water should be reduced. Conventional water treatment, including coagulation and filtration, will remove more than 90 percent of the asbestos fibers in the raw water.[64]

Asbestos cement pipe was found to behave much like other piping materials, except polyvinyl chloride (PVC), that are commonly used for the distribution of drinking water. It has been concluded that, where "aggressive water conditions exist, the pipe will corrode and deteriorate; if aggressive water conditions do not exist, the pipe will not corrode and deteriorate."[65] Aggressive water can leach calcium hydroxide from the cement in A/C pipe. The American Water Works Association (AWWA) Standard C400-77 establishes criteria for the type of pipe to use for nonaggressive water (≥ 12.0), moderately aggressive water (10.9–11.9), and highly aggressive water (≤ 10.0), based on the sum of the pH plus the log of the alkalinity times the calcium hardness, as calcium carbonate. Remedial measures, in addition to pH adjustment and control of corrosion, include chemical addition to build up a protective film, elimination of hydrogen sulfide, rehabilitation and lining of existing pipe, pipe replacement, and a flushing program. Asbestos cement pipe should not be used to carry aggressive water.

If the water is heavily contaminated, its use for humidifiers, showers, food preparation, clothes laundering, and drinking is not advised since the asbestos fibers can become airborne and be inhaled. The EPA has recommended a maximum contaminant level of 7.1×10^6 asbestos fibers longer than 10 μm/1 from all sources, including naturally occurring asbestos. On July 6, 1989, the EPA ruled to prohibit manufacture, importation, and processing of asbestos in certain products and to phase out the use of asbestos in all other products. This action was meant to reduce airborne asbestos in the workplace and ambient air and thereby the carcinogenic health risk associated with the inhalation of asbestos fibers.

Barium   Barium may be found naturally in groundwater (usually in concentrations less than 0.1 mg/1) and in surface water receiving industrial wastes; it is also found in air. It is a muscle stimulant and in large quantities may be harmful to the nervous system and heart. The fatal dose is 550 to 600 mg. The MCL is 2 mg/1 in drinking water. A WHO guideline has not been established; concentrations of 10 mg/1 are not considered significant. Barium can be removed by weak-acid ion exchange.

Benzene   This chemical is used as a solvent and degreaser of metals.[66] It is also a major component of gasoline. Drinking water contamination generally results from leaking underground gasoline and petroleum tanks or improper waste disposal. Benzene has been associated with significantly increased risks of leukemia among certain industrial workers exposed to relatively large amounts of this chemical during their working careers. This chemical has also been shown to cause cancer in laboratory animals when the animals are exposed to high levels over their lifetimes. Chemicals that cause increased risk of cancer among exposed industrial workers and in laboratory animals also may increase the risk of cancer in humans who are exposed at lower levels over long periods of time. The EPA has set the enforceable drinking water standard for benzene at 0.005 mg/1 to reduce the risk of cancer or other adverse health effects observed in humans and laboratory animals. The OSHA standard is 1 mg/1 with 5 mg/1 for short-term (15-minutes) exposure.[67]

Cadmium   The federal drinking water MCL for cadmium is 0.005 mg/1. The WHO guideline is 0.005 mg/1.[68] Common sources of cadmium are water mains and galvanized iron pipes, tanks, metal roofs where cistern water is collected, industrial wastes (electroplating), tailings, pesticides, nickel plating, solder, incandescent light filaments, photography wastes, paints, plastics, inks, nickel–cadmium batteries, and cadmium-plated utensils. It is also found in zinc and lead ores. Cadmium vaporizes when burned; salts of cadmium readily dissolve in water and can, therefore, be found in air pollutants, wastewater, wastewater sludge, fertilizer, land runoff, some food crops, tobacco, and drinking water. Beef liver and shellfish are very high in cadmium. Large concentrations may be related to kidney damage, hypertension (high blood pressure), chronic bronchitis, and emphysema. Cadmium builds up in the human body, plants, and food animals. It has a biological half-life of about 20 years.[69] The direct relationship between cardiovascular death rates in the United States, Great Britain, Sweden, Canada, and Japan and the degree of softness or acidity of water point to cadmium as the suspect.[70] In 1972, the Joint WHO Food and Agriculture Organization Expert Committee on Food Additives established a provisional tolerable weekly cadmium intake of 400 to 500 μg. Cadmium removal from water is discussed in Chapter 2.

Carbon–Chloroform Extract (CCE) and Carbon–Alcohol Extract (CAE) (Tests No Longer Routinely Used)   Carbon–chloroform extract may include chlorinated hydrocarbon pesticides, nitrates, nitrobenzenes, aromatic ethers, and many others adsorbed on an activated carbon cartridge. Water from uninhabited and nonindustrial watersheds usually show CCE concentrations of less than 0.04 mg/1. The taste and odor of drinking water can be expected to be poor when the concentration of CCE reaches 0.2 mg/1. Carbon–alcohol extract measures gross organic chemicals including synthetics. A goal of less than 0.04 mg/1 CCE and 0.10 mg/1 CAE has been proposed.

Carbon Dioxide   The only limitation on carbon dioxide is that pertaining to corrosion. It should be less than 10 mg/1, but when the alkalinity is less than 100 mg/1, the CO_2 concentration should not exceed 5.0 mg/1.

Carbon Tetrachloride   This chemical was once a popular household cleaning fluid.[71] It generally gets into drinking water by improper disposal. This chemical has been shown to cause cancer in laboratory animals such as rats and mice when exposed at high levels over their lifetimes. Chemicals that cause cancer in laboratory animals may also increase the risk of cancer in humans exposed at lower levels over long periods of time. The EPA has set the enforceable drinking water standard for carbon tetrachloride at 0.005 mg/1 to reduce the risk of cancer or other adverse health effects observed in laboratory animals. The WHO *tentative* guideline value is 3 μg/1.

Chlorides of Intestinal Origin   Natural waters remote from the influence of ocean or salt deposits and not influenced by local sources of pollution have a low chloride content, usually less than 4.0 mg/1. Due to the extensive salt deposits in certain parts of the country, it is impractical to assign chloride concentrations that, when exceeded, indicate the presence of sewage, agricultural, or industrial pollution, unless a chloride record over an extended period of time is kept on each water supply. In view of the fact that chlorides are soluble, they will pass through pervious soil and rock for great distances without diminution in concentration, and thus the chloride content must be interpreted with considerable discretion in connection with other constituents in the water. The concentration of chlorides in urine is about 5000 mg/1, in septic tank effluent about 80 mg/1, and in sewage from a residential community 50 mg/1 depending on the water source.

Chlorides of Mineral Origin   The WHO guideline for chloride ion is 250 mg/1.[72] A goal of less than 100 mg/1 is recommended. The permissible chloride content of water depends on the sensitivity of the consumer. Many people notice a brackish taste imparted by 125 mg/1 of chlorides in combination with sodium, potassium, or calcium, whereas others are satisfied with concentrations as high as 250 mg/1. Irrigation waters should contain less than 200 mg/1. When the chloride is in the form of sodium chloride, use of the water for drinking may be inadvisable for persons who are under medical care for certain forms of heart disease. The main intake of chlorides is with foods. Hard water softened by the ion exchange or lime-soda process (with Na_2CO_3) will increase sodium concentrations in the water. Salt used for highway deicing may contaminate groundwater and surface-water supplies. Its use should be curtailed and storage depots covered. Chlorides can be removed from water by distillation, reverse osmosis, or electrodialysis and minimized by proper aquifer selection and well construction. Water sources near oceans or in the vicinity of underground salt deposits may contain high salt concentrations. Well waters from sedimentary rock are likely to contain chlorides. The corrosivity of water is increased by high concentrations of chlorides, particularly if the water has a low alkalinity.

Chromium   The total chromium MCL and WHO guideline[73] is 0.1 mg/l in drinking water. Chromium is found in cigarettes, some foods, the air, and industrial plating, paint, and leather tanning wastes. Chromium deficiency is associated with atherosclerosis. Hexavalent chromium dust can cause cancer of the lungs and kidney damage.[74]

Copper   The EPA action level for copper is 1.3 mg/l; the WHO guideline is 1.0 mg/l.[75] The goal is less than 0.2 mg/l. Concentrations of this magnitude are not present in natural waters but may be due to the corrosion of copper or brass piping; 0.5 to 1.0 mg/l in soft water stains laundry and plumbing fixtures blue–green. A concentration in excess of 0.2 to 0.3 mg/l will cause an "off" flavor in coffee and tea; 5 mg/l or less results in a bitter metallic taste; 1 mg/l may affect film and reacts with soap to produce a green color in water; 0.25 to 1.0 mg/l is toxic to fish. Corrosion of galvanized iron and steel fittings is reported to be enhanced by copper in public water supplies. Copper appears to be essential for all forms of life, but excessive amounts are toxic to fish. The estimated adult daily requirement is 2.0 mg, coming mostly from food. Copper deficiency is associated with anemia. Copper salts are commonly used to control algal growths in reservoirs and slime growths in water systems. Copper can be removed by ion exchange, conventional coagulation, sedimentation, filtration, softening, or reverse osmosis; when caused by corrosion of copper pipes, it can be controlled by proper water treatment and pH control. Copper sulfate treatment of the water source for algae control may contribute copper to the finished water. Electrical grounding to copper water pipe can add to the copper dissolution.

Corrosivity   Water should be noncorrosive. Corrosivity of water is related to its pH, alkalinity, hardness, temperature, dissolved oxygen, carbon dioxide, total dissolved solids, and other factors. Waters high in chlorides and low in alkalinity are particularly corrosive. Since a simple, rapid test for corrosivity is not available, test pipe sections or metal coupons (90-day test) are used, supplemented, where possible, by water analyses such as calcium carbonate saturation, alkalinity, pH, and dissolved solids and gases. Incrustation on stainless steel test pipe or metal coupon should not exceed 0.05 mg/cm^2; loss by corrosion of galvanized iron should not exceed 5.00 mg/cm^2 (AWWA). The corrosion of copper tubing increases particularly when carrying water above 140°F (60°C). Schroeder[76] reports that pewter, britannia metal, water pipes, and cisterns may contain antimony, lead, cadmium, and tin, which leach out in the presence of soft water or acidic fluids. Soft water flowing over galvanized iron roofs or through galvanized iron pipes or stored in galvanized tanks contains cadmium and zinc. Ceramic vessels contain antimony, beryllium, barium, nickel, and zirconium; pottery glazes contain lead, all of which may be leached out if improper firing and glazing are used. Corrosivity is controlled by pH, alkalinity, and calcium carbonate adjustment, including use of lime, sodium carbonate, and/or sodium hydroxide. Other means include the addition of polyphosphate, orthophosphate, and silicates and pH control. In any case, corrosion-resistant pipe should be used where possible.

Cyanide   Cyanide is found naturally and in industrial wastes. Cyanide concentrations as low as 10 μg/l have been reported to cause adverse effects in fish. Long-term consumption of up to 4.7 mg/day has shown no injurious effects (ref. 45, pp. 128–136). The cyanide concentration in drinking water should not exceed 0.2 mg/l. The probable oral lethal dose is 1.0 mg/kg. The WHO guideline is 0.1 mg/l. An MCL and MCLG of 0.2 mg/l has been established by the EPA. Cyanates can ultimately decompose to carbon dioxide and nitrogen gas.[77] Cyanide is readily destroyed by conventional treatment processes.

1,1-Dichloroethylene   This chemical is used in industry and is found in drinking water as a result of the breakdown of related solvents.[78] The solvents are used as cleaners and degreasers of metals and generally get into drinking water by improper waste disposal. This chemical has been shown to cause liver and kidney damage in laboratory animals such as rats and mice when exposed at high levels over their lifetimes. Chemicals that cause adverse effects in laboratory animals may also cause adverse health effects in humans exposed at lower levels over long periods of time. The EPA has set the enforceable drinking water standard for 1,1-dichloroethylene at 0.007 mg/l to reduce the risk of the adverse health effects observed in laboratory animals.

1,2-Dichloroethane   This chemical is used as a cleaning fluid for fats, oils, waxes, and resins.[79] It generally gets into drinking water from improper waste disposal. This chemical has been shown to cause cancer in laboratory animals such as rats and mice when exposed at high levels over their lifetimes. Chemicals that cause cancer in laboratory animals may also increase the risk of cancer in humans exposed at lower levels over long periods of time. The EPA has set the enforceable drinking water standard for 1,2-dichloroethane at 0.005 mg/l to reduce the risk of cancer or other adverse health effects observed in laboratory animals. The WHO guideline is 10 μg/l.

Dissolved Oxygen   Water devoid of dissolved oxygen frequently has a "flat" taste, although many attractive well waters are devoid of oxygen. In general, it is preferable for the dissolved-oxygen content to exceed 2.5 to 3.0 mg/l to prevent secondary tastes and odors from developing and to support fish life. Game fish require a dissolved oxygen of at least 5.0 mg/l to reproduce and either die off or migrate when the dissolved oxygen falls below 3.0 mg/l. The concentration of dissolved oxygen in potable water may be related to problems associated with iron, manganese, copper, and nitrogen and sulfur compounds.

Fluorides   Fluorides are found in many groundwaters as a natural constituent, ranging from a trace to 5 mg/1 or more, and in some foods. Fluorides in concentrations greater than 4 mg/1 can cause the teeth of children to become mottled and discolored, depending on the concentration and amount of water consumed. Mottling of teeth has been reported very occasionally above 1.5 mg/1 according to WHO guidelines. Drinking water containing 0.7 to 1.2 mg/1 natural or added fluoride is beneficial to children during the time they are developing permanent teeth. An optimum level is 1.0 mg/1 in temperate climates. The Centers for Disease Control and Prevention (CDC) estimates that in 2006, approximately 69.2 percent of the United States' population (or 184 million people) had access to optimum levels of fluoridated water (0.7 mg/1 to 1.2 mg/l). More than 65 percent of the nation's nine-year-old children are free of tooth decay and the CDC also considers fluoridation of community water systems one of the 10 great public health achievements of the 20th century.[80]

The maximum contaminant level in drinking water has been established in the National Drinking Water Regulations at 4 mg/1. The probable oral lethal dose for sodium fluoride is 70 to 140 mg/kg. Fluoride removal methods include reverse osmosis, lime softening, ion exchange using bone char or activated alumina, and tricalcium phosphate adsorption. It is not possible to reduce the fluoride level to 1 mg/1 using only lime.[81] The WHO and CDC reports show no evidence to support any association between fluoridation of drinking water and the occurrence of cancer (1982).

Free Ammonia   Free ammonia represents the first product of the decomposition of organic matter; thus, appreciable concentrations of free ammonia usually indicate "fresh pollution" of sanitary significance. The exception is when ammonium sulfate of mineral origin is involved. The following values may be of general significance in appraising free ammonia content in groundwater: low—0.015 to 0.03 mg/1; moderate—0.03 to 0.10 mg/1; high—0.10 mg/1 or greater. In treated drinking water, the goal is less than 0.1 mg/1, but less than 0.5 mg/1 is acceptable. Special care must be exercised to allow for ammonia added if the "chlorine–ammonia" treatment of water is used or if crenothrix organisms are present. If ammonia is present or added, chloramines are formed when chlorine is added to the water. Ammonia in the range of 0.2 to 2.0 mg/1 is toxic to many fish. A recommended maximum is 0.5 mg/1 to 0.2 mg/1 for rainbow trout. Chloramines are also toxic to other aquatic life. Ammonia serves as a plant nutrient, accelerating eutrophication in receiving waters. It is converted to nitrite and then to nitrate, first by *Nitrosomonas* and then by *Nitrobacter* organisms. Ammonia can be removed by breakpoint or superchlorination.

Hardness   Hardness is due primarily to calcium and secondarily to magnesium carbonates and bicarbonates (carbonate or temporary hardness that can be removed by heating) and calcium sulfate, calcium chloride, magnesium sulfate, and magnesium chloride (noncarbonate or permanent hardness, which cannot be removed by heating); the sum is the total hardness expressed as calcium carbonate. In general, water softer than 50 mg/1, as $CaCO_3$ is corrosive, whereas waters harder than about 80 mg/1 lead to the use of more soap and above 200 mg/1 may cause incrustation in pipes. Lead, cadmium, zinc, and copper in solution are usually caused by pipe corrosion associated with soft water. Desirable hardness values, therefore, should be 50 to 80 mg/1, with 80 to 150 mg/1 as passable, over 150 mg/1 as undesirable, and greater than 500 as unacceptable. The U.S. Geological Survey (USGS) and WHO[82] classify hardness, in milligrams per liter as $CaCO_3$, as 0 to 60 soft, 61 to 120 moderately hard, 121 to 180 hard, and more than 180 very hard. Waters high in sulfates (above 600 to 800 mg/1 calcium sulfate, 300 mg/1 sodium sulfate, or 390 mg/1 magnesium sulfate) are laxative to those not accustomed to the water. Depending on alkalinity, pH, and other factors, hardness above 200 mg/1 may cause the buildup of scale and flow reduction in pipes. In addition to being objectionable for laundry and other washing purposes due to soap curdling, excessive hardness contributes to the deterioration of fabrics. Hard water is not suitable for the production of ice, soft drinks, felts, or textiles. Satisfactory cleansing of laundry, dishes, and utensils is made difficult or impractical. When heated, bicarbonates precipitate as carbonates and adhere to the pipe or vessel. In boiler and hot-water tanks, the scale resulting from hardness reduces the thermal efficiency and eventually causes restriction of the flow or plugging in the pipes. Calcium chloride, when heated, becomes acidic and pits boiler tubes. Hardness can be reduced by lime-soda ash chemical treatment or the ion exchange process, but the sodium concentration will be increased. See "Water Softening," in Chapter 2. Desalination will also remove water hardness.

There seem to be higher mortality rates from cardiovascular diseases in people provided with soft water than in those provided with hard water. Water softened by the ion exchange process increases the sodium content of the finished water. The high concentration of sodium and the low concentration of magnesium have been implicated, but low concentrations of chromium and high concentrations of copper have also been suggested as being responsible. High concentrations of cadmium are believed to be associated with hypertension. Cause and effect for any of these is not firm.

Hydrogen Sulfide   Hydrogen sulfide is most frequently found in groundwaters as a natural constituent and is easily identified by a rotten-egg odor. It is caused by microbial action on organic matter or the reduction of sulfate ions to sulfide. A concentration of 70 mg/1 is an irritant, but 700 mg/1 is highly poisonous. In high concentration, it paralyzes the sense of smell, thereby making it more dangerous. Black stains on laundered clothes and black deposits in piping and on plumbing fixtures are caused by hydrogen sulfide in the presence of soluble iron. Hydrogen sulfide in drinking water should not be detectable by smell or exceed 0.05 mg/1. Hydrogen sulfide predominates at pH of 7.0 or less. It is removed by aeration or chemical oxidation followed by filtration.

Iron   Iron is found naturally in groundwaters and in some surface waters and as the result of corrosion of iron pipe. Iron deposits and mining operations and distribution systems may be a source of iron and manganese. Water should have a soluble iron content of less than 0.1 mg/l to prevent reddish-brown staining of laundry, fountains, and plumbing fixtures and to prevent pipe deposits. The secondary MCL and WHO guideline level is 0.3 mg/l; the goal should be less than 0.05 mg/l. Some staining of plumbing fixtures may occur at 0.05 mg/l. Precipitated ferric hydroxide may cause a slight turbidity in water that can be objectionable and cause clogging of filters and softener resin beds. In combination with manganese, concentrations in excess of 0.3 mg/l cause complaints. Precipitated iron may cause some turbidity. Iron in excess of 1.0 mg/l will cause an unpleasant taste. A concentration of about 1 mg/l is noticeable in the taste of coffee or tea. Conventional water treatment or ion exchange will remove iron. Chlorine or oxygen will precipitate soluble iron. Iron is an essential element for human health. See "Iron and Manganese Occurrence and Removal," in Chapter 2.

Lead   The EPA requires that when more than 10 percent of tap water samples exceed 15 μg/l, the utility must institute corrosion control treatment. Concentrations exceeding this value occur when corrosive waters of low mineral content and softened waters are piped through lead pipe and old lead house services. Zinc-galvanized iron pipe, copper pipe with lead-based solder joints, and brass pipe, faucets, and fittings may also contribute lead. The lead should not exceed 5 μg/l in the distribution system.

Lead, as well as cadmium, zinc, and copper, is dissolved by carbonated beverages, which are highly charged with carbon dioxide. Limestone, galena, water, and food are natural sources of lead. Other sources are motor vehicle exhaust, certain industrial wastes, mines and smelters, lead paints, glazes, car battery salvage operations, soil, dust, tobacco, cosmetics, and agricultural sprays. Fallout from airborne pollutants also contributes significant concentrations of lead to water supply reservoirs and drainage basins. About one-fifth of the lead ingested in water is absorbed. The EPA estimates that in young children about 20 percent of lead exposure comes from drinking water; dust contributes at least 30 percent, air 5 to 20 percent, and food 30 to 45 percent.[83]

The Safe Drinking Water Amendments of 1986 require that any pipe, solder, or flux used in the installation or repair of any public water system or any plumbing connected to a public water system shall be lead free. Acceptable substitutes for lead solder are tin–silver, tin–antimony, and tin–copper. Solder and flux containing not more than 0.2 percent lead and pipes and pipe fittings containing not more than 8.0 percent lead are considered to be lead free. Lead-free solder may contain trace amounts of lead, tin, silver, and copper. (Leaded joints necessary for the repair of cast-iron water mains are excluded from the prohibition.) Exposure to lead in tap water is more likely in new homes, less than 5 years old, where plumbing contains lead solder or flux. A survey by the AWWA showed an average lead concentration of 193.3 μg/l in first-draw samples from homes less than 2 years old, 45.7 μg/l from homes 2 to 5 years old, 16 μg/l from homes 5 to 10 years old, and 8.2 μg/1 from homes older than 10 years.[84] Hot water would normally contain higher concentrations of lead. Lead flux is reported to dissolve at about 140° to 150°F (60°–66°C). Hot-water flushing is an economical method for removing residual flux from piping in newer buildings.[85] Galvanic corrosion due to dissimilar metals—copper and lead–tin solder—will also contribute lead. Electric water cooler piping, water contact surfaces, and fittings have also been implicated as sources of lead in drinking water. Defective coolers are being replaced.

Water containing lead in excess of the standard should not be used for baby formula or for cooking or drinking. Flushing the standing water out of a faucet for about 1 minute will minimize the lead concentration, but it does not solve the problem. The Secretary of Housing and Urban Development and the Administrator of the Veterans' Administration may not ensure or guarantee a mortgage or furnish assistance with respect to newly constructed residential property, which contains a potable water system, unless such system uses only lead-free pipe, solder, and flux.

The EPA requires the following measures and standards to control lead in community and noncommunity nontransient water systems:

1. Corrosion control when tap water sample average exceeds 0.01 mg/1, when the pH level is less than 8.0 in more than 5 percent of samples, and when the copper level exceeds 1.3 mg/1 (pH not greater than 9.0, alkalinity of 25–100 mg/1 as calcium carbonate)
2. An MCL for lead of 0.005 mg/1 and a MCLG of zero leaving the treatment plant
3. An MCL and an MCLG for copper of 1.3 mg/1
4. Tap water lead "action level" of 0.015 mg/1 in not more than 10 percent of samples of tap water that has been allowed to stand at least 6 hours (usually the first draw in the morning) from dwelling units that contain copper pipes with lead solder installed after 1982

Water treatment or use of a corrosion inhibitor is advised where indicated. Conventional water treatment, including coagulation, will partially remove natural or manmade lead in raw water. Measures to prevent or minimize lead dissolution include maintenance of pH $\geq$ 8.0 and use of zinc orthophosphate or polyphosphates. Silicates may have a long-term beneficial effect. No apparent relationship was found between lead solubility and free chlorine residual, hardness, or calcium level. Electrical grounding to plumbing increased lead levels. Alkalinity level control was not of value at pH 7.0 to <8.0.[86] However, since only 3 to 5 percent of the free chlorine is in the active hypochlorous acid form at pH 9.0, whereas 23 to 32 percent is in the hypochlorous acid form at pH 8.0, pH level control is critical for corrosion control and the maintenance of disinfection efficiency.

Removal of lead service lines is required if treatment is not adequate to reduce lead level.

Manganese Manganese is found in gneisses, quartzites, marbles, and other metamorphic rocks and, hence, in well waters from these formations. It is also found in many soils and sediments, such as in deep lakes and reservoirs, and in surface water. Manganese concentrations (MCL) should be not greater than 0.05 mg/1, and preferably less than 0.01, to avoid the black-brown staining of plumbing fixtures and laundry when chlorine bleach is added. The WHO guideline value for manganese is 0.1 mg/1.

Concentrations greater than 0.5 to 1.0 mg/1 may give a metallic taste to water. Concentrations above 0.05 mg/1 or less can sometimes build up coatings on sand filter media, glass parts of chlorinators, and concrete structures and in piping, which may reduce pipe capacity. When manganous manganese in solution comes in contact with air or chlorine, it is converted to the insoluble manganic state, which is very difficult to remove from materials on which it precipitates. Excess polyphosphate for sequestering manganese may prevent absorption of essential trace elements from the diet[87] it is also a source of sodium. See "Iron and Manganese Occurrence and Removal," in Chapter 2.

Mercury Episodes associated with the consumption of methylmercury-contaminated fish, bread, pork, and seed have called attention to the possible contamination of drinking water. Mercury is found in nature in the elemental and organic forms. Concentrations in unpolluted waters are normally less than 1.0 μg/1. The organic methylmercury and other alkylmercury compounds are highly toxic, affecting the central nervous system and kidneys. It is taken up by the aquatic food chain. The maximum permissible contaminant level in drinking water is 0.002 mg/1 as total mercury. The WHO guideline is 0.001 mg/1.

Methylene Blue Active Substances (MBASs) The test for MBASs also shows the presence of alkyl benzene sulfonate (ABS), linear alkylate sulfonate (LAS), and related materials that react with methylene blue. It is a measure of the apparent detergent or foaming agent and hence sewage presence. The composition of detergents varies. Household washwater in which ABS is the active agent in the detergent may contain 200 to 1,000 mg/1. Alkyl benzene sulfonate has been largely replaced by LAS, which can be degraded under aerobic conditions; if not degraded, it too will foam at greater than 1 mg/1 concentration. Both ABS and LAS detergents contain phosphates that may, if allowed to enter, fertilize plant life in lakes and streams. The decay of plants will use oxygen, leaving less for fish life and wastewater oxidation. Because of these effects, detergents containing phosphates have been banned in some areas. In any case, the presence of MBAS in well-water supply is objectionable and an indication of sewage pollution, the source of which should be identified and removed, even though it has not been found to be of health significance in the concentrations found in drinking water. The level of MBAS in a surface water is also an indicator of sewage pollution. Carbon adsorption can be used to remove MBAS from drinking water. Foaming agents should be less than 0.5 mg/1; 1.0 mg/1 is detectable by taste. Anionic (nondegradable) detergents should not exceed 0.2 mg/1.

Nitrates   Nitrates represent the final product of the biochemical oxidation of ammonia. Its presence is probably due to the presence of nitrogenous organic matter of animal and, to some extent, vegetable origin, for only small quantities are naturally present in water. Septic tank systems may contribute nitrates to the groundwater if free oxygen is present. Manure and fertilizer contain large concentrations of nitrates. However, careful management practices of efficient utilization of applied manure and fertilizer by crops will reduce nitrates leaching below the root zone. Shallow (18–24-in.) septic tank absorption trenches will also permit nitrate utilization by vegetation. The existence of fertilized fields, barnyards, or cattle feedlots near supply sources must be carefully considered in appraising the significance of nitrate content. Furthermore, a cesspool may be relatively close to a well and contributing pollution without a resulting high nitrate content because the anaerobic conditions in the cesspool would prevent biochemical oxidation of ammonia to nitrites and then nitrates. In fact, nitrates may be reduced to nitrites under such conditions. In general, however, nitrates disclose the evidence of "previous" pollution of water that has been modified by self-purification processes to a final mineral form. Allowing for these important controlling factors, the following ranges in concentration may be used as a guide: low, less than 0.1 mg/1; moderate, 0.1 to 1.0 mg/1; high, greater than 1.0 mg/1. Concentrations greater than 3.0 mg/1 indicate significant manmade contribution.

The presence of more than 10 mg/1 of nitrate expressed as nitrogen, the maximum contaminant level in drinking water, appears to be the cause of methemoglobinemia, or "blue babies." The standard has also been expressed as 45 mg/1 as nitrate ion (10 mg/1 as nitrogen). Methemoglobinemia is largely a disease confined to infants less than three months old but may affect children up to age six. Boiling water containing nitrates increases the concentration of nitrates in the water. The recommended maximum for livestock is 100 mg/1.

Nitrate is corrosive to tin and should be kept at less than 2 mg/1 in water used in food canning. There is a possibility that some forms of cancer might be associated with very high nitrate levels.

Nitrates may stimulate the growth of water plants, particularly algae if other nutrients such as phosphorus and carbon are present. Nitrates seem to serve no useful purpose, other than as a fertilizer. Gould points out that

> a more objective review of literature would perhaps indicate that without any sewage additions most of our waterways would contain enough nitrogen and phosphorous (due to nonpoint pollution source) to support massive algal blooms and that the removal of these particular elements would have little effect on existing conditions.[88]

The feasible methods for the removal of nitrates are anion exchange, reverse osmosis, distillation, and electrodialysis. See "Nitrate Removal" in Chapter 2.

Nitrites   Nitrites represent the first product of the oxidation of free ammonia by biochemical activity. Free oxygen must be present. Unpolluted natural waters contain practically no nitrites, so concentrations exceeding the very low value of 0.001 mg/1 are of sanitary significance, indicating water subject to pollution that is in the process of change associated with natural purification. The nitrite concentration present is due to sewage and the organic matter in the soil through which the water passes. Nitrites in concentrations greater than 1 mg/1 in drinking water are hazardous to infants and should not be used for infant feeding.

Oxidation–Reduction Potential (ORP, Also Redox)   Oxidation–reduction potential is the potential required to transfer electrons from the oxidant to the reductant and is used as a qualitative measure of the state of oxidation in water treatment systems.[89] An ORP meter is used to measure in millivolts the oxidation–loss of electrons or reduction–gain of electrons.

Oxygen-Consumed Value   This represents organic matter that is oxidized by potassium permanganate under the test conditions. Pollution significant from a bacteriological examination standpoint is accompanied by so little organic matter as not to significantly raise the oxygen-consumed value. For example, natural waters containing swamp drainage have much higher oxygen-consumed values than water of low original organic content that are subject to bacterial pollution. This test is of limited significance.

Para-Dichlorobenzene   This chemical is a component of deodorizers, moth-balls, and pesticides.[90] It generally gets into drinking water by improper waste disposal. This chemical has been shown to cause liver and kidney damage in laboratory animals such as rats and mice exposed to high levels over their lifetimes. Chemicals that cause adverse effects in laboratory animals also may cause adverse health effects in humans exposed at lower levels over long periods of time. The EPA has set the enforceable drinking water standard for *para*-dichlorobenzene at 0.075 mg/1 to reduce the risk of the adverse health effects observed in laboratory animals.

Pesticides   Pesticides include insecticides, herbicides, fungicides, rodenticides, regulators of plant growth, defoliants, or desiccants. Sources of pesticides in drinking water are industrial wastes, spills and dumping of pesticides, and runoff from fields, inhabited areas, farms, or orchards treated with pesticides. Surface and groundwater may be contaminated. Conventional water treatment does not adequately remove pesticides. Powdered or granular activated carbon treatment may also be necessary. Maximum permissible contaminant levels of certain pesticides in drinking water and their uses and health effects are given in Table 1.4.

pH*   The pH values of natural water range from about 5.0 to 8.5 and are acceptable except when viewed from the standpoint of corrosion. A guideline value of

---

*pH is defined as the logarithm of the reciprocal of the hydrogen ion concentration. The concentration increases and the solution becomes more acidic as the pH value decreases below 7.0; the solution becomes more alkaline as the concentration decreases and the pH value increases above 7.0.

6.5 to 8.5 is suggested. The pH is a measure of acidity or alkalinity using a scale of 0.0 to 14.0, with 7.0 being the neutral point, a higher value being alkaline and lower value acidic. The bactericidal, virucidal, and cysticidal efficiency of chlorine as a disinfectant increases with a decrease in pH. The pH determination in water having an alkalinity of less than 20 mg/1 by using color indicators is inaccurate. The electrometric method is preferred in any case. The ranges of pH color indicator solutions, if used, are as follows: bromphenol blue, 3.0 to 4.6; bromcresol green, 4.0 to 5.6; methyl red, 4.4 to 6.0; bromcresol purple, 5.0 to 6.6; bromthymol blue, 6.0 to 7.6; phenol red, 6.8 to 8.4; cresol red, 7.2 to 8.8; thymol blue, 8.0 to 9.6; and phenol phthalein, 8.6 to 10.2. Waters containing more than 1.0 mg/1 chlorine in any form must be dechlorinated with one or two drops of 1/4 percent sodium thiosulfate before adding the pH indicator solution. This is necessary to prevent the indicator solution from being bleached or decolorized by the chlorine and giving an erroneous reading. The germicidal activity is greatly reduced at a pH level above 8.0. Corrosion is associated with pH levels below 6.5 to 7.0 and with carbon dioxide, alkalinity, hardness, and temperature.[91]

Phenols   The WHO guideline for individual phenols, chlorophenols, and 2,4,6-trichlorophenol is not greater than 0.1 μg/1 (0.1 ppb), as the taste and odor can be detected at or above that level after chlorination. The odor of some chlorophenols is detected at 1 μg/1. In addition, 2,4,6-trichlorophenol, found in biocides and chlorinated water containing phenol, is considered a chemical carcinogen based on animal studies.[92] The guideline for pentachlorophenol in drinking water, a wood preservative, is 0.001 mg/1 based on its toxicity. It also causes objectionable taste and odor. If the water is not chlorinated, phenols up to 100 μg/1 are acceptable.[93] Phenols are a group of organic compounds that are byproducts of steel, coke distillation, petroleum refining, and chemical operations. They should be removed prior to discharge to drinking water sources. Phenols are also associated with the natural decay of wood products, biocides, and municipal wastewater discharges. The presence of phenols in process water can cause serious problems in the food and beverage industries and can taint fish. Chlorophenols can be removed by chlorine dioxide and ozone treatment and by activated carbon. The AWWA advises that phenol concentrations be less than 2.0 μg/1 at the point of chlorination. Chlorine dioxide, ozone, or potassium permanganate pretreatment is preferred, where possible, to remove phenolic compounds.

Phosphorus   High phosphorus concentrations, as phosphates, together with nitrates and organic carbon are often associated with heavy aquatic plant growth, although other substances in water also have an effect. Fertilizers and some detergents are major sources of phosphates. Uncontaminated waters contain 10 to 30 μg/1 total phosphorus, although higher concentrations of phosphorus are also found in "clean" waters. Concentrations associated with nuisances in lakes would not normally cause problems in flowing streams. About 100 μg/1 complex phosphate interferes with coagulation. Phosphorus from septic tank subsurface absorption system effluents is not readily transmitted through sandy soil and groundwater.[94] Most waterways naturally contain sufficient nitrogen and phosphorus to support massive algal blooms.

Polychlorinated Biphenyls (PCBs) Polychlorinated biphenyls give an indication of the presence of industrial wastes containing mixtures of chlorinated byphenyl compounds having various percentages of chlorine. Organochlorine pesticides have a similar chemical structure. The PCBs cause skin disorders in humans and cancer in rats. They are stable and fire resistant and have good electrical insulation capabilities. They have been used in transformers, capacitors, brake linings, plasticizers, pumps, hydraulic fluids, inks, heat exchange fluids, canvas waterproofing, ceiling tiles, fluorescent light ballasts, and other products. They are not soluble in water but are soluble in fat. They cumulate in bottom sediment and in fish, birds, ducks, and other animals on a steady diet of food contaminated with the chemical. Concentrations up to several hundred and several thousand milligrams per liter have been found in fish, snapping turtles, and other aquatic life. Polybrominated biphenyl, a derivative of PCB, is more toxic than PCB. Aroclor is the trade name for a PCB mixture used in a pesticide. The manufacture of PCBs was prohibited in the United States in 1979 under the Toxic Substances Control Act of 1976. The use in transformers and electromagnets was banned after October 1985 if they pose an exposure risk to food or animal feed. Continued surveillance of existing equipment and its disposal is necessary for the life of the equipment. The toxicity of PCB and its derivatives appears to be due to its contamination with dioxins. The Food and Drug Administration (FDA) action levels are 1.5 mg/l in fat of milk and dairy products; 3 mg/l in poultry and 0.3 mg/l in eggs; and 2 mg/l in fish and shellfish. The MCL for drinking water is 0.0005 mg/l with zero as the EPA MCLG. The OSHA permissible 8-hour time-weighted average (TWA) airborne exposure limit is 0.5 mg/m^3 for PCBs containing 42 percent chlorine.[95] The National Institute of Occupational Safety and Health (NIOSH) recommended that the 8-hour TWA exposure by inhalation be limited to 1.0 μg/m^3 or less.[96] A level not exceeding 0.002 μg/l is suggested to protect aquatic life.[97] The PCBs are destroyed at 2000°F (1093°C) and 3 percent excess oxygen for 2 seconds contact time. They are vaporized at 1584°F (862°C). The PCB contamination of well water has been associated with leakage from old submersible well pumps containing PCB in capacitors. These pumps were manufactured between 1960 and 1978, are oil cooled rather than water cooled, and have a two-wire lead rather than three-wire. Pumps using 220-volt service would not be involved.[98] Activated carbon adsorption and ozonation plus UV are possible water treatments to remove PCBs.

Polynuclear Aromatic Hydrocarbons Polynuclear aromatic hydrocarbons such as fluoranthene, 3,4-benzfluoranthene, 11,12-benzfluoranthene, 3,4-benzpyrene, 1,12-benzperyline, and indeno [1,2,3-*cd*] pyrene are known carcinogens and are potentially hazardous to humans. The WHO set a limit of 0.2 μg/l for the sum of these chemicals in drinking water, comparable in quality with unpolluted groundwater. Because of its carcinogenicity, a guideline value of 0.01 $\mu g/l$ is proposed for benzo[a]pyrene in drinking water. It is also recommended that the use of cool-tar-based pipe linings be discontinued. [99]

Polysaccharides In soft drink manufacturing, polysaccharides* in surface waters may be found in the water used. In waters of low pH, the polysaccharides come out of solution to form a white precipitate. The CO_2 in carbonated water is also sufficient to cause this. Coagulation and sedimentation or reverse osmosis treatment can remove polysaccharides.

Brewing water should ideally be low in alkalinity and soft but high in sulfates.[100]

Radioactivity The maximum contaminant levels for radioactivity in drinking water are given in Table 1.4. The exposure to radioactivity from drinking water is not likely to result in a total intake greater than recommended by the Federal Radiation Council. Naturally occurring radionuclides include Th-232, U-235, and U-238 and their decay series, including radon and radium 226 and 228. They may be found in well waters, especially those near uranium deposits. (Radium is sometimes found in certain spring and well supplies.) Since these radionuclides emit alpha and beta radiation (as well as gamma), their ingestion or inhalation may introduce a serious health hazard, if found in well-water supplies.[101] Possible manmade sources of radionuclides in surface waters include fallout (in soluble form and with particulate matter) from nuclear explosions in precipitation and runoff, releases from nuclear reactors and waste facilities, and manufacturers. Radon is the major natural source of radionuclides.

Radon Radon is a natural decay product of uranium and is a byproduct of uranium used in industry and the manufacture of luminescent faces of clocks and instruments. It is also found in soil, rock, and well water and is readily released when water is agitated such as in a washing machine (clothes and dish), when water flows out of a faucet, and when water is sprayed from a shower head. Radon is particularly dangerous when released and inhaled in an enclosed space such as indoors. Radon-222 is emitted from tailings at uranium mill sites.

The EPA estimates that 10,000 pCi/l in water will result in a radon air concentration of about 1 pCi/l. The EPA has proposed a maximum contaminant level of 300 pCi/l for drinking water supplies.

Radon can be removed from water by aeration—packed tower or diffused air, filtration through granular activated carbon, ion exchange, and reverse osmosis. The concentration of radon in removal raises a disposal problem.

Selenium Selenium is associated with industrial pollution (copper smelting) and vegetation grown in soil containing selenium. It is found in meat and other foods. Selenium causes cancers and sarcomas in rats fed heavy doses. Chronic exposure to excess selenium results in gastroenteritis, dermatitis, and central

---

*One of a group of carbohydrates.

nervous system disturbance.[102] Selenium is considered an essential nutrient and may provide protection against certain types of cancer. Selenium in drinking water should not exceed the MCL of 0.05 mg/l. An intake of 25 or 50 μg/day is not considered harmful.

Silver   The secondary MCL for silver in drinking water is 0.10 mg/l. Silver is sometimes used to disinfect small quantities of water and in home faucet purifiers. Colloidal silver may cause permanent discoloration of the skin, eyes, and mucous membranes. A continuous daily dose of 400 μg of silver may produce the discoloration (argyria). Only about 10 percent of the ingested silver is absorbed.[103]

Sodium   Persons on a low-sodium diet because of heart, kidney, or circulatory (hypertension) disease or pregnancy should use distilled water if the water supply contains more than 20 mg/l of sodium and be guided by a physician's advice. The consumption of 2.0 liters of water per day is assumed. Water containing more than 200 mg/l sodium should not be used for drinking by those on a moderately restricted sodium diet. It can be tasted at this concentration when combined with other anions. Many groundwater supplies and most home-softened (using ion exchange) well waters contain too much sodium for persons on sodium-restricted diets. If the well water is low in sodium (less than 20 mg/l) but the water is softened by the ion exchange process because of excessive hardness, the cold-water system can be supplied by a line from the well that bypasses the softener and low-sodium water can be made available at cold-water taps. A home water softener adds 0.46 times the hardness removed as $CaCO_3$. Sodium can be removed by reverse osmosis, distillation, and cation exchange, but it is costly. A laboratory analysis is necessary to determine the exact amount of sodium in water. The WHO guideline for sodium in drinking water is 200 mg/l. Common sources of sodium, in addition to food, are certain well waters, ion exchange water-softening units, water treatment chemicals (sodium aluminate, lime-soda ash in softening, sodium hydroxide, sodium bisulfite, and sodium hypochlorite), road salt, and possibly industrial wastes. Sodium added in fluoridation and corrosion control is not significant.

Specific Electrical Conductance   Specific electrical conductance is a measure of the ability of a water to conduct an electrical current and is expressed in micromhos per cubic centimeters of water at 77°F (25°C). Because the specific conductance is related to the number and specific chemical types of ions in solution, it can be used for approximating the dissolved-solids content in the water, particularly the mineral salts in solution if present. The higher the conductance, the more mineralized the water and its corrosivity. Different minerals in solutions give different specific conductance. Commonly, the amount of dissolved solids (in milligrams per liter) is about 65 percent of the specific conductance. This relationship is not constant from stream to stream from well to well, and it may even vary in the same source with changes in the composition of the water.

Specific conductance is used for the classification of irrigation waters. In general, waters of less than 200 μmho/cm^3 are considered acceptable, and conductance in excess of 300 μmho/cm^3 unsuitable. Good fresh waters for fish in the United States are reportedly under 1100 μmho/cm^3.[104] Wastewater with a conductivity up to 1,200 to 4,000 μmho/cm^3 may be acceptable for desert reclamation. Electrical conductivity measurements give a rapid approximation of the concentration of dissolved solids in milligrams per liter.

Sulfates　The sulfate content should not exceed the secondary MCL of 250 mg/l. The WHO guideline is 400 mg/l.[105] With zeolite softening, calcium sulfate or gypsum is replaced by an equal concentration of sodium sulfate. Sodium sulfate (or Glauber salts) in excess of 200 mg/l, magnesium sulfate (or Epsom salts) in excess of 390 to 1,000 mg/l, and calcium sulfate in excess of 600 to 800 mg/l are laxative to those not accustomed to the water. Magnesium sulfate causes hardness; sodium sulfate causes foaming in steam boilers. Sulfate is increased when aluminum sulfate is used in coagulation. High sulfates also contribute to the formation of scale in boilers and heat exchangers. Concentrations of 300 to 400 mg/l cause a taste. Sulfates can be removed by ion exchange, distillation, reverse osmosis, or electrodialysis. Sulfates are found in surface waters receiving industrial wastes such as those from sulfate pulp mills, tanneries, and textile plants. Sulfates also occur in many waters as a result of leaching from gypsum-bearing rock.

Total Dissolved Solids (TDS)　The total solid content should be less than 500 mg/l; however, this is based on the industrial uses of public water supplies and not on public health factors. Higher concentrations cause physiological effects and make drinking water less palatable. Dissolved solids, such as calcium, bicarbonates, magnesium, sodium, sulfates, and chlorides, cause scaling in plumbing above 200 mg/l. The TDS can be reduced by distillation, reverse osmosis, electrodialysis, evaporation, ion exchange, and, in some cases, chemical precipitation. Water with more than 1000 mg/l of dissolved solids is classified as *saline*, irrespective of the nature of the minerals present.[106] The USGS classifies water with less than 1000 mg/l as fresh, 1,000 to 3,000 as slightly saline, 3,000 to 10,000 as moderately saline, 10,000 to 35,000 as very saline, and more than 35,000 as briny.

1,1,1-Trichloroethane　This chemical is used as a cleaner and degreaser of metals.[107] It generally gets into drinking water by improper waste disposal. This chemical has been shown to damage the liver, nervous system, and circulatory system of laboratory animals such as rats and mice exposed at high levels over their lifetimes. Some industrial workers who were exposed to relatively large amounts of this chemical during their working careers also suffered damage to the liver, nervous system, and circulatory system. Chemicals that cause adverse effects among exposed industrial workers and in laboratory animals may also cause adverse health effects in humans exposed at lower levels over long periods of time. The EPA has set the enforceable drinking water standard for 1,1,1-trichloroethane at 0.2 mg/l to protect against the risk of adverse health effects observed in humans and laboratory animals.

Trichloroethylene This chemical is a common metal-cleaning and dry-cleaning fluid.[108] It generally gets into drinking water by improper waste disposal. This chemical has been shown to cause cancer in laboratory animals such as rats and mice when exposed at high levels over their lifetimes. Chemicals that cause cancer in laboratory animals may also increase the risk of cancer in humans exposed at lower levels over long periods of time. The EPA has set forth the enforceable drinking water standard for trichloroethylene at 0.005 mg/l to reduce the risk of cancer or other adverse health effects observed in laboratory animals.

Trihalomethanes Trihalomethanes (THMs) and other nonvolatile, higher molecular weight compounds are formed by the interaction of free chlorine with humic and fulvic substances and other organic precursors produced either by normal organic decomposition or by metabolism of aquatic biota. The precursor level is determined through testing by prechlorination of a sample and then analyzing the sample after seven days storage under controlled temperature and pH. A rapid surrogate THM measurement can be made using UV absorbent measurement. Two gas chromatographic analytic techniques are acceptable by the EPA for THM analysis. The THMs include chloroform (trichloromethane), bromoform (tribromomethane), dibromochloromethane, bromodichloromethane, and iodoform (dichloroiodomethane). Toxicity, mutagenicity, and carcinogenicity have been suspected as being associated with the ingestion of trihalomethanes. The EPA has stated that:

> epidemiological evidence relating THM concentrations or other drinking water quality factors and cancer morbidity-mortality is not conclusive but suggestive. Positive statistical correlations have been found in several studies,* but causal relationships cannot be established on the basis of epidemiological studies. The correlation is stronger between cancer and the brominated THMs than for chloroform.[109]

Chloroform is reported to be carcinogenic to rats and mice in high doses and hence is a suspected human carcinogen. The Epidemiology Subcommittee of the National Research Council (NRC) says that cancer and THM should not be linked.[110] The Report on Drinking Water and Health, NRC Safe Drinking Water and Health, states: "A review of 12 epidemiological studies failed either to support or refute the results of positive animal bioassays suggesting that certain trihalomethanes, chloroform for example, may cause cancer in humans."[111] However, the National Drinking Water Advisory Council, based on studies in the review and evaluation by the National Academy of Sciences, the work done by

---

*The reliability and accuracy of studies such as these are often subject to question.

the National Cancer Institute, and other research institutions within the EPA, has accepted the regulation of trihalomethanes on "the belief that chloroform in water does impose a health threat to the consumer."[112] The EPA has established a standard of 80 μg/l for total THMs for public water supplies. The WHO guideline for chloroform is 30 μg/l[113] and 35 μg/l for THM in Canada.

Uranyl Ion This ion may cause damage to the kidneys. Objectionable taste and color occur at about 10 mg/l. It does not occur naturally in most waters above a few micrograms per liter. The taste, color, and gross alpha MCL will restrict uranium concentrations to below toxic levels; hence, no specific limit is proposed.[114]

Vinyl Chloride This chemical is used in industry and is found in drinking water as a result of the breakdown of related solvents.[115] The solvents are used as cleaners and degreasers of metals and generally get into drinking water by improper waste disposal. This chemical has been associated with significantly increased risks of cancer among certain industrial workers who were exposed to relatively large amounts of this chemical during their working careers. This chemical has also been shown to cause cancer in laboratory animals when exposed at high levels over their lifetimes. Chemicals that cause increased risk of cancer among exposed industrial workers and in laboratory animals also may increase the risk of cancer in humans exposed at lower levels over long periods of time. The EPA has set the enforceable drinking water standard for vinyl chloride at 0.002 mg/l to reduce the risk of cancer or other adverse health effects observed in humans and laboratory animals. Packed-tower aeration removes vinyl chloride.

Zinc The concentration of zinc in drinking water (goal) should be less than 1.0 mg/l. The MCL and the WHO guideline is 5.0 mg/l.[116]Zinc is dissolved by surface water. A greasy film forms in surface water containing 5 mg/l or more zinc upon boiling. More than 5.0 mg/l causes a bitter metallic taste and 25 to 40 mg/l may cause nausea and vomiting. At high concentrations, zinc salts impart a milky appearance to water. Zinc may contribute to the corrosiveness of water. Common sources of zinc in drinking water are brass and galvanized pipe and natural waters where zinc has been mined. Zinc from zinc oxide in automobile tires is a significant pollutant in urban runoff.[117] The ratio of zinc to cadmium may also be of public health importance. Zinc deficiency is associated with dwarfism and hypogonadism.[118] Zinc is an essential nutrient. It can be reduced by ion exchange, softening, reverse osmosis, and electrodialysis.

Drinking Water Additives

Potentially hazardous chemicals or contaminants may inadvertently be added directly or indirectly to drinking water in treatment, well drilling, and distribution. Other contaminants potentially may leach from paints, coatings, pumps, storage tanks, distribution system pipe and plumbing systems, valves, pipe fittings, and other equipment and products.

Chemicals (direct additives) used in water treatment for coagulation, corrosion control, and other purposes may contain contaminants such as heavy metals or organic substances that may pose a health hazard. In addition, significant concentrations of organic and inorganic contaminants (indirect additives) may leach or be extracted from various drinking water system components.

Since its inception, the EPA has maintained an advisory list of acceptable products for drinking water contact, but this function was transferred to the private sector on April 7, 1990. In 1985, the EPA provided seed funding for a consortium to establish a program for setting standards and for the testing, evaluation, inspection, and certification to control potentially hazardous additives. The consortium included the AWWA, the American Water Works Association Research Foundation (AWWARF), the Association of State Drinking Water Administrators (ASDWA), and the National Sanitation Foundation (NSF).

In 1988, the NSF published American National Standards Institute (ANSI)/NSF Standard 60, Drinking Water Treatment Chemicals—Health Effects, and ANSI/NSF Standard 61, Drinking Water System Components—Health Effects.[119] The ANSI approved NSF Standards 60 and 61 in May of 1989.

Third-party certification organizations, like the NSF, Underwriters Laboratories (UL), and the Safe Water Additives Institute,[120]* can certify products for compliance with the ANSI/NSF standards. In addition to the NSF listing of certified products, the AWWA plans to maintain and make available a directory of all products certified as meeting the ANSI/NSF standards.

In mid-1990, the ANSI announced a program to "certify the certifiers." Because each state regulates drinking water additives products, the ANSI program is expected to provide the basis for state acceptance of independent certification organizations to test and evaluate equipment and products for compliance with the standards. The ANSI program includes minimum requirements for certification agencies that address chemical and microbiological testing, toxicology review and evaluation, factory audits, follow-up evaluations, marking, contracts and policies, and quality assurance. Many state drinking water regulations and rules require independent third-party certification of additives products.

Water Quantity

The quantity of water used for domestic purposes will generally vary directly with the availability of the water, habits of the people, cost of water, number and type of plumbing fixtures provided, water pressure, air temperature, newness of a community, type of establishment, metering, and other factors. Wherever possible, the actual water consumption under existing or similar circumstances and the number of persons served should be the basis for the design of a water and sewage system. Special adjustment must be made for unaccounted-for water

---

*The NSF is accredited by the ANSI, UL has applied for accrediation, and the Safe Water Additives Institute is developing a program for ANSI review (*AWWA MainStream*, May 1991).

and for public, industrial, and commercial uses. The average per-capita municipal water use has increased from 150 gpd in 1960 to 168 gpd in 1975 to 183 gpd in 1980 and remained relatively steady at 179 gpd in 1995. Approximately 70 gal is residential use, 50 gal industrial, 35 gal commercial, and 10 gal public and 14 gal is lost.[121] Included is water lost in the distribution system and water supplied for firefighting, street washing, municipal parks, and swimming pools. USGS estimated rural water use at 68 gpd in 1975 and 79 gpd in 1980.[122]

Table 1.14 gives estimates of water consumption at different types of places and in developing areas of the world. Additions should be made for car washing, lawn sprinkling, and miscellaneous uses. If provision is made for firefighting requirements, then the quantity of water provided for this purpose to meet fire underwriters' standards will be in addition to that required for normal domestic needs in small communities.

Developing Areas of the World   Piped water delivery to individual homes and waterborne sewage disposal are not affordable in many developing countries. This calls for sequential or incremental improvements from centrally located hand pumps to water distribution systems. Social, cultural, and economic conditions, hygiene education, and community participation must be taken into account in project selection and design.[123] Community perception of needs, provision of local financial management, operation, and maintenance must be taken into consideration and assured before a project is started. The annual cost of water purchased from a water vendor may equal or exceed the cost of piped metered water. In addition, much time is saved where water must be hauled from a stream. Hand pumps, where used, should be reliable, made of corrosion-resistant materials, with moving parts resistant to abrasion, including sand, and readily maintained at the local level. A detailed analysis of hand pump tests and ratings has been made by Arlosoroff et al.[124] It is important to keep mechanical equipment to a minimum and to train local technicians. Preference should be given to drilled wells where possible. For surface-water supplies, slow sand filters are generally preferred over the more complex rapid sand filters.

Water Conservation

Water conservation can effect considerable saving of water with resultant reduction in water treatment and pumping costs and wastewater treatment. With water conservation, development of new sources of water and treatment facilities and their costs can be postponed or perhaps made unnecessary, and low-distribution system water pressure situations are less likely. However, the unit cost of water to the consumer may not be reduced; it may actually increase because the fixed cost will remain substantially the same. The revenue must still be adjusted to meet the cost of water production and distribution.

Water conservation can be accomplished, where needed, by a continuing program of leak detection and repair in the community distribution system and in buildings; use of low water-use valves and plumbing fixtures; water pressure and flow control in the distribution system and in building services (orifices);

TABLE 1.14 Guides for Water Use in Design

Type of Establishment	gpd[a]
Residential	
Dwellings and apartments (per bedroom)	150
Rural	60
Suburban	75
Urban	180
Temporary Quarters	
Boarding houses	65
Additional (or nonresident boarders)	10
Campsites (per site), recreation vehicle with individual connection	100
Campsites, recreational vehicle, with comfort station	40–50
Camps without toilets, baths, or showers	5
Camps with toilets, without baths or showers	25
Camps with toilets and bathhouses	35–50
Cottages, seasonal with private bath	50
Day camps	15–20
Hotels	65–75
Mobile home parks (per unit)	125–150
Motels	50–75
Public Establishments	
Restaurants (toilets and kitchens)	7–10
Without public toilet facilities	$2\frac{1}{2}$–3
With bar or cocktail lounge, additional	2
Schools, boarding	75–100
Day with cafeteria, gymnasium, and showers	25
Day with cafeteria, without gymnasium and shower	15
Hospitals (per bed)	175–400
Institutions other than hospitals (per bed)	75–125
Places of public assembly	3–10
Turnpike rest areas	5
Turnpike service areas (per 10% of cars passing)	15–20
Prisons	120
Amusement and Commercial	
Airports (per passenger), add for employees and special uses	3–5
Car wash (per vehicle)	40
Country clubs, excluding residents	25
Day workers (per shift)	15–35
Drive-in theaters (per car space)	5
Gas station (per vehicle serviced)	10
Milk plant, pasteurization (per 100 lb of milk)	11–25
Movie theaters (per seat)	3
Picnic parks with flush toilets	5–10
Picnic parks with bathhouse, showers, bathrooms	20
Self-service laundries (per machine) (or 50 gal per customer)	400–500

(continues)

TABLE 1.14   (*continued*)

Type of Establishment	gpd[a]
Shopping center (per 1,000 ft² floor area), add for employees, restaurants, etc.	250
Stores (per toilet room)	400
Swimming pools and beaches with bathhouses	10
Fairgrounds (based on daily attendance), also sports arenas	5
Farming (per Animal)	
Cattle or steer	12
Milking cow, including servicing	35
Goat or sheep	2
Hog	4
Horse or mule	12
Cleaning milk bulk tank, per wash	30–60
Milking parlor, per station	20–30
Liquid manure handling, cow	1–3
Poultry (per 100)	
Chickens	5–10
Turkeys	10–18
Cleaning and sanitizing equipment	4

Miscellaneous Home Water Use	Estimated (gal)
Toilet, tank, per use[b]	1.6–3.5
Toilet, flush valve 25 psi (pounds per square inch), per use[b]	1.6–3.5
Washbasin, gpm[b]	2–3
Bathtub	30/use
Shower, gpm[b]	2.5–3
Dishwashing machine, domestic, 15.5/load	9.5–
Garbage grinder, 2/day	1–
Automatic laundry machine, domestic	
34–57/load, top load	
22–33/load, front load	
Garden hose	
$\frac{5}{8}$ in., 25-ft head	200/hr
$\frac{3}{4}$ in., $\frac{1}{4}$ in. nozzle, 25-ft head	300/hr
Lawn sprinkler, 3,000-ft² lawn, 1 in. per week	120/hr
Air conditioner, water-cooled, 3-ton, 8 hr per day	1,850/week
	2,880/day

Household Water Use	Percent	Municipal Water Use	Percent
Toilet flushing	36	Residential	38
Bathing	26	Industrial: factories	27
Drinking and cooking	5	Commercial: hospitals, restaurants	19
Dishwashing	6	Public: fires, parks	6
Clothes washing	15	Waste: leaks	10
Cleaning and miscellaneous	12		

TABLE 1.14 (*continued*)

Water Demand per Dwelling Unit: Surburban, Three-Bedrooms (BR)	Water Use (gpd)
Average day	300
Maximum day	600
Maximum hourly rate	1500
Maximum hourly rate with appreciable lawn watering	1800

Home Water System (Minimums)	2 BR	3 BR	4 BR	5 BR
Pump capacity, gal/hr	250	300	360	450
Pressure tank, gal minimum	42	82	82	120
Service line from pump, diameter (in.)[c]	$\frac{3}{4}$	$\frac{3}{4}$	1	$1\frac{1}{4}$

Other Water Use	Gallons
Fire hose, $1\frac{1}{2}$ in., $\frac{1}{2}$ in. nozzle, 70-ft head	2400/hr
Drinking fountain, continuous flowing	75/hr
Dishwashing machine, commercial	
Stationary rack type, 15 psi	6–9/min
Conveyor type, 15 psi	4–6/min
Fire hose, home, 10 gpm at 60 psi for 2 hr, $\frac{3}{4}$ in.	600/hr
Restaurant, average	35/seat
Restaurant, 24-hr	50/seat
Restaurant, tavern	20/seat
Gas station	500/set of pumps

Developing Areas of the World
One well or tap/200 persons; controlled tap or hydrant:
Fordilla or Robovalve type
Average consumption, 5 gal/capita/day at well or tap, water carried
Water system design, 30 gal/capita/day (10 gal/capita is common)
(50 gal is recommended)
Pipe size, 2 in. and preferably larger (1 and $1\frac{1}{2}$ in. common)
Drilled well, cased, 6–8 in. diameter
Water system pressure, 20 lb/in.^{2}
(Keep mechanical equipment to a minimum.)

Developing Country[d]	Liters	Gallons
China	80	21
Africa	15–35	4–9
Southeast Asia	30–70	8–19
Western Pacific	30–90	8–24
Eastern Mediterranean	40–85	11–23
Europe (Algeria, Morocco, Turkey)	20–65	5–17
Latin America and Caribbean	70–190	19–51
World average	35–90	9–24

[a] Per person unless otherwise stated.
[b] Water conservation fixtures. See text.
[c] Service lines less than 50 ft long, brass or copper. Use next larger size if iron pipe is used. Use minimum $\frac{1}{3}$-in. service with flush valves. Minimum well yield, 5 gal/min.
[d] Assumes hydrant or hand pump available within 200 m; 70 liters per capita per day (Lpcd) or more could mean house or central courtyard outlet.) Mechanical equipment kept at a minimum.

universal metering and price adjustment; conservation practices by the consumer; and a rate structure that encourages conservation.

Leak detection activities would include metering water use and water production balance studies; routine leak detection surveys of the distribution system; investigation of water ponding or seepage reports and complaints; and reporting and prompt follow-up on leaking faucets, running flushometer valves and water closet ball floats, and other valves. Universal metering will make possible water balance studies to help detect lost water and provide a basis for charging for water use. Meters must be periodically tested for accuracy and read. However, centralized remote meter reading can simplify this task. Reduction in water use, perhaps 20 percent, may be temporary in some instances; many users may not economize.

Low water-use plumbing fixtures and accessories would include the low-flush water closets; water-saving shower-head flow controls, spray taps, and faucet aerators; and water-saving clothes washers and dishwashers. In a dormitory study at a state university, the use of flow control devices (pressure level) on shower heads effected a 40 to 60 percent reduction in water use as a result of reducing the shower-head flow rates from 5.5 gpm to 2.0 to 2.5 gpm.[125] Plumbing codes should require water-saving fixtures and pressure control in new structures and rehabilitation projects. For example, only water-efficient plumbing fixtures meeting the following standards are permitted to be sold or installed in New York State*:

Sink 3 gpm, lavatory faucet not greater than 2 gpm;

Shower heads not greater than 3 gpm;

Urinals and associated flush valve, if any, not greater than 1 gal of water per flush;

Toilets and associated flush valve, if any, not greater than 1.6 gal of water per flush

Drinking fountains, sinks, and lavatories in public restrooms with self-closing faucets[126]

Special fixtures such as safety showers and aspirator faucets are exempt, and the commissioner may permit use of fixtures not meeting standards if necessary for proper operation of the existing plumbing or sewer system.

On March 1, 1989, Massachusetts became the first state to require ultra-low-flow toilets using 1.6 gal per flush. The federal government adopted (effective January 1991) the following standards[127]:

Toilets 1.6 gal per flush

Urinals 1.0 gal per flush

---

*The Washington Suburban Sanitary District plumbing code has similar requirements. (R. S. McGarry and J. M. Brusnighan, "Increasing Water and Sewer Rate Schedules: A Tool for Conservation," *J. Am. Water Works Assoc.* (September 1979): 474–479.) The National Small Flows Clearinghouse, West Virginia University, reported in *Small Flows*, July 1991, that 12 states have adopted low-flow plumbing fixture regulations.

Showerheads 2.5 gpm

Lavatory faucets 2.0 gpm

Kitchen faucets 2.5 gpm

An ultra-low-flush toilet using 0.8 gal per flush was found to perform equal to or better than the conventional toilet.[128] One might also add to the list of water conservation possibilities, where appropriate, use of the compost toilet, recirculating toilet, chemical toilet, incinerator toilet, and various privies. Air-assisted half-gallon flush toilets are also available.[129]

Pressure-reducing valves in the distribution system (pressure zones) to maintain a water pressure of 20 to 40 psi at fixtures will also reduce water use. A water saving of 6 percent can be expected at new single-family homes where water pressure in the distribution system is reduced from 80 to 30 to 40 psi based on HUD studies.[130] The potential water saving through pressure control is apparent from the basic hydraulic formulas:

$$Q = VA \quad Q = (2gpw)^{1/2} \times A \quad Q = (2gh)^{1/2} \times A$$

where

Q = cfs

V = fps

A = ft^2

g = 32.2 ft/sec/sec

p = lb/ft^2

w = lb/ft^3 (62.4)

h = ft of water

which show that the quantity of water flowing through a pipe varies with the velocity or the square root of the pressure head. For example, a pressure reduction from 80 to 40 psi will result in a flow reduction of 29 percent, but the actual water savings would probably be 6 percent, as previously noted.

The success of water-use conservation also depends largely on the extent to which consumers are motivated. They can be encouraged to repair leaking faucets and running toilets immediately; to not waste water; to understand that a leak causing a 1/8-inch-diameter stream adds up to 400 gal in 24 hours, which is about the amount of water used by a family of five or six in one day; to purchase a water-saving clothes washer and dishwasher; to add 1-liter bottles or a "dam" to the flush tank to see if the closet still flushes properly; to install water-saving shower heads and not use the tub; to install mixing faucets with single-lever control; and to install aerators on faucets. Consumer education and motivation must be a continuing activity. In some instances, reuse of shower, sink, and laundry wastewater for gardens is feasible.[131]

Water Reuse

An additional way of conserving drinking water and avoiding or minimizing large capital expenditures is to reduce or eliminate its use for nonpotable purposes by substituting treated municipal wastewater. This could increase the available supply for potable purposes at least cost and reduce the wastewater disposal problem. However, a distinctly separate nonpotable water system and monitoring protocol would be required.

Discussion of wastewater reuse should clearly distinguish between direct reuse and indirect reuse. In *direct reuse*, the additional wastewater treatment (such as storage, coagulation, flocculation, sedimentation, sand or anthracite filtration or granular activated-carbon filtration, and disinfection) is usually determined by the specific reuse. The wastewater is reclaimed for *nonpotable* purposes such as industrial process or cooling water, agricultural irrigation, groundwater recharge, desert reclamation, and fish farming; lawn, road median, tree farm, and park irrigation; landscape and golf-course watering; and toilet flushing. The treated wastewater must *not* be used for drinking, culinary, bathing, or laundry purposes. The long-term health effects of using treated wastewater for potable purposes are not fully understood at this time, and fail-safe, cost-effective treatment technology for the removal of all possible contaminants is not currently available.[132] In *indirect reuse*, wastewater receiving various degrees of treatment is discharged to a surface water or a groundwater aquifer, where it is diluted and after varying detention periods and treatment may become a source of water for potable purposes. *Recycling* is the reuse of wastewater, usually by the original user.

Direct municipal wastewater reuse, where permitted, would require a clearly marked dual water system, one carrying potable water and the other reclaimed wastewater. It has been estimated that the average person uses only about 25 to 55 gal of water per day for potable purposes.[133] The reclaimed water is usually bacteriologically safe but questionable insofar as other biological or organic and inorganic chemical content is concerned. A dye added to the reclaimed water would help avoid its inadvertent use for potable purposes. Okun emphasizes that the reclaimed or nonpotable water should.

> equal the quality of the potable systems that many communities now provide—the health hazard that results from the continuous ingestion of low levels of toxic substances over a period of years would not be present.[134]

Advanced wastewater treatment, monitoring, and surveillance cannot yet in practice guarantee removal of all harmful substances (microcontaminants) from wastewater at all times. However, numerous projects are now investigating reuse of water for potable purposes.[135]* More knowledge is needed concerning acute and long-term effects on human health of wastewater reuse.[136] In Windhoek, Namibia, Southwest Africa, reclaimed sewage, which is reported to contain no industrial wastes, blended with water from conventional sources has occasionally been used for drinking for many years without any apparent problems. The sewage is given very elaborate treatment involving some 18 unit processes.[137]

---

*The July 1985 issue of the *Journal of the American Water Works Association* is devoted to wastewater reuse.

Monitoring is done for *Salmonella, Shigella*, enteropathogenic *E. coli, Vibrio*, enterovirus, *Schistosoma*, viral hepatitis, meningitis, and nonbacterial gastroenteritis, in addition to turbidity and organic and inorganic chemicals. None of the pathogens was associated with the reclaimed wastewater.

More emphasis is needed on the removal of hazardous substances at the source and on adequate wastewater treatment prior to its discharge to surface and underground water supply sources. This will at least reduce the concentrations of contaminants discharged from urban and industrial areas and, it is hoped, the associated risks.

In any case, it is axiomatic that, in general, the cleanest surface and underground water source available should be used as a source of drinking water, and water conservation practiced, before a polluted raw water source is even considered, with cost being secondary.

SOURCE AND PROTECTION OF WATER SUPPLY

General

The sources of water supply are divided into two major classifications: groundwater and surface water. To these should be added rainwater and demineralized water. The groundwater supplies include dug, bored, driven and drilled wells, rock and sand or earth springs, and infiltration galleries. The surface-water supplies include lake, reservoir, stream, pond, river, and creek supplies.

The location of groundwater supplies should take into consideration the recharge tributary wellhead area,[138] the probable sources and travel of pollution through the ground, the well construction practices and standards actually followed, depth of well casing and grouting, and the type of sanitary seal provided at the point where the pump line(s) pass out of the casing.

Wellhead area has been defined under the 1986 Amendments to the Safe Drinking Water Act as "the surface and subsurface area surrounding a water well or wellfield, supplying a public water system, through which contaminants are reasonably likely to move toward and reach such water well or wellfield." The time of travel of a potential contaminant, distance, drawdown, flow boundaries, and assimilative capacity are critical factors in determining the wellhead protection area.[139] Some of the other hydrogeological considerations, in addition to well drawdown, radius of influence,* withdrawal rate, recharge area, and aquifer formation, are the hydraulic gradient, natural dilution, filtration, attenuation, and degradation of the contaminant in its movement through the zone of aeration (unsaturated zone) to the saturated zone and into the water table of the wellhead drainage area. These factors must be evaluated in the light of available topographic, geologic, and engineering information and the practicality of land-use controls, conservation easements, and dedication of land to parks to

---

*Circular only with flat water table, when drawdown cone of depression is 99 percent stabilized.

effectively prevent or adequately minimize the potential effects of contaminants on the recharge area. See earlier discussion under "Sanitary Survey and Water Sampling."

The chemical quality of shallow groundwater (8–20 ft) and its quantity can be expected to vary substantially throughout the year and after heavy rains, depending on the soil depth and characteristics in the unsaturated zone above the water table.

It is sometimes suggested that the top of a well casing should terminate below the ground level or in a pit. This is not considered good practice except when the pit can be drained above flood level to the surface by gravity or to a drained basement. Frost-proof sanitary seals with pump lines passing out horizontally from the well casing are generally available. Some are illustrated later in Figures 1.7 through 1.10.

In order that the basic data on a new well may be recorded, a form such as the well driller's log and report shown in Figure 1.4 should be completed by the well driller and kept on file by the owner for future reference. A well for a private home should preferably have a capacity (well yield) of at least 500 gal/hr, but 300 gal/hr is usually specified as a minimum for domestic water use in serving a three-bedroom home. The long-term yield of a well is dependent on the seasonal static water level, other withdrawals from the aquifer, the recharge area and storage in the aquifer, and the hydraulic characteristics of the aquifer. Because of this and the uncertainty of when stabilized drawdown is reached, the determined well yield should be reduced to compensate for long-term use and possible decline of aquifer yield. Pumping tests should therefore ensure that the water level in the well returns to the original static level. See Tables 1.14 and 1.15.

Surface-water supplies are all subject to continuous or intermittent pollution and must be treated to make them safe to drink. One never knows when the organisms causing typhoid fever, gastroenteritis, giardiasis, infectious hepatitis A, or dysentery, in addition to organic and inorganic pollutants, may be discharged or washed into the water source. The extent of the treatment required will depend on the results of a sanitary survey made by an experienced professional, including physical, chemical, and microbiological analyses. The minimum required treatments are coagulation, flocculation, sedimentation, filtration, and chlorination, unless a conditional waiver is obtained from the regulatory agency. If more elaborate treatment is needed, it would be best to abandon the idea of using a surface-water supply and resort to a protected groundwater supply if possible and practical. Where a surface supply must be used, a reservoir or a lake that provides at least 30 days *actual* detention, that does not receive sewage, industrial, or agricultural pollution, and that can be controlled through ownership or watershed rules and regulations would be preferred to a stream or creek, the pollution of which cannot from a practical standpoint be controlled. There are many situations where there is no practical alternative to the use of polluted streams for water supply. In such cases, carefully designed water-treatment plants providing multiple barriers must be provided.

Well at _____ in _____ County of _____
    Name of place   City, village or town

Owner _____ P.O. Address _____

Depth of well _____ Diameter _____ Yield _____ Was well disinfected? _____
        ft   in.   gpm   yes or no

Amt. of casing above ground _____ Below ground _____ Well seal _____
                   in.   ft   cement grout

Draw a well diagram in the space provided below and show the depth and type of casing, the well seal, kind and thickness of formations penetrated, water-bearing formations, diameter of drill holes with dotted lines, and casing(s) with solid lines.

Well Diagram		Formations Penetrated	Remarks			
Diameter, in.	Depth in ft	Kind, thickness, and if water bearing	Type of well _____ Drilling method _____			
⏐ ⏐ ⏐ ⏐ ⏐ ⏐ ⏐	Grade		Was well dynamited? _____			
			Pumping Tests			
			Details	#1	#2	#3
	25		Static water level, in feet below grade			
	50		Pumping rate in gpm			
	75		Pumping level in feet below grade			
			Duration of test, in hours			
	100		*Water at end of test:* Clear ___ Cloudy ___ Turbid ___			
	150		Recommended depth of pump in well, ft below grade ___ Capacity ___ gpm			
	200		*Wells in sand & gravel:* Sand Eff. size ___ mm Unif. Coef. _____			
	250		Length of screen ___ ft Diam. of screen ___ in. Type of screen _____ Screen openings ___ x ___			
			Comments:			

Show cross-section of well & formations penetrated above. Draw a sketch of the property on the back of this sheet locating the well and sewage disposal systems within 200 ft, also land uses.	Drilling started _____ Completed _____ Well driller _____ *Signature*

FIGURE 1.4 Well driller's log and report. *Well yield* is the volume of water per unit of time, such as gallons per minute, discharged from a well either by pumping to a stabilized drawdown or by free flow. The *specific capacity* of a well is the yield at a stabilized drawdown and given pumping rate, expressed as gallons per minute per foot of drawdown. Chalked tape, electric probe, or known length of air line is used with pressure gauge. Test run is usually 4 to 8 hours for small wells; 24 to 72 hours for wells serving the public, or for 6 hours at a stabilized drawdown when pumping at 1.5 times the design pumping rate.

TABLE 1.15 Standards for Construction of Wells[a]

| Water-Bearing Formation | Overburden | Oversize Drill Hole | | Cased Portion |
		Diameter	Depth[b]	
1. Sand or gravel	Unconsolidated caving material; sand or sand and gravel	None required	None	2 in. minimum, 5 in. or more preferred
2. Sand or gravel	Clay, hardpan, silt, or similar material to depth of more than 20 ft	Casing size plus 4 in.	Minimum 20 ft	2 in. maximum, 5 in. or more preferred
3. Sand or gravel	Clay, hardpan, silt, or similar material containing layers of sand or gravel within 15 ft of ground surface	Casing size plus 4 in.	Minimum 20 ft	2 in. minimum, 5 in or more preferred
4. Sand or gravel	Creviced or fractured rock, such as limestone, granite, quartzite	Casing size plus 4 in.	Through rock formation	4 in. minimum
5. Creviced, shattered, or otherwise fractured limestone, granite, quartzite, or similar rock types	Unconsolidated caving material, chiefly sand or sand and gravel to a depth of 40 ft or more and extending at least 2000 ft in all directions from the well site	None required	None required	6 in. minimum

| Well Diameter | | | | | |
Uncased Portion	Well Screen Diameter[c]	Minimum Casing Length or Depth[b]	Liner Diameter (If Required)	Construction Conditions[b]	Miscellaneous Requirements
Does not apply	2 ft minimum	20 ft minimum; but 5 ft below pumping level[d]	2 in. minimum		
Does not apply	2 ft minimum	5 ft below pumping level[d]	2 in. minimum	Upper drill hole shall be kept at least one-third filled with clay slurry while driving permanent casing; after casing is in permanent position annular space shall be filled with clay slurry or cement grout.	An adequate well screen shall be provided where necessary to permit pumping sand-free water from the well.
Does not apply	2 ft minimum	5 ft below pumping level[d]	2 in. minimum	Annular space around casing shall be filled with cement grout.	
Does not apply	2 in. minimum	5 ft below overburden of rock	2 in. minimum	Annular space around casing shall be filled with cement grout.	
6 in. preferred	Does not apply	Through caving overburden	4 in. minimum	Casing shall be firmly seated in the rock.	

(continues)

TABLE 1.15 (*continued*)

| Water-Bearing Formation | Overburden | Oversize Drill Hole | | Cased Portion |
		Diameter	Depth[b]	
6. Creviced, shattered, or otherwise fractured limestone, granite, quartzite, or similar rock types	Clay, hardpan, shale, or similar material to a depth of 40 ft or more and extending at least 2000 ft in all directions from well site	Casing size plus 4 in.	Minimum 20 ft	6 in. minimum
7. Creviced, shattered, or otherwise fractured limestone, granite, quartzite, or similar rock	Unconsolidated materials to a depth of less than 40 ft and extending at least 2,000 ft in all directions	Casing size plus 4 in.	Minimum 40 ft	6 in. minimum
8. Sandstone	Any material except creviced rock to a depth of 25 ft or more	Casing size plus 4 in.	15 ft into firm sandstone or to 30 ft depth, whichever is greater	4 in. minimum
9. Sandstone	Mixed deposits mainly sand and gravel, to a depth of 25 ft or more	None required	None required	4 in. minimum

Well Diameter		Minimum Casing Length or Depth[b]	Liner Diameter (If Required)	Construction Conditions[b]	Miscellaneous Requirements
Uncased Portion	Well Screen Diameter[c]				
6 in. preferred	Does not apply	Through overburden	4 in. minimum	Annular space around casing shall be grouted. Casing shall be firmly seated in rock.	
6 in. preferred	Does not apply	40 ft minimum	4 in. minimum	Casing shall be firmly seated in rock. Annular space around casing shall be grouted.	If grout is placed through casing pipe and forced into annular space from the bottom of the casing, the oversize drill hole may be only 2 in. larger than the casing pipe.
4 in. preferred		Same as oversize drill hole or greater	2 in. minimum	Annular space around casing shall be grouted. Casing shall be firmly seated in sandstone.	Pipe 2 in. smaller than the drill hole and liner pipe 2 in. smaller than casing shall be assembled without couplings.
4 in. preferred		Through overburden into firm sandstone	2 in. minimum	Casing shall be effectively seated into firm sandstone.	

(*continues*)

TABLE 1.15   (*continued*)

| Water-Bearing Formation | Overburden | Oversize Drill Hole | | Cased Portion |
		Diameter	Depth[b]	
10. Sandstone	Clay, hardpan, or shale to a depth of 25 ft or more	Casing size plus 4 in.	Minimum 20 ft	4 in. minimum

Well Diameter		Minimum Casing Length or Depth[b]	Liner Diameter (If Required)	Construction Conditions[b]	Miscellaneous Requirements
Uncased Portion	Well Screen Diameter[c]				
4 in. preferred		Through overburden into sandstone	2 in. minimum	Casing shall be effectively seated into firm sandstone. Oversized drill hole shall be kept at least one-third filled with clay slurry while driving permanent casing; after the casing is in the permanent position, annular space shall be filled with clay slurry or cement grout.	Pipe 2 in. smaller than the oversize drill hole and liner pipe 2 in. smaller than casing shall be assembled without couplings.
	2 in. minimum, if well screen required to permit pumping sand-free water from partially cemented sandstone				

(continues)

TABLE 1.15 (*continued*)

| Water-Bearing Formation | Overburden | Oversize Drill Hole | | Cased Portion |
		Diameter	Depth[b]	
11. Sandstone	Creviced rock at variable depth	Casing size plus 4 in.	15 ft or more into firm sandstone	6 in. minimum

Note: For wells in creviced, shattered, or otherwise fractured limestone, granite, quartzite, or similar rock in which the overburden is less than 40 ft and extends less than 2,000 ft in all directions and no other practical acceptable water supply is available, the well construction described in line 7 of this table is applicable.

[a]Requirements for the proper construction of wells vary with the character of subsurface formations, and provisions applicable under all circumstances cannot be fixed. The construction details of this table may be adjusted, as conditions warrant, under the procedure provided by the Health Department and in the Note above.

[b]In the case of a flowing artesian well, the annular space between the soil and rock and the well casing shall be tightly sealed with cement grout from within 5 ft of the top of the aquifer to the ground surface in accordance with good construction practice.

[c]These diameters shall be applicable in circumstances where the use of perforated casing is deemed practicable. Well points commonly designated in the trade as $1\frac{1}{4}$-in. pipe shall be considered as being 2 ft nominal diameter well screens for purposes of these regulations.

[d]As used herein, the term *pumping level* shall refer to the lowest elevation of the surface of the water in a well during pumping, determined to the best knowledge of the water well contractor, taking into consideration usual seasonal fluctuations in the static water level and drawdown level.

Source: Recommended State Legislation and Regulations, Public Health Service, Department of Health, Education, and Welfare, Washington, DC, July 1965.

Well Diameter					
Uncased Portion	Well Screen Diameter[c]	Minimum Casing Length or Depth[b]	Liner Diameter (If Required)	Construction Conditions[b]	Miscellaneous Requirements
6 in. preferred		15 ft into firm sandstone	4 in. minimum	Annular space around casing shall be filled with cement grout.	If grout is placed through casing pipe and forced into annular space from the bottom of the casing, the oversize drill hole may be only 2 in. larger than the casing pipe. Pipe 2 in. smaller than the drill hole and liner pipe 2 in. smaller than casing shall be assembled without couplings.

Groundwater

About one-half of the U.S. population depends on groundwater for drinking and domestic purposes; 98 percent of the rural population is almost entirely dependent on groundwater. Some 43.5 million people are served by individual, on-site well-water systems (2000 USGS). These are not protected or regulated under the Safe Drinking Water Act. In view of this, protection of our groundwater resources must receive the highest priority. Elimination of groundwater pollution and protection of aquifers and their drainage areas by land-use and other controls require state and local regulations and enforcement.

It is estimated that there is more than 100 times more water stored underground than in all the surface streams, lakes, and rivers. Protection and development of groundwater sources can significantly help meet the increasing water needs. Exploration techniques include use of data from USGS and state agencies, previous studies, existing well logs, gains or losses in stream flow, hydrogeologic mapping using aerial photographs, surface resistivity surveys electromagnetic induction surveys or other geophysical prospecting, and exploratory test wells.

A technique for well location called *fracture-trace mapping* is reported to be a highly effective method for increasing the ratio of successful to unsuccessful well-water drilling operations and to greatly improve water yields (up to 50 times). Aerial photographs give the skilled hydrogeologist clues of the presence of a zone of fractures underneath the earth's surface. Clues are abrupt changes in the alignment of valleys, the presence of taller or more lush vegetation, the alignment of sink holes or other depressions in the surface, or the existence of shallow, longitudinal depressions in the surface overtop of the fracture zone. The soil over fracture zones is often wetter and, hence, shows up darker in recently plowed fields. The aerial photograph survey is then followed by a field investigation and actual ground location of the fractures and potential well-drilling sites.[140]

It has been suggested that all groundwater supplies be chlorinated. Exceptions may be properly located, constructed, and protected wells *not* in limestone or other channeled or fractured rock and where the highest water table level is at least 10 feet below ground level; where sources of pollution are more than 5,000 feet from the well; and where there is a satisfactory microbiological history. Other criteria include soil permeability, rate and direction of groundwater flow, and underground drainage area to the well. Chlorination should be considered a factor of safety and not reason to permit poor well construction and protection.

Dug Well

A dug well is one usually excavated by hand, although it may be dug by mechanical equipment. It may be 3 to 6 feet in diameter and 15 to 35 feet deep, depending on where the water-bearing formation or groundwater table is encountered. Wider and deeper wells are less common. Hand pumps over wells and pump lines entering wells should form watertight connections, as shown in Figure 1.5 and Figure 1.6. Since dug wells have a relatively large diameter, they have large storage capacity. The level of the water in dug wells will lower at times of

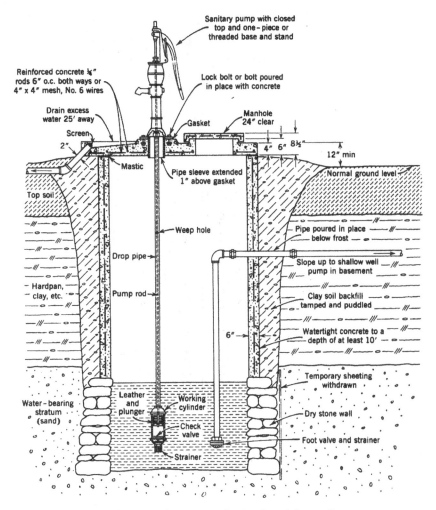

FIGURE 1.5   A properly developed dug well.

drought and the well may go dry. Dug wells are not usually dependable sources of water supply, particularly where modern plumbing is provided. In some areas, properly developed dug wells provide an adequate and satisfactory water supply. However, dug wells are susceptible to contamination deposited on or naturally present in the soil when subjected to heavy rains, particularly if improperly constructed. This potential hazard also applies to shallow bored, driven, and jetted wells. Water quality can be expected to change significantly.

Bored Well

A bored well is constructed with a hand- or machine-driven auger. Bored wells vary in diameter from 2 to 30 inches and in depth from 25 to 60 feet. A casing

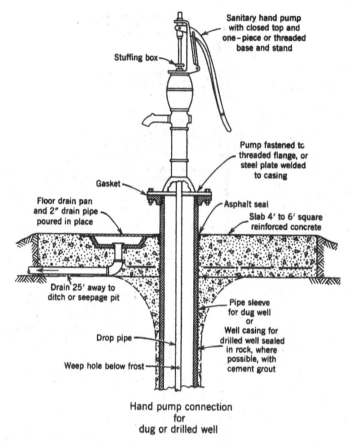

Hand pump connection
for
dug or drilled well

FIGURE 1.6 Sanitary hand pump and well attachment. Place 2 feet of gravel under slab where frost is expected.

of concrete pipe, vitrified clay pipe, metal pipe, or plastic pipe is necessary to prevent the relatively soft formation penetrated from caving into the well. Bored wells have characteristics similar to dug wells in that they have small yields, are easily polluted, and are affected by droughts.

Driven and Jetted Well

These types of wells consist of a well point with a screen attached, or a screen with the bottom open, which is driven or jetted into a water-bearing formation found at a comparatively shallow depth. A series of pipe lengths are attached to the point or screen as it is forced into position. The driven well is constructed by driving the well point, preferably through at least 10 to 20 feet of casing, with the aid of a maul or sledge, pneumatic tamper, sheet pile driver, drive monkey, hand-operated driver, or similar equipment. In many instances, the casing is omitted, but then less protection is afforded the driven well, which also serves as the pump suction line. The jetted well is constructed by directing a stream of water at the bottom of the open screen, thereby loosening and flushing the soil up the casing to the surface as the screen is lowered. Driven wells are commonly between 1-1/4 and 2 inches in diameter and less than 50 feet in depth; jetted wells may be 2 to 12 inches in diameter and up to 100 feet deep, although larger and deeper wells can be constructed. In the small-diameter wells, a shallow well hand or mechanical suction pump is connected directly to the well. Large-diameter driven wells facilitate installation of the pump cylinder close to or below the water surface in the well at greater depth, in which case the hand pump must be located directly over the well. In all cases, however, care must be taken to see that the top of the well is tightly capped, the concrete pump platform extends 2 or 3 feet around the well pipe or casing, and the annular space between the well casing and drop pipe(s) is tightly sealed. This is necessary to prevent the entrance of unpurified water or other pollution from close to the surface.

A radial well is a combination dug-and-driven well in which horizontally driven well collectors radiate out from a central sump or core and penetrate into a water-bearing stratum.

Drilled Well

Studies have shown that, in general, drilled wells are superior to dug, bored, or driven wells and springs. But there are some exceptions. Drilled wells are less likely to become contaminated and are usually more dependable sources of water. When a well is drilled, a hole is made in the ground, usually with a percussion (cable tool) or rotary (air or mud) drilling machine. Drilled wells are usually 4 to 12 inches in diameter or larger and may reach 750 to 1,000 feet in depth or more. Test wells are usually 2 to 5 inches in diameter with a steel casing. A steel or wrought-iron casing is lowered as the well is drilled to prevent the hole from caving in and to seal off water of doubtful quality. Special plastic pipe is also used if approved. Lengths of casing should be threaded and coupled or properly field welded. The drill hole must, of course, be larger than the casing, thereby leaving an irregular space around the outside length of the casing. Unless this space or channel is closed by cement grout or naturally by formations that conform to the casing almost as soon as it is placed, pollution from the surface or crevices close to the surface or from polluted formations penetrated will flow down the side of the casing and into the water source. Water can also move up and down this annular space in an artesian well and as the groundwater and pumping water level changes.

The required well diameter is usually determined by the size of the discharge piping, fittings, pump, and motor placed inside the well casing. In general, for well yields of less than 100 gpm, a 6-inch-inside-diameter casing should be used; for 75 to 175 gpm an 8-inch casing; for 150 to 400 gpm a 10-inch casing; for 350 to 650 gpm a 12-inch casing; for 600 to 900 gpm a 14-inch-outside-diameter casing; for 850 to 1,300 gpm a 16-inch casing; for 1,200 to 1,800 gpm a 20-inch casing; and for 1,600 to 3,000 gpm a 24-inch casing.[141] Doubling the diameter of a casing increases the yield up to only 10 to 12 percent.

When the source of water is water-bearing sand and gravel, a gravel well or gravel-packed well with screen may be constructed. Such a well will usually yield more water than the ordinary drilled well with a screen of the same diameter and with the same drawdown. A slotted or perforated casing in a water-bearing sand will yield only a fraction of the water obtainable through the use of a proper screen selected for the water-bearing material. On completion, the well should be developed and tested, as noted previously. A completed well driller's log should be provided to the owner on each well drilled. See Figure 1.4.

Only water well casing of clean steel or wrought iron should be used. Plastic pipe may be permitted. Used pipe is unsatisfactory. Standards for well casing are available from the American Society for Testing Materials, the American Iron and Steel Institute, and state health or environmental protection agencies.

Extending the casing at least 5 feet below the pumping water level in the well—or if the well is less than 30 feet deep, 10 feet below the pumping level—will afford an additional measure of protection. In this way, the water is drawn from a depth that is less likely to be contaminated. In some sand and gravel areas, extending the casing 5 to 10 feet below the pumping level may shut off the water-bearing sand or gravel. A lesser casing depth would then be indicated, but in no instance should the casing be less than 10 feet, provided sources of pollution are remote and provision is made for chlorination. The recommended depth of casing, cement grouting, and need for double-casing construction or the equivalent are given in Table 1.15.

A vent is necessary on a well because, if not vented, the fluctuation in the water level will cause a change in air pressure above and below atmospheric pressure in a well, resulting in the drawing in of contaminated water from around the pump base over the well or from around the casing if not properly sealed. Reduced pressure in the well will also increase lift or total head and reduce volume of water pumped.

It must be remembered that well construction is a very specialized field. Most well drillers are desirous of doing a proper job, for they know that a good well is their best advertisement. However, in the absence of a state or local law dealing with well construction, the enforcement of standards, and the licensing of well drillers, price alone frequently determines the type of well constructed. Individuals proposing to have wells drilled should therefore carefully analyze bids received. Such matters as water quality, well diameter, type and length of casing, minimum well yield, type of pump and sanitary seal where the pump line(s) passes through the casing, provision of a satisfactory well log, method used to seal off undesirable formations and cement grouting of the well, plans to pump the well until clear, and disinfection following construction should all be taken into consideration. See Figures 1.6 through 1.12.

Recommended water well protection and construction practices and standards are given in this text. More detailed information, including well construction and development, contracts, and specifications, is available in federal, state, and other

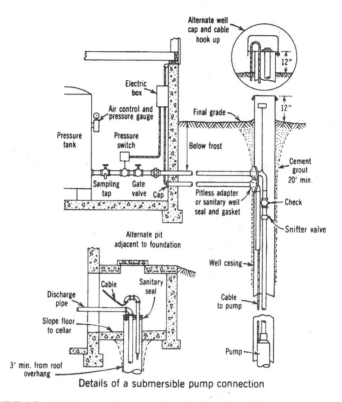

Details of a submersible pump connection

FIGURE 1.7   Sanitary well caps and seal and submersible pump connection.

publications.[142] A hydrogeologist or professional engineer can help assure proper location, construction, and development of a well, particularly for a public water supply. It has been estimated that the radius of the cone of depression of a well in fine sand is 100 to 300 feet, in coarse sand 600 to 1,000 feet, and in gravel 1,000 to 2,000 feet. In a consolidated formation, determination of the radius of the cone of depression requires a careful hydrogeological analysis. Remember, the cheapest well is not necessarily the best buy.

Well Development

Practically all well-drilling methods, and especially the rotary drill method, cause smearing and compaction or cementing of clay, mud, and fine material on the bore hole wall and in the crevices of consolidated formations penetrated. This will reduce the sidewall flow of water into the well and, hence, the well yield. Various methods are used to remove adhering mud, clay, and fines and to develop a well to its full capacity. These include pumping, surging (valved surge device, solid surge device, pumping with surge device, air surge), and fracturing (explosives, high-pressure jetting, backwashing). Adding a polyphosphate or a nonfoaming

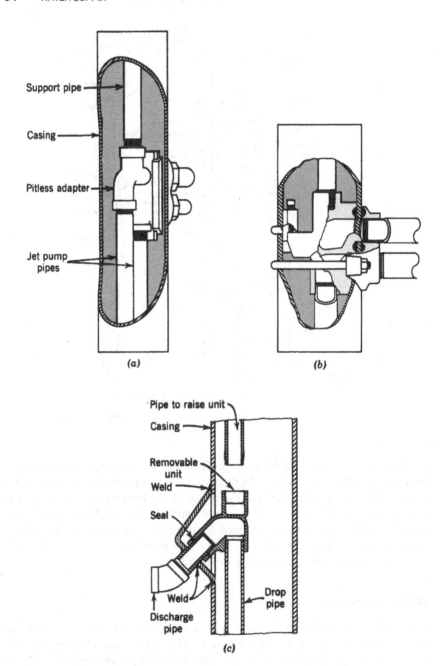

FIGURE 1.8 Pitless adapters. ((*a*) Courtesy Martin Manufacturing Co., Ramsey, NJ. (*b*) Courtesy Williams Products Co., Joliet, IL. (*c*) Courtesy Herb Maass Service, Milwaukee, WI.)

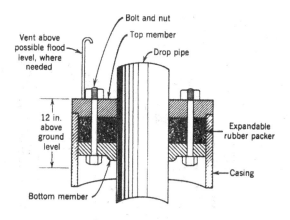

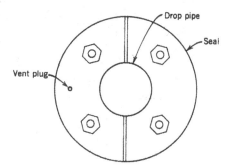

FIGURE 1.9   Sanitary expansion well cap.

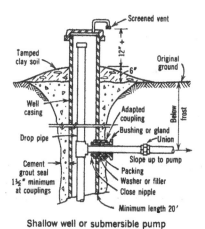

Shallow well or submersible pump
underground connection

FIGURE 1.10   Improved well seal.

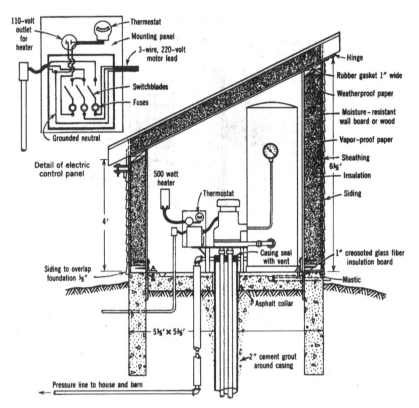

FIGURE 1.11 Insulated pumphouse. (*Source: Sewage Disposal and Water Systems on the Farm*, Extension Bulletin 247, University of Minnesota Extension Service, revised 1956. Reproduced with permission.)

detergent can also aid in removing adhering materials. The well development operation is continued until the discharge becomes practically clear of sand (5 ppm or less). Following development, the well should be tested to determine the dependable well yield. The well is then disinfected and the log completed.

Grouting

One of the most common reasons for contamination of wells drilled through rock, clay, or hardpan is failure to properly seal the annular space around the well casing. A proper seal is needed to prevent water movement between aquifers, protect the aquifers, and prevent entry of contaminated water from the surface or near the surface.

A contaminated well supply causes the homeowner or municipality considerable inconvenience and extra expense, for it is difficult to seal off contamination after the well is drilled. In some cases, the only practical answer is to build a new well.

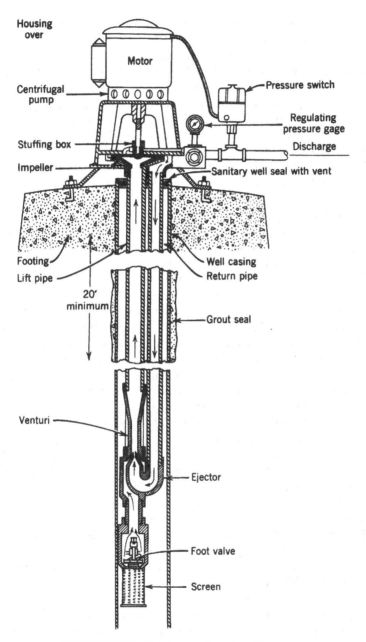

FIGURE 1.12   Sanitary well seal and jet pump.

Proper *cement grouting* of the space between the drill hole and well casing, the annulus, where the overburden over the water-bearing formation is clay, hardpan, or rock, can prevent this common cause of contamination. (See Table 1.15.)

There are many ways to seal well casings. The best material is neat cement grout.* However, to be effective, the grout must be properly prepared (a proper mixture is 5-1/2 to 6 gal of clean water to a bag of cement), pumped as one continuous mass, and placed upward from the bottom of the space to be grouted. An additive such as bentonite may be used to minimize shrinkage and increase fluidity, if approved.

The clear annular space around the outside of the casing couplings and the drill hole must be at least 1-1/2 in. on all sides to prevent bridging of the grout. Guides must be welded to the casing.

Cement grouting of a well casing along its entire length of 50 to 100 feet or more is good practice but expensive for the average farm or rural dwelling. An alternative is grouting to at least 20 feet below ground level. This provides protection for most installations, except in limestone and fractured formations. It also protects the casing from corrosion.

For a 6-inch-diameter well, a 10-inch hole is drilled, if 6-inch welded pipe is used, to at least 20 feet or to solid rock if the rock is deeper than 20 feet. If 6-inch coupled pipe is used, a 12-inch hole will be required. From this depth the 6-inch hole is drilled deeper until it reaches a satisfactory water supply. A temporary outer casing, carried down to rock, prevents cave-in until the cement grout is placed.

Upon completion of the well, the annular space between the 6-inch casing and temporary casing or drill hole is filled from the bottom up to the grade with cement grout. The temporary pipe is withdrawn as the cement grout is placed—it is not practical to pull the casing after all the grout is in position.

The extra cost of the temporary casing and larger drill hole is small compared to the protection obtained. The casing can be reused as often as needed. In view of this, well drillers who are not equipped should consider adding larger casing and equipment to their apparatus.

A temporary casing or larger drill hole and cement grouting are not required where the entire earth overburden is 40 feet or more of silt or sand and gravel, which immediately close in on the total length of casing to form a seal around the casing; however, this condition is not common.

Drilled wells serving public places are usually constructed and cement grouted as explained in Table 1.15.

In some areas, limestone and shale beneath a shallow overburden represent the only source of water. Acceptance of a well in shale or limestone might be conditioned on an extended observation period to determine the sanitary quality of the water. Continuous chlorination should be required on satisfactory supplies serving the public and should be recommended to private individuals. However,

---

*Sand–cement grout, two parts sand to one part Portland cement by weight, with not more than 6 gal of water per sack of cement, may also be used. The curing time for neat cement is 72 hr; for high early strength cement, at least 36 hr.

chlorination should not be relied on to make a heavily contaminated well-water supply satisfactory. Such supplies should be abandoned and filled in with concrete or puddled clay unless the source of contamination can be eliminated.

Well drillers may have other sealing methods suitable for particular local conditions, but the methods just described utilizing a neat cement or sand-cement grout will give reasonably dependable assurance that an effective seal is provided, whereas this cannot be said of some of the other methods used. Driving the casing, a lead packer, drive shoe, rubber sleeves, and similar devices do not provide reliable annular space seals for the length of the casing.

Well Contamination — Cause and Removal

Well-water supplies are all too often improperly constructed, protected, or located, with the result that microbiological examinations show the water to be contaminated. Under such conditions, all water used for drinking or culinary purposes should first be boiled or adequately treated. Boiling will not remove chemical contaminants other than volatiles; treatment may remove some. If practical, abandonment of the well and connection to a public water supply would be the best solution. A second alternative would be investigation to find and remove the cause of pollution; however, if the aquifer is badly polluted, this may take considerable time. A third choice would be a new, properly constructed and located drilled well in a clean aquifer. See "Travel of Pollution through the Ground," earlier in this chapter.

When a well shows the presence of bacterial contamination, it is usually due to one or more of four probable causes: lack of or improper disinfection of a well following repair or construction; failure to seal the annular space between the drill hole and the outside of the casing; failure to provide a tight sanitary seal at the place where the pump line(s) passes through the casing; and wastewater pollution of the well through polluted strata or a fissured or channeled formation. On some occasions, the casing is found to be only a few feet in length and completely inadequate. Chemical contamination usually means the aquifer has been polluted.

If a new well is constructed or if repairs are made to the well, pump, or piping, contamination from the work is probable. The well, pump, storage tank, and piping should be disinfected, as explained in this chapter.

If a sewage disposal system is suspected of contamination, a dye such as water-soluble sodium or potassium fluorescein or ordinary salt can be used as a tracer. A solution flushed into the disposal system or suspected source may appear in the well water within 12 to 24 hours. It can be detected by sight, taste, or analysis if a connection exists. Samples should be collected every few hours and set aside for comparison. If the connection is indirect, fluoroscopic or chemical examination for the dye or chlorides is more sensitive. One part of fluorescein in 10 to 40 million parts of water is visible to the naked eye, and in 10 billion parts if viewed in a long glass tube or if concentrated in the laboratory. The chlorides in the well before adding salt should, of course, be known. Where chloride determinations are routinely made on water samples, sewage pollution may be apparent without making the salt test. Dye is not decolorized by passage through sand, gravel, or manure; it is slightly decomposed by calcareous soils and entirely decolorized by peaty formations and free acids, except carbonic acid.[143] A copper sulfate solution (300 mg/1), nonpathogenic bacteria and spores, radionuclides, strong electrolytes, and nonfluorescent dyes have also been used. Dyes include congo red, malachite green, rhodamine, pyranine, and photine.[144]

If the cause of pollution is suspected to be an underground seal where the pump line(s) passes through the side of the casing, a dye or salt solution or even plain water can be poured around the casing. Samples of the water can be collected for visual or taste test or chemical examination. The seal might also be excavated for inspection. Where the upper part of the casing can be inspected, a mirror or strong light can be used to direct a light beam inside the casing to see if water is entering the well from close to the surface. Sometimes it is possible to hear the water dripping into the well. Inspection of the top of the well will also show if the top of the casing is provided with a sanitary seal and whether the well is subject to flooding. See Figures 1.7 to 1.12.

The path of pollution entry can also be holes in the side of the casing, channels along the length of the casing leading to the well source, crevices or channels connecting surface pollution with the water-bearing stratum, or the annular space around the casing. A solution of dye, salt, or plain water can be used to trace the pollution, as previously explained.

The steps taken to provide a satisfactory water supply would depend on the results of the investigation. If a sanitary seal is needed at the top or side of the casing where the pump lines pass through, then the solution is relatively simple. On the other hand, an unsealed annular space is more difficult to correct. A competent well driller could be engaged to investigate the possibility of grouting the annular space and installing an inner casing or a new casing carefully sealed in solid rock. If the casing is found tight, it would be assumed that pollution is finding its way into the water-bearing stratum through sewage-saturated soil or creviced or channeled rock at a greater depth. It is sometimes possible, but costly, to seal off the polluted stratum and, if necessary, drill deeper.

Once a stratum is contaminated, it is very difficult to prevent future pollution of the well unless all water from such a stratum is effectively sealed off. Moving the offending sewage disposal system to a safe distance or replacing a leaking oil or gasoline tank is possible, but evidence of the pollution may persist for some time.

If a dug well shows evidence of contamination, the well sidewalls may be found to consist of stone or brick lining, which is far from being watertight. In such cases, the upper 6 to 10 feet should be removed and replaced with a poured concrete lining and platform. As an alternative, a concrete collar 6 to 12 inches thick, 6 to 10 feet deep could be poured around the *outside* of the stone or brick lining (see Figure 1.5). Take safety precautions (see *Safety* in Index).

Chemical contamination of a well and the groundwater aquifer can result from spills, leaking gasoline and oil tanks, or improper disposal of chemical wastes such as by dumping—on the ground in landfills—lagooning, or similar methods. Gasoline and oil tanks typically have a useful life of about 20 years, depending on the type of soils and tank coatings. Since many tanks have been in the ground 20 to 30 years or longer, their integrity must be uncertain and they are probably leaking to a greater or lesser degree. New tanks are not necessarily immune from leakage. If not already being done, oil, gasoline, and other buried tanks containing hazardous chemicals should be tested periodically and, of course, at the first sign of leakage promptly replaced with approved tanks. The number of tanks, surreptitious dumpings, discharges to leaching pits, and other improper disposals make control a formidable task. This subject is discussed further in this chapter; see "Groundwater Pollution Hazard" and "Travel of Pollution through the Ground."

Unless all the sources of pollution can be found and removed, it is recommended that the well be abandoned and filled with neat cement grout, puddled clay, or concrete to prevent the pollution from traveling to other aquifers or wells. In some special cases and under controlled conditions, use of a slightly contaminated water supply may be permitted provided approved treatment facilities are installed. Such equipment is expensive and requires constant attention. If a public water supply is not available and a new well is drilled, it should be located and constructed as previously explained.

Spring

Springs are broadly classified as either rock springs or earth springs, depending on the source of water. To obtain satisfactory water, it is necessary to *find the source*, properly develop it, eliminate surface water, and prevent animals from gaining access to the spring area.

Protection and development of a source of water are shown in Figure 1.13. A combination of methods may also be possible under certain ground conditions and would yield a greater supply of water than either alone.

In all cases, the spring should be protected from surface-water pollution by constructing a deep diverting ditch or the equivalent above and around the spring. The spring and collecting basin should have a watertight top, preferably concrete, and water obtained by gravity flow or by means of a properly installed sanitary hand or mechanical pump. Access or inspection manholes, when provided, should be tightly fitted (as shown) and kept locked. Water from limestone or similar type channeled or fissured rock springs is not purified to any appreciable extent when traveling through the formation and hence may carry pollution from nearby or distant places. Under these circumstances, it is advisable to have periodic bacteriological examinations made and chlorinate the water.

Infiltration Gallery

An infiltration gallery consists of a system of porous, perforated, or open-joint pipe or other conduit draining to a receiving well. The pipe is surrounded by gravel and located in a porous formation such as sand and gravel below the water table. The collecting system should be located 20 feet or more from a lake or stream or under the bed of a stream or lake if installed under expert supervision. It

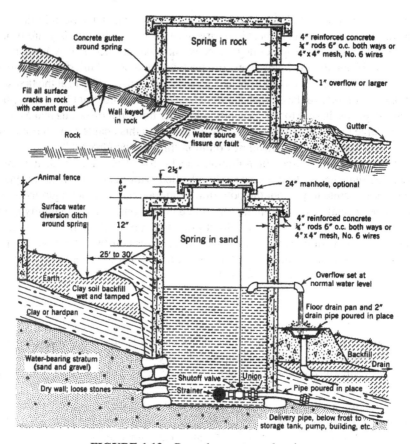

FIGURE 1.13 Properly constructed springs.

is sometimes found desirable, where possible, to intercept the flow of groundwater to the stream or lake. In such cases, a cofferdam, cutoff wall, or puddled clay dam is carefully placed between the collecting conduit and the lake or stream to form an impervious wall. It is not advisable to construct an infiltration gallery unless the water table is relatively stable and the water intercepted is free of pollution. The water-bearing strata should not contain cementing material or yield a very hard water, as it may clog the strata or cause incrustation of the pipe, thereby reducing the flow. An infiltration gallery is constructed similar to that shown in Figure 1.14. The depth of the collecting tile should be about 10 feet below the normal ground level, and below the lowest known water table, to assure a greater and more constant yield. An infiltration gallery may also be located at a shallow depth, above a highly mineralized groundwater, such as saline water, to collect the fresh or less mineralized water. An infiltration system consisting of horizontally perforated or porous radial collectors draining to a collecting well can also be designed and constructed where hydrogeological conditions are suitable, usually under a stream bed or lake, or where a thin water-bearing stratum exists. The

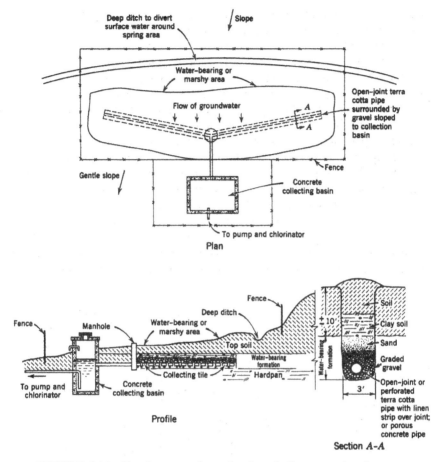

FIGURE 1.14 Development of a spring in a shallow water-bearing area.

infiltration area should be controlled and protected from pollution by sewage and other wastewater and animals. Water derived from infiltration galleries should, at the minimum, be given chlorination treatment.

Cistern

A cistern is a watertight tank in which rainwater collected from roof runoff or other catchment area is stored. When the quantity of groundwater or surface water is inadequate or the quality objectionable and where an adequate municipal water supply is not available, a cistern supply may be acceptable as a limited source of water. On the one hand, because rainwater is soft, little soap is needed when used for laundry purposes. On the other hand, rain will wash air pollutants, dust, dirt, bird and animal droppings, leaves, paint, and other material on the roof or in roofing materials or catchment area into the cistern unless special provision is made to bypass the first rainwater and filter the water. The bypass may consist of a simple manually or float-operated damper or switch placed in the leader drain. When in one position, all water will be diverted to a float control tank or to waste away from the building foundation and cistern; when in the other position, water will be run into the cistern. The filter will not remove chemical pollutants. If the water is to be used for drinking or food preparation, it should also be pointed out that because rainwater is soft and acidic, and therefore corrosive, hazardous concentrations of zinc from galvanized iron sheet roofing, gutters, and pipe and lead and copper from soldered copper pipe may also be released, in addition to cadmium.

The capacity of the cistern is determined by the size of the roof or catchment area, the probable water consumption, the maximum 24-hour rainfall, the average annual rainfall, and maximum length of dry periods. Suggested rainwater cistern sizes are shown in Figure 1.15. The cistern storage capacity given allows for a reserve supply, plus a possible heavy rainfall of 3-1/2 inches in 24 hours. The calculations assume that 25 percent of the precipitation is lost. Weather bureaus, the *World Almanac*, airports, water departments, and other agencies give rainfall figures for different parts of the country. Adjustment should therefore be made in

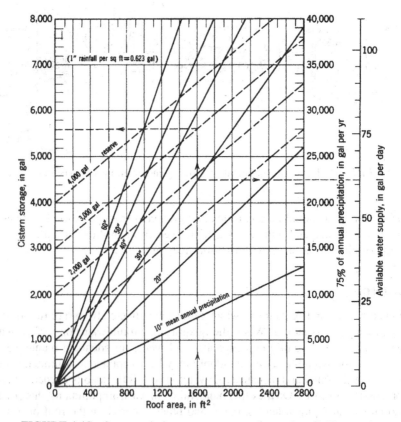

FIGURE 1.15 Suggested cistern storage capacity and available supply.

TABLE 1.16   Quantity and Type of Chlorine to Treat 1,000 gal of Clean Water at Rate of 1 mg/l

Chlorine Compound	Quantity
High test, 70% chlorine	1/5 oz or 1/4 heaping tablespoon
Chlorinated lime, 25% chlorine	1/2 oz or 1 heaping tablespoon
Sodium hypochlorite	
14% chlorine	1 oz
10% chlorine	1-1/3 oz
Bleach, 5-1/4 chlorine	2-3/5 oz

the required cistern capacity to fit local conditions. The cistern capacity will be determined largely by the volume of water one wishes to have available for some designated period of time, the total volume of which must be within the limits of the volume of water that the roof or catchment area and annual rainfall can safely yield. Monthly average rainfall data can be expected to depart from the true values by 50 percent or more on occasion. The drawing of a mass diagram is a more accurate method of estimating the storage capacity, since it is based on past actual rainfall in a given area.

It is recommended that the cistern water be treated after every rain with a chlorine compound of at least 5 mg/l chlorine. This may be accomplished by adding five times the quantities of chlorine shown in Table 1.16, mixed in 5 gallons of water to each 1,000 gallon of water in the cistern. A stack or tablet chlorinator and carbonate (limestone) contact tank on the inlet to the cistern is advised for disinfection and acidity neutralization. In areas affected by air pollution, fallout on the roof or catchment area will contribute chemical pollutants that may not be neutralized by limestone or chlorine treatment. Soft water flowing over galvanized iron roofs or through galvanized iron pipe or stored in galvanized tanks contains cadmium and zinc.[145]

Example With a roof area of 1,600 ft^2, in a location where the mean annual precipitation is 30 inches and it is desired to have a reserve supply of 3,000 gallon, the cistern storage capacity should be about 5,600 gallons. This should yield an average annual supply of about 62 gallons per day.

In some parts of the world, large natural catch basins are lined to collect rainwater. The water is settled and chlorinated before distribution. The amount of water is of course limited and may supplement groundwater, individual home cisterns, and desalinated water.

Domestic Well-Water Supplies — Special Problems*

Domestic well-water supply problems are discussed in this section.[146] The local health department and commercial water-conditioning companies may be of assistance to a homeowner.

---

*This section is adapted from ref.146.

Hard Water   Hard water makes it difficult to produce suds or rinse laundry, dishes, or food equipment. Water hardness is caused by dissolved calcium and magnesium bicarbonates, sulfates, and chlorides in well water. Pipes clog and after a time equipment and water heaters become coated with a hard mineral deposit, sometimes referred to as lime scale. A commercial zeolite or synthetic resin water softener is used to soften water. The media must be regenerated periodically and disinfected with chlorine to remove contamination after each regeneration. Softeners do not remove contamination in the water supply. A filter should be placed ahead of a softener if the water is turbid. See also "Water Softening," in Chapter 2.

The sodium content of the water passing through a home water softener will be increased. Individuals who are on a sodium-restricted diet should advise their physician that they are using home-softened water since such water is a continual source of dietary sodium. A cold-water bypass line can be installed around the softener to supply drinking water and water for toilet flushing.

Turbidity or Muddiness   This usually occurs in water from a pond, creek, or other surface source. Such water is polluted and requires coagulation, flocculation, sedimentation, filtration, and chlorination treatment. Wells sometimes become cloudy from cave-in or seepage from a clay or silt stratum but usually clear up with prolonged pumping. If the clay is in the colloidal state, coagulation, such as with aluminum sulfate (alum), is needed.

Sand filters can remove mud, dirt, leaves, foreign matter, and most bacteria, viruses, and protozoa if properly operated, but they may clog rapidly. Chlorination is also required to ensure destruction of pathogens. Charcoal, zeolite, or carbon filters are not suitable for this purpose, and, in addition, they clog. Iron and iron growths that sometimes cause turbidity in well water are discussed next. See also "Filtration", in Chapter 2.

Iron and Manganese in Well Water   Iron and manganese may be found in water from deep wells and springs. In high concentrations it causes a bitter taste in tea or coffee. When exposed to the air, iron, and manganese are oxidized and settle out. Red to brown or black (manganese) stains form on plumbing fixtures, equipment, and laundry. Chlorine bleach exacerbates the staining problem. Iron and manganese in solution (colloidal) form may be found in shallow wells, springs, and surface waters. In this form, the water has a faint red or black color.

A commercial home zeolite water softener removes 1.5 to 2.0 mg/l, and an iron removal filter removes up to 10 mg/l iron from well water devoid of oxygen. The water should *not* be aerated prior to zeolite filtration, as this will cause precipitation of oxidized (ferric) iron rather than the exchange of sodium by ferrous iron, which is washed out as ferrous chloride when regenerated. An iron removal filter will also remove some hydrogen sulfide. The water softener is regenerated with salt water. The iron removal filter is backwashed to remove the precipitated iron and regenerated with potassium permanganate. Since potassium permanganate is toxic, it must all be flushed out before the treated water is used. The controlled addition of a polyphosphate can keep 1.0 to 2.0 mg/l iron in solution, but, as with the zeolite softener, sodium is also added to the water. Heating of water to 140° to 150°F (60°–66°C) nullifies the effectiveness of polyphosphate.

With higher concentrations of iron, the water is chlorinated to oxidize the iron in solution and allowed a short contact period, but the water should then be filtered to remove the iron precipitate before it enters the distribution system. The pH of the water should be raised to above 7.0 if the water is acid; soda ash, added to the chlorine solution, is usually used for this purpose. Hydrogen peroxide or potassium permanganate will also oxidize the iron.

Another approach is to discharge the water to the air chamber of a pressure tank, or to a sprinkler over a cascade above a tank, but this will require double pumping. It is necessary to flush out the iron that settles in the tank and filter out the remainder. Air control is needed in a pressure tank. Air is admitted with the well water entering and air is vented from the tank. Manganese is also removed with iron treatment.

Injecting a chlorine solution into the water at its source, where possible, controls the growth of iron bacteria, if this is a problem. See also "Iron and Manganese Occurrence and Removal" and "Iron Bacteria Control," in Chapter 2. Before purchasing any equipment, seek expert advice and a proper demonstration should be sought.

Corrosive Water Water having a low pH or alkalinity and dissolved oxygen or carbon dioxide tends to be corrosive. Corrosive water dissolves metal, shortens the life of water tanks, discolors water, and clogs pipes. Iron corrosion causes rusty water; copper or brass pipe causes blue-green stains. Water can be made noncorrosive by passing it through a filter containing broken limestone, marble chips, or other acid neutralizers. The controlled addition of a polyphosphate, silicate, or soda ash to raise the water pH (commercial units are available) usually prevents metal from going into solution. The water remains clear and staining is prevented. However, bear in mind that a sodium polyphosphate would add sodium to the water, making it undesirable for individuals on a low-sodium diet. The use of low-lead solder (95:5 tinantimony solder), plastic pipe, maintenance of water temperature below 140°F (60°C), and a glass-lined hot water storage tank will minimize the problems associated with corrosion in home plumbing.

Taste and Odors Activated-carbon filters or cartridges are normally used to remove undesirable tastes and odors from domestic water supplies, but they do not remove microbiological contamination. Hydrogen sulfide in water causes a rotten-egg odor; corrosion of iron, steel, and copper; and black stains on laundry and crockery. It can also be eliminated by aeration and chlorination followed by filtration. An activated-carbon filter is not efficient. The activated carbon will have to be replaced when its capacity has been exhausted. Filtration alone, through a pressure filter containing a special synthetic resin, also removes up to 5 mg/l hydrogen sulfide in most cases. The water in question should be used to check the effectiveness of a process before any equipment is purchased. See also "Hydrogen Sulfide, Sources and Removal" in Chapter 2.

Detergents   Detergents in water can be detected visually, by taste, or by laboratory examination. When some detergents exceed 1 mg/l, foam appears in a glass of water drawn from a faucet. Detergents themselves have not been shown to be harmful, but their presence is evidence that wastewater from one's own sewage disposal system or from a neighbor's system is entering the water supply source. In such circumstances, the sewage disposal system may be moved, a well constructed in a new area, or the well extended and sealed into a deeper water-bearing formation not subject to pollution. There is no guarantee that the new water-bearing formation will not be or become polluted later. The solution to this problem is connection to a public water supply and/or a public sewer. A granular activated-carbon (GAC) filter may be used to remove detergent, but its effectiveness and cost should first be demonstrated. See also "Methylene Blue Active Substances (MBAS)," previously in this chapter.

Salty Water   In some parts of the country, salty water may be encountered. Since the salt water generally is overlain by fresh water, the lower part of the well in the salt water zone can be sealed off. But when this is done, the yield of the well is decreased.

Sometimes, waste salt water resulting from the backwashing of a home ion exchange water softener is discharged close to the well. Since salt water is not filtered out in seeping through the soil, it may find its way into the well. The best thing to do is to discharge the wastewater as far as possible and downgrade from the well or utilize a commercial water softener service. Salt water is corrosive; it will damage grass and plants and sterilize soil. Road salting or salt storage areas may also contribute to well pollution.

Special desalting units (using distillation, deionization, and reverse osmosis) are available for residential use, but they are of limited capacity and are relatively expensive, and pretreatment of the water may be needed. Complete information, including effectiveness with the water in question and annual cost, should be obtained before purchase. See "Desalination," this chapter, for additional information.

Radon in Well Water   See "Radon," in this chapter.

Gasoline or Fuel Oil in Water   See "Removal of Gasoline, Fuel Oil, and Other Organics in an Aquifer" and "Travel of Pollution through the Ground," in Chapter 2.

Household Treatment Units (Point-of-Use and Point-of-Entry)

Sometimes a chlorinator, faucet filter (point-of-use unit), dwelling filter (point-of-entry unit), or UV light disinfection unit is suggested to make an on-site polluted water supply safe for drinking without regard to the type, amount, or cause of pollution. This is hazardous. Instead, every effort should first be made to identify the pollutant and remove the source. This failing, every effort should be made to obtain water from a public water system. As a last resort, a household treatment unit or bottled water may have to be used. But the treatment units do not remove all microbiological, chemical, and physical pollutants. Careful selection of the proper treatment unit, which will resolve the particular pollution problem, in addition to cost, required maintenance and operation control, must be considered.

Household treatment unit processes include filtration, UV light radiation, chlorination, granular or powdered activated-carbon filtration, reverse osmosis, cartridge filters, cation exchange, anion exchange, distillation, pasteurization,[147] and activated-alumina filtration, as well as sand, porous stone, and ceramic filters. Each has limitations.

Ultraviolet light radiation and chlorination units are not considered satisfactory for the purification of surface-water supplies such as from ponds, lakes, and streams, which usually vary widely in physical, chemical, and microbiological quality, or for well or spring supplies, which may contain turbidity, color, iron, or organic matter. Pretreatment, usually including coagulation, flocculation, sedimentation, filtration, and disinfection or the equivalent, would be required to remove organic and inorganic contaminants that interfere with the effectiveness of the treatment. Chlorination and UV radiation treatment may be considered microbiologically acceptable only if the water supply is always clean, clear, and not subject to chemical or organic pollution and the units are operated as intended.* Certain controls are needed to ensure that the efficiency of the UV unit is not impaired by changes in light intensity, loss of power, rate of water flow, short circuiting, condition of the lamp, slime accumulation, turbidity, color, and temperature of the water.[148] Public Health Service 1974 standards state that acceptable UV units must have a flow rate of less than 0.2 gpm/effective inch of lamp, which must emit 2437 Å at an intensity of 4.85 UV watts/ft^2 at a distance of 2 inches, or an equivalent ratio of lamp intensity to flow, with a minimum retention time of 15 seconds at the maximum flow rate.[149] A flow control device, UV light-sensing device, alarm, and shutdown device are also needed.[150] Ultraviolet radiation units have application in the dairy, beverage, pharmaceutical, cosmetic, electronic, and food industries for the treatment of wash and cooling waters and for lowering the bacterial count in potable water used for soft drinks and bottled water. A chlorination unit requires inspection, solution replacement, and daily residual chlorine tests to ensure the unit operates as intended.

Most household filters contain activated carbon for the removal of organic substances. Taste and odor compounds are reduced, including chlorine, radon, and volatile halogenated organics such as trichloroethylene and carbon

---

*Normal chlorination treatment and UV radiation treatment do not inactivate the *Giardia lamblia* and *Cryptosporidium* protozoan cysts.

tetrachloride.[151] Sediment is trapped in the filter, and organic compounds, such as trihalomethanes resulting from chlorination, are removed to some extent. The activated-carbon filter cartridge needs periodic replacement, as recommended by the manufacturer. Microorganisms may grow in the filter and be released, but no harmful effects have been reported.[152] Many volatile organic compounds and radon are also removed by boiling and aeration. A cartridge filter to remove particulates should precede the carbon filter if the raw water is turbid. It should be understood that the water to be filtered must be potable. Microbiological and inorganic contaminants in solution are not removed.

A reverse-osmosis filter can reduce the concentrations of fluoride, mercury, lead, nitrates, sodium, iron, sulfate, alkalinity, total dissolved solids, and similar substances that might be present in drinking water, but not radon (GAC is effective). Sediment and many organic compounds are also removed, but prefiltration through a filter that removes particulates is indicated if sediment is present to prevent premature membrane clogging, followed by an activated-carbon filter to remove taste and odor compounds and other organics.[153] Arsenic and uranium are also removed under certain operating conditions.[154] The unit should have an automatic shut-off valve. The filter membrane requires backwashing.*

An activated-alumina unit can reduce the fluorides, arsenic, barium, and nitrates if sulfates are not too high. Uranium is also reduced.[155] The unit requires periodic regeneration. The activated-alumina lead removal cartridge is effective in removing lead.[156]

Electric distillation units that boil and condense water are also available. These units remove most microorganisms and inorganic compounds, including lead, salt, and nitrates, but not volatile organic compounds like benzene and chloroform—their capacity is limited.

Special ion exchange cartridge filters can remove inorganic contaminants from drinking water, including fluoride, uranium, and arsenic.[157] Ion exchange units can be regenerated with sodium chloride.

Porous stone "candles" and unglazed porcelain Pasteur or Berkefield filters for microbiological control are available and can be attached to a faucet spigot. They may develop hairline cracks and become unreliable for the removal of pathogenic microorganisms. They should be scrubbed, cleaned, and sterilized in boiling water once a week. Portable pressure-type ceramic microfiltration units, with single or multiple candles, having a capacity to remove 0.2-μm particles (bacteria, protozoa, helminths, and fungi), but not all viruses or chemical contaminants, are also available.[158]

Environmental Protection Agency studies of home water treatment filter devices showed THM removals of 6 to 93 percent and total organic carbon removals of 2 to 41 percent, depending on the unit. In some cases, higher bacterial counts were found in the water that had passed through the filter.[159] A subsequent study showed similar results.[160] Another study of halogenated organic

---

*Typically, about 75 percent of the tap water put into the reverse-osmosis system is wasted. ("FACTS for Consumers," Federal Trade Commission, Washington, DC, August 1989, p. 2.)

removal showed reductions ranging from 76 percent for a faucet-mount unit to 99 percent for several line bypass units.[161] These filter units do not remove nitrates, fluorides, or chlorides; do not soften water; remove little dissolved lead, iron, manganese, and copper; and do not remove microorganisms. They should not be used on any water supply that does not otherwise meet drinking water standards. The ability of a unit to remove the particular deleterious contaminants in the raw water should be confirmed with the manufacturer and the health department before purchase.* In general, reverse osmosis and distillation are most effective for inorganic contaminant reduction and granular activated carbon for organic contaminant removal.

Household treatment units have a limited flow capacity, which can be compensated for in part by incorporating a storage tank in the water system. Provision must be made for replacement or washing and disinfection of the filter element on a planned basis.

The satisfactory operation of a large number of household point-of-entry units in an area requires an effective management system, including monitoring, maintenance, and timely replacement of units or components and, in some instances, pre–and post–water treatment such as preclarification and postdisinfection.[162]

Desalination

Desalination or desalting is the conversion of seawater or brackish water to fresh water for potable and industrial purposes. The conversion of treated wastewater to potable water using multiple desalination processes is also being utilized in water scarce areas of the world. This conversion uses a variety of technologies to separate the dissolved solids from a source water. Desalination technology is being used to remove contaminants from surface and underground waters, including inorganics, radionuclides, emerging contaminants (such as pharmaceuticals), and THM precursors.

Many countries have used desalination technology for decades having either exhausted all of their primary sources of freshwater or to supplement and diversify their portfolio of water supplies. Considered by many as a drought-proof and inexhaustible supply of *new* water, municipal planners in the United States are also turning to desalination treatment plants as a means to ensure water supply in times of extended drought or as a back-up supply during an emergency. California for example, has 16 desalination plants with a combined capacity exceeding 400 mgd, in either planning, design, piloting, or construction as of 2008.

About seven-tenths of our globe is covered by seawater. The world's oceans have a surface area of $139,500,000\,mi^2$ and a volume of $317,000,000\ mi^3$.[163] The oceans contain about 97 percent of the world's water; brackish inland sites

---

*The National Sanitation Foundation, 3475 Plymouth Road, Ann Arbor, MI 48106, can provide a list of units certified for specific purposes. Also, The Water Quality Research Council, 4151 Naperville Road, Lisle, IL 60532.

and polar ice make up 2.5 percent, leaving less than 0.5 percent fresh water to be used and reused for municipal, industrial, agricultural, recreational, and energy-producing purposes.[164] In addition, more than half of the earth's surface is desert or semidesert. Under circumstances where adequate and satisfactory groundwater, surface water, or rainwater is not available and a high-quality water is required but where seawater or brackish water is available, desalination may provide an answer to the water problem. For seawater applications however, construction of intakes and the discharge of brine concentrate make siting new seawater plants a challenge. Prior to 2000, the high amounts of energy used in a desalination plant rendered plants feasible only where: 1) energy was plentiful and cheap; 2) where there were absolutely no other choices in water supply, or 3) the application was low volumes of high value product water such as for beverages, pharmaceuticals, or the electronics industry. Today however and with technology advances in both membrane materials and energy recovery, desalination costs are now affordable—even for high volumes of low-value water such as a municipal application.

Having begun in the arid Middle East 40 years ago with thermal (or distillation processes), desalting plants are now in use all over the world. Global Water Intelligence and the International Desalination Association reports 12,791 plants worldwide with capacity exceeding 11,000 mgd (over 42 million cubic meters per day) in operation as of 2006.[165]

Seawater has a total dissolved solids (TDS) concentration of about 35,000 mg/l. About 78 percent is sodium chloride, 11 percent magnesium chloride, 6 percent magnesium sulfate, 4 percent calcium sulfate, with the remainder primarily potassium sulfate, calcium carbonate, and magnesium bromide, in addition to suspended solids and microbiological organisms. The U.S. Geological Survey classifies water with less than 1,000 mg/l TDS as fresh, 1,000 to 3,000 mg/l as slightly saline, 3,000 to 10,000 mg/l as moderately saline, 10,000 to 35,000 mg/l as very saline, and more than 35,000 mg/l as brine. The U.S. Office of Technology Assessment defines potable water as generally having less than 500 ppm TDS (salt and/or dissolved solids), less brackish water as 500 to 3,000 ppm, moderately brackish water as 3,000 to 10,000 ppm, and highly brackish water as 10,000 to 35,000 ppm.[166] The source of brackish water may be groundwater or surface-water sources such as oceans, estuaries, saline rivers, and lakes. Its composition can be extremely variable, containing different concentrations of sodium, magnesium, sulfate, calcium, chloride, bicarbonate, fluoride, potassium, and nitrate. Iron, manganese, carbon dioxide, and hydrogen sulfide might also contribute to the variability of brackish water quality.

Desalting will remove dissolved salts and minerals such as chlorides, sulfates, and sodium, in addition to hardness. Nitrates, nitrites, phosphates, fluorides, ammonia, and heavy metals are also removed to some degree, depending on the process. Very hard brackish water will require prior softening to make reverse osmosis or electrodialysis very effective.[167] Desalination is not normally used to remove iron, manganese, fluorides, calcium, or magnesium.

Some known methods for desalting water are as follows[168]:

- *Membrane:* Reverse osmosis; electrodialysis and electrodialysis reversal; transport depletion; piezodialysis
- *Distillation or Thermal:* Multistage flash distillation; multieffect multistage distillation; vapor compression; vertical tube distillation; solar humidification
- *Crystallization:* Vacuum freezing–vapor compression; secondary refrigerant freezing; eutectic freezing; hydrate formation
- *Chemical:* Ion exchange

Distillation   In distillation or thermal desalination, seawater is heated to the boiling point and then into steam, usually under pressure, at a starting temperature of 250°F (121°C). The steam is collected and condensed in a chamber by coming into contact with tubes (condenser–heat exchanger) containing cool seawater. The heated saline water is passed through a series of distillation chambers in which the pressure is incrementally reduced and the water boils (made to "flash"), again at reduced temperature, with the production of steam, which is collected as fresh water. The remaining, more concentrated, seawater (brine) flows to waste. In each step, the temperature of the incoming seawater is increased by the condenser–heat exchangers as it flows to the final heater. The wastewater (brine) and distilled water are also used to preheat the incoming seawater. This process is referred to as multistage flash distillation (MSF). There may be as many as 15 to 25 stages. A major problem is the formation of scale (calcium carbonate, calcium sulfate, and magnesium hydroxide) on the heat transfer surfaces of the pipe or vessel in which the seawater is permitted to boil. This occurs at a temperature of about 160°F (71°C), but scale can be greatly minimized by pretreating the seawater to remove either the calcium or the carbon dioxide. Distilled sea water normally has 5 to 50 mg/l salt. Most volatile substances are removed.

Vertical-tube distillation, multieffect multistage distillation, vapor compression distillation, and solar distillation are distillation variations. Solar humidification (distillation) depends on water evaporation at a rate determined by the temperature of the water and the prevailing humidity. The unit is covered with a peaked glass or plastic roof from which the condensate is collected. Distilled water is tasteless and low in pH if not aerated and adjusted before distribution.

Reverse Osmosis   Normally, if salt water and fresh water are separated by a semipermeable membrane, the fresh water diffuses through to the salt water as if under pressure, actually osmotic pressure. The process is known as osmosis. In reverse osmosis, hydraulic pressures of 200 to 500 psi for brackish water and 800 to 1,200 psi for seawater[169] are applied to the concentrated salt water on one side of a special flat or cylindrical supported membrane, a spiral wound, or hollow-fiber unit. The life of the membrane decreases with increasing pressure. In the process, fresh water is separated out from the salt water into a porous or hollow channel from which the fresh water is collected. The concentration of

TDS in the salt water flowing through the unit must be kept below the point at which calcium sulfate precipitation takes place. Some of the dissolved solids, 5 to 10 percent, will pass through the membrane, including total hardness, sulfates, chlorides, ammonium, chemical oxygen demand (COD) materials, color, bacteria, and viruses. Chlorinated methanes and ethanes, which are common solvents, are not removed by reverse osmosis; however, air stripping is effective.[170] An increase in the TDS will result in a small increase of solids in the fresh water.

In reverse osmosis, the salt water to be treated must be relatively clear and free of excessive hardness, iron, manganese, and organic matter to prevent fouling of the system membranes. The maximum water temperature must be between 86° and 122°F (30° and 50°C), depending on membrane type.[171] Since the RO elements are designed to remove only the dissolved material in the source water, all suspended particles must be removed before entering the RO membranes, or the elements will become fouled prematurely. The pretreatment of the source water is a critical component of a well-designed plant and may consist of:

1. Softening to remove hardness;
2. Coagulation and filtration (sand, anthracite, multimedia; cartridge, or diatomaceous earth) to remove turbidity, suspended matter, iron, and manganese;
3. Low pressure micro filtration (MF) or ultra filtration (UF) membranes for turbidity and suspended particle removal; and
4. Filtration through activated-carbon columns to remove dissolved organic chemicals.

If the pretreatment design uses conventional process (coagulation/ sedimentation/sand filtration), a 1 micron cartridge or bag filter is installed between the sand filter and the RO elements. This disposable filter is an insurance step to prevent an accidental loading of unwanted foulant material onto the RO elements. Acid is used if necessary to lower the pH and prevent calcium carbonate and magnesium hydroxide scale. Citric acid is used to clean membranes of inorganic and chlorine bleach for organics removal. Special cleaners may be needed to remove silicates, sulfates, hydroxides, and sulfides. Chlorine might also be used to control biological growths on the membranes,[172] but prior filtration of water through GAC is necessary to protect membranes not resistant to chlorine and prevent the formation of trihalomethanes (bromoform). Salt, dissolved solids, some microorganisms, organic and colloidal materials, and other contaminants, including radiologic, are removed. Reverse-osmosis treated water usually requires posttreatment for pH adjustment, degasification (H_2S and CO_2), corrosion adjustment, and disinfection, possibly further demineralization by ion exchange, and UV radiation disinfection for certain industrial waters. Other membrane processes include nanofiltration, ultrafiltration, and microfiltration.

Electrodialysis In electrodialysis, the dissolved solids in the brackish water (less than 10,000 mg/l TDS) are removed by passage through a cell in which a direct electric current is imposed. Dissolved solids in the water contain positively charged ions (cations) and negatively charged ion (anions). The cations migrate to and pass through a special membrane allowing passage of the positive ions. Another special membrane allows the negative ions to pass through. The concentration of dissolved solids determines the amount of current needed. The process removes salt, other inorganic materials, and certain low-molecular-weight organics.[173] Operating pressures vary from 70 to 90 psi. The partially desalted–demineralized water is collected and the wastewater is discharged to waste. Maximum water operating temperature is 113°F (45°C).[174]

The plant size is determined in part by the desired amount of salt removal. However, a change in the TDS in the brackish water will result in an equal change in the treated water.[175] As in reverse osmosis, pretreatment of the brackish water is necessary to prevent fouling of the membranes and scale formation. Scaling or fouling of membranes is reported to be prevented in most units by reversing the electric current at 15- to 30-minute intervals.[176] The cost of electricity limits the use of electrodialysis.

Transport depletion is a variation of the electrodialysis process. Piezodialysis is in the research stage; it uses a new membrane desalting process.

Ion Exchange   In the deionization process, salts are removed from brackish water (2,000 to 3,000 mg/l TDS). Raw water passes through beds of special synthetic resins that have the capacity to exchange ions held in the resins with those in the raw water.

In the two-step process, at the first bed (acidic resin) sodium ions and other cations in the water are exchanged for cations (cation exchange) in the resin bed. Hydrogen ions are released and, together with the chloride ions in the raw water, pass through to the second resin bed as a weak hydrochloric acid solution. In the second resin bed, the chloride ions and other anions are taken up (anion exchange) from the water, are exchanged for hydroxide ions in the resin bed that are released, combine with the hydrogen ions to form water, and pass through with the treated water. The ion exchange beds may be in a series or in the same shell.

When the resins lose their exchange capacity and become saturated, the treatment of water is interrupted and the beds are regenerated, with acids or bases. The resins may become coated or fouled if the raw water contains excessive turbidity, microorganisms, sediment, color, or organic matter, including dissolved organics, hardness, iron, or manganese. In such cases, pretreatment to remove the offending contaminant is necessary. Chlorine in water would attack the cation resin and must also be removed prior to deionization.

Waste Disposal   The design of a desalting plant must make provision for the disposal of waste sludge from pretreatment and also of the concentrated salts and minerals in a solution removed in the desalting process. The amount or volume of waste is dependent on the concentration of salts and minerals in the raw water and the amount of water desalted. The percent disposed as waste concentrate from a reverse osmosis unit treating brackish water may be 20 to 50 percent, from a seawater unit 60 to 80 percent, from an electrodialysis unit 10 to 20 percent, and from a distillation unit 5 to 75 percent.[177]

TABLE 1.17   Country Inventory of Global Desalination Treatment Plants

Country	No. of Plants	Treatment Technology		User Category		Source Water		
		Membrane	Thermal	Municipal	Industrial[a]	Seawater	Brackish	Other[b]
Algeria	147	75%	25%	68%	32%	73%	19%	8%
Australia	181	74%	26	15%	85%	18%	46%	36%
Bahrain	140	26%	74%	77%	23%	91%	9%	—
China	189	84	16	48	52	45	13	42
India	193	57	43	9	91	68	18	14
Israel	50	97	3	98	2	87	11	3
Japan	1457	95	5	19	81	17	16	67
Kuwait	84	16	84	85	15	84	2	14
Libya	295	18	82	74	26	87	13	—
Oman	133	39	61	92	8	96	3	1
Qatar	87	2	98	96	4	99	1	—
Saudi Arabia	2086	41	59	84	16	79	20	1
Spain	760	95	5	83	17	72	21	7
UAE	351	20	80	95	5	98	2	—
USA	2174	95	5	63	37	12	49	39
Total[c]	8327							

[a]Industrial includes other categories such as power, irrigation, military, tourism
[b]Other source waters include: river water, wastewater, pure water
[c]These 15 countries contain 8327 of the total number of 12,791 global plants
Source: IDA Desalination Yearbook 2007-2008, T. Pankratz and E. Yell; Media Analytics, Ltd., Oxford, UK 2008.

The waste from mildly brackish water (1000 to 3000 mg/l TDS) will contain from 5,000 to 10,000 mg/l (TDS). The waste from a seawater desalting plant can contain as much as 70,000 mg/l (TDS).[178]

The waste disposal method will usually be determined by the location of the plant and the site geography. Methods that would be considered include disposal to the ocean, inland saline lakes and rivers, existing sewer outfalls, injection wells or sink holes where suitable rock formations exist, solar evaporation ponds, lined or tight-bottom holding ponds, or artificially created lakes. In all cases, prior approval of federal (EPA) and state regulatory (water pollution and water supply) agencies having jurisdiction must be obtained. Surface and underground sources of drinking water and irrigation water must not be endangered.

Table 1.17 presents an inventory of the top 15 countries incorporating desalination technology as of 2006. In addition to the number of installed plants in each country, the table presents the technology used, the application categories, and the source of the water supply.

Costs   The Office of Water Research and Technology reported that the cost of desalted water from global desal plants varies from upward from 85 cents per 1,000 gallons, except where fuel is available at very low cost.[179]. In 2007, the costs in the United States for seawater ranged from $2.00 to $6.00 per 1000 gallons depending on the size of the plant. Commissioned in late 2007 for example, the total cost of water from the Tampa Bay Water desalination plant cost approximately $3.18 per 1000 gallons. This is the largest operating plant in the United States at 28 mgd, although two 50 mgd plants are scheduled to begin construction in Carlsbad and Huntington Beach California in 2009. [180] Costs in 1985 (capital and operating costs) were estimated to be $2 to $2.50 per 1,000 gallons for brackish water, reverse osmosis, and electrodialysis treatment, with conventional treatment at approximately $0.40 to $2 per 1,000 gallon.[181]

An analysis was made by Miller[182] of 15 municipalities in the western United States demineralizing brackish water by reverse osmosis, electrodialysis, or ion exchange and combinations thereof. Flows varied from 0.13 to 7.18 mgd and TDS from 941 to 3,236 mg/l. The demineralization cost varied from $0.37 to $1.56 per 1,000 gal. Reverse osmosis was found to be the least costly process by most of the communities. Reverse osmosis plant construction and operating costs for seawater desalting were reported to be usually less than for distillation.[183] This may not be the case, however, where large volumes of seawater are to be distilled and where a convenient source of heat energy is available,[184] such as from a power plant or incinerator or where fuel costs are low. In another report, the energy break-even point of the reverse osmosis and electrodialysis treatment of brackish water and wastewater was approximately 1,200 mg/l. Electrodialysis was more energy efficient below 1,200 mg/l and reverse osmosis above that level.[185]

Construction and operating cost comparisons must be made with care. They are greatly influenced by location; material, labor, and energy costs; size; TDS concentration; and amount of pollutants such as suspended and other dissolved solids in the water to be desalted. Waste disposal and water distribution are additional factors usually considered separately.

General   The use of desalted water usually implies a dual water distribution and plumbing system, one carrying the potable desalted water and the other carrying nonpotable brackish water or seawater. Obviously, special precaution must be taken to prevent interconnections between these two water systems. The brackish water or seawater may be used for firefighting, street flushing, and possibly toilet flushing.

The finished desalted water requires pH adjustment for corrosion control (lime, sodium hydroxide) and disinfection prior to distribution. It must contain not more than 500 mg/l total dissolved solids to meet drinking water standards. Up to 1,000 mg/l dissolved solids might be acceptable in certain circumstances. Other standards would apply if the desalted water is used for industrial purposes. The EPA considers a groundwater containing less than 10,000 mg/l TDS as a potential source of drinking water.[186]

Indirect benefits of desalting brackish water may include the purchase of less bottled water, use of less soap and detergents, no need for home water softeners and water-conditioning agents, and fewer plumbing and fixture repairs and replacements due to corrosion and scale buildup.[187]

REFERENCES

1. *Code of Federal Regulations*, Title 40, Protection of Environment, Part 141, National Primary Drinking Water Regulations, July 2002; 2002 and *Factoids: Drinking Water and Groundwater Statistics for 2007*, EPA, Office of Water, Washington, DC, March 2008.

2. *Estimated Water Use in the United States in 2000*, U.S. Geological Survey Circular 1268, March 2004 (Revised February 2005).

3. A. Levin, "The Rural Water Survey", *J. Am. Water Works Assoc.* (August 1978): 446–452.

4. S. Arlosoroff, *Community Water Supply*, World Bank, Washington, DC, May 1987, p.1.

5. A. Wolman, "Water Supply and Environmental Health", *J. Am. Water Works Assoc.* (December 1970): 746–749.

6. D. R. Hopkins, "Guinea Worm: The Next To Go"? *World Health* (April 1988): 27.

7. *WHO Report*, Global Water Supply and Sanitation Assessment 2000.

8. *Aid Highlights*, 3(1), USAID, Washington, DC, Winter 1986.

9. G. E. Arnold, "Water Supply Projects in Developing Countries", *J. Am. Water Works Assoc.* (December 1970): 750–753.

10. H. R. Acuña, "Health for All by the Year 2000", *Bull. PAHO*, **16**(1), 1982.

11. USEPA, *Needs Survey Report to Congress*, Environmental Protection Agency, Washington, DC, 1992.

12. *Waste Disposal Practices and Their Effects on Ground Water*, Report to Congress, Executive Summary, EPA, Washington, DC, January 1977.

13. R. W. Schowengerdt, "An Overview of Ground Water Techniques." In N. Dee, W. F. McTernan, and E. Kaplan (Eds.), *Detection, Control, and Renovation of Contaminated Ground Water*, ASCE, New York, 1987, pp. 35–50.

14. C. W. Stiles, H. R. Crohurst, and G. E. Thomson, *Experimental Bacterial and Chemical Pollution of Wells via Ground Water, and the Factors Involved*, PHS Bull. 147, DHEW, Washington, DC, June 1927.

15. *Report on the Investigation of Travel of Pollution*, Pub. No. 11, State Water Pollution Control Board, Sacramento, CA, 1954.

16. R. G. Butler, G. T. Orlob, and P. H. McGauhey, "Underground Movement of Bacterial and Chemical Pollutants", *J. Am. Water Works Assoc.*, **46**(2) (February 1954): 97–111.

17. Stiles, Crohurst, and Thomson.

18. Butler, Orlob, and McGauhey, p. 97.

19. B. H. Keswick and C. P. Gerba, "Viruses in Groundwater," *Environ. Sci. Technol.* (November 1980): 1290–1297.

20. S. Buchan and A. Key, "Pollution of Ground Water in Europe," *WHO Bull.*, **14**(5–6): 949–1006 (1956).

21. Protection of Public Water Supplies from Ground-Water Contamination, Seminar Publication, EPA/625/4-85/016, Center for Environmental Research Information, Cincinnati, OH, September 1985.

22. *Hydrology Handbook*, Manual 28, ASCE, New York, 1949.

23. P.H. Gleick, Water Resources, *Encyclopedia of Climate and Weather*, Oxford University Press, New York, vol. **2**. 1996, pp. 817-823.

24. *Handbook—Ground Water*, EPA/625/6-87/016, Center for Environmental Research Information, Cincinnati, OH, March 1987, p. 76.

25. T. M. Schad, *"Ground Water Classification: Goals and Basis,"* in P. Churchill and R. Patrick (Eds.), *Ground Water Contamination: Sources, Effects and Options to Deal with the Problem*, p. 335, Proceedings of the Third National Water Conference, Philadelphia, January 13–15, 1987, Academy of Natural Sciences, Philadelphia, PA, 1987.

26. "Disinfection—Committee Report," *J. Am. Water Works Assoc.* (April 1978): 219–222.

27. T. R. Hauser, "Quality Assurance Update," *Environ. Sci. Technol.* (November 1979): 1356–1366.

28. *Guidelines for Drinking-Water Quality*, Vols. **1, 2, 3**, WHO, Geneva, 1984, 1985; *Manual for Evaluating Public Drinking Water Supplies*, DHEW, Bureau of Water Hygiene, Cincinnati, OH, 1971.

29. A. D. Eaton, L. S. Clesceri, E. W. Rice, and A. E. Greenberg (Eds.), *Standard Methods for the Examination of Water and Wastewater*, 21st ed., APHA, AWWA, WEF, 2005.

30. *Analytical Handbook, Laboratory of Organic Analytical Chemistry*, Wadsworth Center for Laboratories and Research, New York State Department of Health, Albany, NY, 1988.

31. J. P. Gibb and M. J. Barcelona, "Sampling for Organic Contaminants in Ground-water," *J. Am. Water Works Assoc.* (May 1984): 48–51.

32. Eaton, Clesceri, Rice, and Greenberg.

33. Ibid.

34. E. E. Geldreich, "Is the Total Count Necessary," in *Proceedings—American Water Works Association*, Water Quality Technology Conference, AWWA, Denver, CO, 1974.

35. A. P. Dufour, *Escherichia coli:* The Fecal Coliform, in A.W. Hoadley and B. J. Dutka (Eds.), *Bacterial Indicators/Health Hazards Associated with Water*, ASTM STP635, American Society for Testing Materials, Philadelphia, PA, 1977.

36. Eaton, Clesceri, Rice, and Greenberg.

37. Ibid.

38. Geldreich.

39. *Guidelines for Drinking-Water Quality*, Vol. **2**, WHO, Geneva, 1984, p. 34.

40. Eaton, Clesceri, Rice, and Greenberg.

41. *Guidelines for Drinking-Water Quality*, Vol. **1**, WHO, Geneva, 1984, p. 28.

42. J. E. Ongerth, "Giardia Cyst Concentrations in River Water," *J. Am. Water Works Assoc.* (September 1989): 71–80.

43. E. G. Means III and M. J. McGuire, "An Early Warning System for Taste and Odor Control," *J. Am. Water Works Assoc.* (March 1986): 77–83.

44. Eaton, Clesceri, Rice, and Greenberg

45. B. O. Wilen, Options for Controlling Natural Organics, in R. B. Pojasek (Ed.), *Drinking Water Quality Enhancement through Source Protection*, Ann Arbor Science, Ann Arbor, MI; "Research Committee on Color Problems Report for 1966," *J. Am. Water Works Assoc.* (August 1967): 1023–1035.

46. D. F. Kincannon, "Microbiology in Surface Water Sources," *OpFlow*, AWWA, Denver, CO, December 1978, p. 3.

47. C. M. Palmer, *Algae in Water Supplies*, PHS Pub. No. 657, DHEW, Cincinnati, OH, 1959.

48. Eaton, Clesceri, Rice, and Greenberg

49. Ibid.

50. W. M. Ingram and C. G. Prescott, "Toxic Fresh-Water Algae," *Am. Midland Naturalist*, **52**(1), (July 1954): 75–87.

51. *Guidelines for Drinking-Water Quality*, Vol.**2**, p. 54.

52. P. R. Gorham, "Toxic Algae as a Public Health Hazard," *J. Am. Water Works Assoc.*, **56**, 1487 (November 1964).

53. Ibid.

54. *Guidelines for Drinking-Water Quality*, Vol.**1**, p. 107.

55. *Guidelines for Drinking-Water Quality*, Vol.**2**.

56. *Guidelines for Drinking-Water Quality*, Vol.**1**, p. 6–7, Vol. **2**, p. 88.

57. *Water Quality Criteria 1972*, NAS, National Academy of Engineering, Washington, DC, 1972, p. 87.

58. "British Study Links Aluminum to Alzheimer's," *AWWA MainStream*, February 1989, p. 9;   and S. H. Reiber, HDR Engineering, W. A. Kukull, Department of Epidemiology, University of Washington;   *Aluminum, Drinking Water and Alzheimer's Disease*; AWWA Research Foundation, Denver, CO, 1996.

59. "Proposed by State of Washington Department of Ecology," *New York Times*, November 6,1983.

60. M. M. Varma, S. G. Serdahely, and H. M. Katz, "Physiological Effects of Trace Elements and Chemicals in Water," *J. Environ. Health* (September/October 1976): 90–100.

61. J. Malcolm Harrington et al., "An Investigation of the Use of Asbestos Cement Pipe for Public Water Supply and the Incidence of Gastrointestinal Cancer in Connecticut, 1935–1973," *Am. J. Epidemiol.*, **107**(2), 96–103 (1978).

62. J. Wister Meigs et al., "Asbestos Cement Pipe and Cancer in Connecticut, 1955–1974," *J. Environ. Health.*

63. Civil Eng., ASCE, October 1983, p.22.

64. *Asbestos in Drinking Water Fact Sheet*, Document A 0140, Bureau of Toxic Substances Assessment, New York State Department of Health, Albany, NY, October 1985.

65. R. W. Buelow et al., "The Behavior of Asbestos-Cement Pipe under Various Water Quality Conditions: A Progress Report," *J. Am. Water Works Assoc.* (February 1980): 91–102.

66. Code of Federal Regulations, National Drinking Water Regulations, 52 FR, Section 141.32 (e), October 28, 1987.

67. NIOSH Recommendations for Occupational Safety and Health Standards 1988, MMWR, Supplement, August 1988, p. 5.

68. *Guidelines for Drinking-Water Quality*, Vols.**1**, **2**, and **3**.

69. *Guidelines for Drinking-Water Quality*, Vol.**1**, pp. 6–7, Vol. **2**, p. 88.

70. H. A. Schroeder, *New York State Environment*, New York State Department of Environmental Conservation, Albany, NY, April 1976.

71. Code of Federal Regulations.

72. *Guidelines for Drinking-Water Quality*, Vol.1, pp. 80, 81, and 85.

73. Ibid.

74. H. A. Schroeder, "Environmental Metals: The Nature of the Problem," in W. D. McKee (Ed.), *Environmental Problems in Medicine*, Charles C. Thomas, Springfield, IL, 1974.

75. *Guidelines for Drinking-Water Quality*, Vols.1, 2, and 3.

76. Schroeder 1974.

77. *Guidelines for Drinking-Water Quality*, Vol.2, p. 97.

78. Code of Federal Regulations.

79. Ibid.

80. *Community Water Fluridation Fact Sheets*, Centers for Disease Control and Prevention, Atlanta, GA, August, 2006.

81. W.-W. Choi and K. Y. Chen, "The Removal of Fluoride from Waters by Adsorption," *J. Am. Water Works Assoc.* (October 1979): 562–570.

82. *Guidelines for Drinking-Water Quality*, Vol. 1, p. 6–7, Vol. 2, p. 264.

83. *AWWA MainStream*, September 1988, p. 13.

84. "Update," *J. Am. Water Works Assoc.* (February 1989).

85. M. Cousins et al., "Getting the Lead Out—Reducing Lead Concentrations in Water," Public *Works* (November 1987): 68–70.

86. R. G. Lee, W. C. Becker, and D. W. Collins, "Lead at the Tap: Sources and Control," *J. Am. Water Works Assoc.* (July 1989): 52–62.

87. R. J. Bull and G. F. Craun, "Health Effects Associated with Manganese in Drinking Water," *J. Am. Water Works Assoc.* (December 1977): 662–663.

88. R. H. Gould, "Growing Data Disputes Algal Treatment Standards," *Water Wastes Eng.* (May 1978): 78–83.

89. *Glossary Water and Wastewater Control Engineering*, 3rd ed., APHA, ASCE, AWWA, WPCF, 1981, p. 259.

90. Code of Federal Regulations.

91. National Technical Advisory Committee Report, "Raw-Water Quality Criteria for Public Water Supplies," *J. Am. Water Works Assoc.*, **61**, 133 (March 1969).

92. *Guidelines for Drinking-Water Quality*, Vol. 1, pp. 74–76, 85.

93. Ibid.

94. R. A. Jones and G. Fred Lee, "Septic-Tank Disposal Systems as Phosphorus Sources for Surface Waters," Tichardson Institute for Environmental Science, Texas University at Dallas, Dallas, TX, November 1977.

95. Centers for Disease Control, "NIOSH Recommendations for Occupational Safety and Health Standards," MMWR, **35**(Suppl. 1S), 27S (1986).

96. *Criteria for a Recommended Standard: Occupational Exposure to Polychlorinated Biphenyls (PCBs)*, DHEW Pub. No. (NIOSH) 77–225, NIOSH, DHEW, PHS, Cincinnati, OH, 1977.

97. *Water Quality Criteria 1972*, National Academy of Sciences—National Academy of Engineering, EPA-R3-73-033, U.S. Government Printing Office, Washington, DC, March 1973, p. 177.

98. "PCBs in Old Submersible Well Pumps," *Dairy, Food and Environmental Sanitation* (January 1989): 35–36.

99. *Guidelines for Drinking-Water Quality*, Vol. **2**, pp. 185–186.

100. G. S. Solt, "High Purity Water," *J. R. Soc. Health* (February 1984): 6–9.

101. L. J. Kosarek, "Radionuclides Removal from Water," *Environ. Sci. Technol.* (May 1979): 522–525.

102. "Drinking Water and Health, Recommendations of the National Academy of Sciences," Fed. Reg. 42:132: 35764, July 11, 1977.

103. *Guidelines for Drinking Water Quality*, Vol. **2**, p. 144.

104. H. B. Brown, *The Meaning, Significance, and Expression of Commonly Measured Water Quality Criteria and Potential Pollutants*, Louisiana State University and Agricultural and Mechanical College, Baton Rouge, LA, 1957, pp. 35–36.

105. *Guidelines for Drinking-Water Quality*, Vol. **2**, p. 84.

106. C. R. Murray and E. B. Reeves, *Estimated Use of Water in the United States in 1975*, Geological Survey Circular 765, U.S. Department of the Interior, Washington, DC.

107. Code of Federal Regulations.

108. Ibid.

109. U.S. EPA Statement, "Chlorinated and Brominated Compounds are not equal," *J. Am. Water Works Assoc.*, October 1977, p. 12.;   J. A. Cotruvo and Ch. Wu, *J. Am. Water Works Assoc.* (November 1978): 590–594.

110. "Update," *J. Am. Water Works Assoc.*, December 1978, p. 9.

111. A. Wolman, "Reflections, Perceptions, and Projections," *J. Water Pollut. Control Fed.* (December 1983): 1412–1416.

112. M. Gaskie and C. C. Johnson, Jr., "Face to Face," *J. Am. Water Works Assoc.* (August 1979): 16.

113. *Guidelines for Drinking-Water Quality*, Vol. **1**, p. 77.

114. Water Quality Criteria 1972, p. 91.

115. Code of Federal Regulations.

116. *Guidelines for Drinking-Water Quality*, Vols. **1**, **2**, and **3**.

117. E. R. Christensen and V. P. Guinn, "Zinc from Automobile Tires in Urban Runoff," *J. Environ. Eng. Div.*, ASCE, (February 1979): 165–168.

118. Schroeder 1974.

119. N. I. McClelland et al., "The Drinking-Water Additives Program," *Environ. Sci. Technol.* (January 1989): 14–18.

120. J. Bucher, "USEPA Additives Involvement End in April," *AWWA MainStream*, (October 1989): 1.

121. *The Story of Drinking Water*, Catalog No. 70001, AWWA, Denver, CO, p. 13.

122. W. B. Solley, E. B. Chase, and W. B. Mann IV, *Estimated Use of Water in the United States in 1980*, Geological Survey Circular 1001, U.S. Department of the Interior, Washington, DC, 1983.

123. S. Blinco, World Bank Research in Water Supply and Sanitation, The World Bank, Washington, DC, Summer 1982, pp. 3–11;   Village Water Supply, The World Bank, Washington, DC, March 1976.

124. Saul Arlosoroff et al., *Community Water Supply: The Handpump Option*, The World Bank, Washington, DC, May 1987.

125. W. E. Sharpe, "Water and Energy Conservation with Bathing Shower Flow Controls," *J. Am. Water Works Assoc.* (February 1978): 93–97.

126. *Environmental Conservation Law*, Section 15-0107, Albany, NY, 1989.

127. *AWWA MainStream*, April 1989, p. 7.

128. D. L. Anderson and R. L. Siegrist, "The Performance of Ultra-Low-Flush Toilets in Phoenix," *J. Am. Water Works Assoc.* (March 1989): 52–57.

129. Microphor Inc., P.O. Box 1460, Willits, CA 95490.

130. W. O. Maddaus, "The Effectiveness of Residential Water Conservation Measures," *J. Am. Water Works Assoc.* (March 1987): 52–58; Residential Water Conservation Projects—Summary Report, HUD-PDR-903, HUD Washington, DC, June 1984.

131. P. W. Fletcher and W. E. Sharpe, "Water-Conservation Methods to Meet Pennsylvania's Water Needs," *J. Am. Water Works Assoc.* (April 1978): 200–203; Rural Wastewater Management, State of California Water Resources Board, Sacramento, CA, 1979, pp. 11–14.

132. P. Schorr and R. T. Dewling, "Reusing Water," *Civil Eng., ASCE* (August 1988): 69–71.

133. Ibid.

134. D. A. Okun, "The Use of Polluted Sources of Water Supply," APWA *Reporter* (September 1976): 23–25.

135. *Water Environ. Technol.* (January 1991).

136. "Use of Reclaimed Waste Water as a Public Water Supply Source," 1978–79 Officers and Committee Directory Including Policy Statements and Official Documents, *AWWA*, Denver, CO, 1979, p. 78; "Water Should Be Segregated by Use," Water Sewage Works (March 1979): 52–54; D. A. Okun, Wastewater Reuse Dilemma, *ESE Notes*, University of North Carolina at Chapel Hill, November 1975.

137. A. J. Clayton and P. J. Pybus, "Windhoek Reclaiming Sewage for Drinking Water," *Civil Eng., ASCE* (September 1972): 103–106.

138. Ground Water Basin Management, ASCE Manual No. 40, ASCE, New York, 1961; Large-Scale Ground-Water Development, United Nations Water Resources Centre, New York, 1960; Guidelines for Delineation of Wellhead Protection Areas, EPA Office of Ground-Water Protection, Washington, DC, June 1987.

139. R. Hoffer, "The Delineation and Management of Wellhead Protection Areas," in N. Dee, W. F. McTernan, and E. Kaplan (Eds.), *Detection, Control, and Renovation of Contaminated Ground Water*, ASCE, New York, 1987, pp. 143–152.

140. *Water Well Location by Fracture Trace Mapping, Technology Transfer*, Office of Water Research and Technology, Department of the Interior, U.S. Government Printing Office, Washington, DC, 1978.

141. R. C. Heath, *Basic Ground Water Hydrology*, USGS Paper 2220, U.S. Government Printing Office, Washington, DC, 1983.

142. Manual of Water Well Construction Practices, EPA-570/9-75-101, EPA Office of Water Supply, U.S. Government Printing Office, Washington, DC, 1976, pp. 104–116; *Manual of Individual Water Supply Systems*, EPA-570/9-82-004, EPA Office of Drinking Water, U.S. Government Printing Office, Washington, DC, October 1982; Recommended Standards for Water Works, A Report of the Committee of the Great Lakes-Upper Mississippi River Board of State Public Health and Environmental Managers, Health Education Service Division, Albany, NY, 1987, pp. 18–32; Environmental Health Ready Reference, Michigan Environmental Health Association, November 1983, pp. 115–120; F. G. Driscoll, Groundwater and Wells, 2nd ed., Johnson Division, St. Paul, MN, 1986.

143. R. B. Dole, "Use of Fluorescein in the Study of Underground Waters," U.S. Geol. Survey, Water-Supply Paper 160, 1906, pp. 73–85; S. Reznek, W. Hayden, and M. Lee, "Analytical Note—Fluorescein Tracer Technique for Detection of Groundwater Contamination," *J. Am. Water Works Assoc.* (October 1979): 586–587.

144. *Handbook—Ground Water*, EPA/625/6-87/016, EPA, Center for Environmental Research Information, Cincinnati, OH, March 1987, pp. 136–145.

145. Schroeder 1974.

146. J. A. Salvato and A. Handley, *Rural Water Supply*, New York State Department of Health, Albany, 1972, pp. 47–50.

147. Goldstein et al., "Continuous Flow Water Pasteurizer," *J. Am. Water Works Assoc.* (February 1960): 247–254.

148. "Policy Statement on Use of the Ultraviolet Process for Disinfection of Water," HEW, PHS, Washington, DC, April 1, 1966.

149. *Municipal Wastewater Reuse News*, AWWA Research Foundation, Denver, CO, November 1977, p. 4.

150. C. Faust, "Performance and Application of Ultraviolet Light Systems," pp. 69-70, and L. T. Rozelle, "Overview of Point-of-Use and Point-of-Entry Systems," in *Proceedings: Conference on Point-of-Use Treatment of Drinking Water*, EPA/600/9-88/012, Water Engineering Research Laboratory, Cincinnati, OH, June 1988.

151. G. E. Bellen et al., *Point-of-Use Treatment to Control Organic and Inorganic Contaminants in Drinking Water*, EPA/600/S2-85/112, EPA, Water Engineering Research Laboratory, Cincinnati, OH, January 1986.

152. L. T. Rozelle, "Point-of-Use and Point-of-Entry Drinking Water Treatment," *J. Am. Water Works Assoc.* (October 1987): 53–59.

153. M. Anderson et al., "Point-of-Use Treatment Technology to Control Organic and Inorganic Contaminants," in *AWWA Seminar Proceedings on Experiences with Groundwater Contamination*, AWWA, Denver, CO, June 10–14, 1984, pp. 37–56.

154. K. R. Fox and T. J. Sorg, "Controlling Arsenic, Fluoride, and Uranium by Point-of-Use Treatment," *J. Am. Water Works Assoc.* (October 1987): 81–84.

155. Ibid.

156. "The Pollutants That Matter Most: Lead, Radon, Nitrate," *Consumer Reports* (January 1990): 31.

157. Fox and Song.

158. Katadyn USA, Inc., Scottsdale, AZ.

159. "Update," *J. Am. Water Works Assoc.* (July 1979): 8.

160. "Home Filters to 'Purify' Water," *Changing Times* (February 1981): 44–47.

161. Third Phase Update, Home Drinking Water Treatment Units Contract, EPA Criteria and Standards Division, Office of Drinking Water, Washington, DC, March 1982; F. A. Bell, Jr.,"Studies on Home Water Treatment Systems," *J. Am. Water Works Assoc.* (April 1984): 126–133.

162. M. E. Burke and G. A. Stasko, "Organizing Water Quality Districts in New York State," *J. Am. Water Works Assoc.* (October 1987): 39–41.

163. J. H. Feth, *Water Facts and Figures for Planners and Managers*, Geological Survey Circular 601-1, National Center, Reston, VA, 1973, p. 14.

164. *Desalting Water Probably Will Not Solve the Nation's Water Problems, But Can Help*, Report to the Congress, General Accounting Offices, Washington, DC, May 1, 1979, p. 1.

165. T. Pankratz, E. Yell, *IDA Desalination Yearbook 2007-2008*, Media Analytics, Ltd., Oxford, UK, 2008.

166. *Using Desalination Technologies for Water Treatment, Office of Technology Assessment*, U.S. Government Printing Office, Washington, DC, March 1988, pp. 1–2.

167. H. A. Faber, S. A. Bresler, and G. Walton, "Improving Community Water Supplies with Desalting Technology," *J. Am. Water Works Assoc.* (November 1972): 705–710.

168. *The A-B-C of Desalting*, U.S. Department of the Interior, Office of Water Research and Technology, Washington, DC, 1977, p. 2.

169. *Using Desalination Technologies for Water Treatment*, 1988, pp. 14, 57.

170. *Municipal Wastewater Reuse News*, AWWA Research Foundation, Denver, CO, Aug. 1979, p. 7.

171. "Committee Report: Membrane Desalting Technologies, AWWA Water Desalting and Reuse Committee," *J. Am. Water Works Assoc.* (November 1989): 30–37.

172. M. E. Mattson, "Membrane Desalting Gets Big Push," *Water & Wastes Eng.* (April 1975): 35–42.

173. *Using Desalination Technologies for Water Treatment*, pp. 14, 57.

174. Ibid.

175. R. Chambers, "Electrodialysis or RO—How Do You Choose?" *World Water*, March 1979, pp. 36–37.

176. Using Desalination Technologies for Water Treatment, pp. 14, 57.

177. Ibid., pp. 25, 31.

178. W. E. Katz and R. Eliassen, *Saline Water Conversion*, "Water Quality & Treatment," AWWA, McGraw-Hill, New York, 1971, p. 610.

179. *The A-B-C of Desalting*, p. 1.

180. *IDA Desalination Yearbook* 2007–2008.

181. Using Desalination Techniques for Water Treatment, pp. 14, 57.

182. E. F. Miller, "Demineralization of Brackish Municipal Water Supplies—Comparative Costs," *J. Am. Water Works Assoc.* (July 1977): 348–351.

183. R. A. Keller, "Seawater RO Desalting Moving into Big League," World Water (London), March 1979, pp. 44–45.

184. J. D. Sinclair, More Efficient MSF Plants Are There to be Specified, *World Water* (March 1979): 33–34.

185. "Desalting Water Probably Will Not Solve the Nation's Water Problems, But Can Help," pp. i, 10.

186. Fed. Reg., CFR 122, June 14, 1979, p. 34269.

187. S. L. Scheffer, "History of Desalting Operation, Maintenance, and Cost Experience," *J. Am. Water Works Assoc.* (November 1972): 726–734.

BIBLIOGRAPHY

2002 *Code of Federal Regulations*, Title 40, Protection of Environment, Part 141, National Primary Drinking Water Regulations, July 2002; and *Factoids: Drinking Water and Groundwater* Statistics for 2007, EPA, Office of Water, Washington, DC, March 2008.

The A-B-C of Desalting, U.S. Department of the Interior, Office of Water Research and Technology, Washington, DC, 1977, p. 2.

A. Bell, Jr., "Studies on Home Water Treatment Systems," *J. Am. Water Works Assoc.* (April 1984): 126–133.

Acuña, H. R., "Health for All by the Year 2000," *Bull. PAHO*, **16**(1), 1982.

Aid Highlights, **3**(1), USAID, Washington, DC, Winter 1986.

Analytical Handbook, Laboratory of Organic Analytical Chemistry, Wadsworth Center for Laboratories and Research, New York State Department of Health, Albany, 1988.

Anderson, D. L., and R. L. Siegrist, "The Performance of Ultra-Low-Flush Toilets in Phoenix," *J. Am. Water Works Assoc.* (March 1989): 52–57.

Anderson, M., et al., "Point-of-Use Treatment Technology to Control Organic and Inorganic Contaminants," in AWWA Seminar Proceedings on Experiences with Groundwater Contamination, AWWA, Denver, CO, June 10–14, 1984, pp. 37–56.

Arlosoroff, Saul, et al., *Community Water Supply: The Handpump Option*, The World Bank, Washington, DC, May 1987.

Arlosoroff, S., *Community Water Supply*, World Bank, Washington, DC, May 1987, p. 1.

Arnold, G. E., "Water Supply Projects in Developing Countries," *J. Am. Water Works Assoc.* (December 1970): 750–753.

Asbestos in Drinking Water Fact Sheet, *Document A 0140, Bureau of Toxic Substances Assessment*, New York State Department of Health, Albany, NY, October 1985.

AWWA MainStream, April 1989, p. 7.

AWWA MainStream, September 1988, p. 13.

Bellen, G. E., et al., *Point-of-Use Treatment to Control Organic and Inorganic Contaminants in Drinking Water*, EPA/600/S2-85/112, EPA, Water Engineering Research Laboratory, Cincinnati, OH, January 1986.

Blinco, S., World Bank Research in Water Supply and Sanitation, The World Bank, Washington, DC, Summer 1982, pp. 3–11; Village Water Supply, The World Bank, Washington, DC, March 1976.

"British Study Links Aluminum to Alzheimer's," *AWWA MainStream* (February 1989): 9.

Brown, H. B., *The Meaning, Significance, and Expression of Commonly Measured Water Quality Criteria and Potential Pollutants*, Louisiana State University and Agricultural and Mechanical College, Baton Rouge, LA, 1957, pp. 35–36.

Buchan, S., and A. Key, "Pollution of Ground Water in Europe," *WHO Bull.*, **14**(5–6): 949–1006 (1956).

Bucher, J., "USEPA Additives Involvement End in April," *AWWA MainStream*, (October 1989): 1.

Buelow, R. W., et al., "The Behavior of Asbestos-Cement Pipe under Various Water Quality Conditions: A Progress Report," *J. Am. Water Works Assoc.* (February 1980): 91–102.

Bull, R. J., and G. F. Craun, "Health Effects Associated with Manganese in Drinking Water," *J. Am. Water Works Assoc.* (December 1977): 662–663.

Burke, M. E., and G. A. Stasko, "Organizing Water Quality Districts in New York State," *J. Am. Water Works Assoc.* (October 1987): 39–41.

Butler, R. G., G. T. Orlob, and P. H. McGauhey, "Underground Movement of Bacterial and Chemical Pollutants," *J. Am. Water Works Assoc.*, **46**(2) (February 1954): 97–111.

Centers for Disease Control, "NIOSH Recommendations for Occupational Safety and Health Standards," MMWR, **35** (Suppl. 1S), 27S (1986).

Clayton, A. J., and P. J. Pybus, "Windhoek Reclaiming Sewage for Drinking Water," *Civil Eng., ASCE* (September 1972): 103–106.

Chambers, R., "Electrodialysis or RO—How Do You Choose?" *World Water* (March 1979): 36–37.

Choi, W.-W., and K. Y. Chen, "The Removal of Fluoride from Waters by Adsorption," *J. Am. Water Works Assoc.* (October 1979): 562–570.

Christensen, E. R., and V. P. Guinn, "Zinc from Automobile Tires in Urban Runoff," *J. Environ. Eng. Div.*, ASCE, (February 1979): 165–168

Civil Eng., ASCE, October 1983, p. 22.

Code of Federal Regulations.

Code of Federal Regulations, National Drinking Water Regulations, 52 FR, Section 141.32 (e), October 28, 1987.

"Committee Report: Membrane Desalting Technologies, AWWA Water Desalting and Reuse Committee," *J. Am. Water Works Assoc.* (November 1989): 30–37

Cotruvo, J. A., and Ch. Wu, *J. Am. Water Works Assoc.* (November 1978): 590–594.

Cousins, M., et al., "Getting the Lead Out—Reducing Lead Concentrations in Water," *Public Works* (November 1987): 68–70.

Criteria for a Recommended Standard: Occupational Exposure to Polychlorinated Biphenyls (PCBs), DHEW Pub. No. (NIOSH) 77–225, NIOSH, DHEW, PHS, Cincinnati, OH, 1977.

Desalting Water Probably Will Not Solve the Nation's Water Problems, But Can Help, Report to the Congress, General Accounting Offices, Washington, DC, May 1, 1979, p. 1.

"Disinfection—Committee Report," *J. Am. Water Works Assoc.* (April 1978): 219–222

"Drinking Water and Health, Recommendations of the National Academy of Sciences," Fed. Reg. **42:132**: 35764, July 11, 1977.

Eaton, A. D., L. S. Clesceri, E. W. Rice, and A. E. Greenberg (Eds.) *Standard Methods for the Examination of Water and Wastewater*, 21st ed., APHA, AWWA, WEF, 2005.

Environmental Conservation Law, Section 15-0107, Albany, NY, 1989.

Environmental Health Ready Reference, Michigan Environmental Health Association, November 1983, pp. 115–120; F. G. Driscoll, *Groundwater and Wells*, 2nd ed., Johnson Division, St. Paul, MN, 1986.

Estimated Water Use in the United States in 2000, U.S. Geological Survey Circular 1268, March 2004 (Revised February 2005).

Faber, H. A., S. A. Bresler, and G. Walton, "Improving Community Water Supplies with Desalting Technology," *J. Am. Water Works Assoc.* (November 1972): 705–710.

Faust, C., Performance and Application of Ultraviolet Light Systems, pp. 69–70, and L. T. Rozelle, Overview of Point-of-Use and Point-of-Entry Systems, in *Proceedings: Conference on Point-of-Use Treatment of Drinking Water*, EPA/600/9-88/012, Water Engineering Research Laboratory, Cincinnati, OH, June 1988.

Fed. Reg., CFR 122, June 14, 1979, p. 34269.

Feth, J. H., *Water Facts and Figures for Planners and Managers*, Geological Survey Circular 601-1, National Center, Reston, VA, 1973, p. 14.

Fletcher, P. W., and W. E. Sharpe, "Water-Conservation Methods to Meet Pennsylvania's Water Needs," *J. Am. Water Works Assoc.* (April 1978): 200–203.

Fox, K. R., and T. J. Sorg, "Controlling Arsenic, Fluoride, and Uranium by Point-of-Use Treatment," *J. Am. Water Works Assoc.* (October 1987): 81–84.

Gaskie, M., and C. C. Johnson Jr., "Face to Face," *J. Am. Water Works Assoc.* (August 1979): 16.

Geldreich, E. E., Is the Total Count Necessary, in *Proceedings*—American Water Works Association, Water Quality Technology Conference, AWWA, Denver, CO, 1974.

Gibb, J. P., and M. J. Barcelona, "Sampling for Organic Contaminants in Ground-water," *J. Am. Water Works Assoc.* (May 1984): 48–51.

Gleick, P. H., Water Resources, *Encyclopedia of Climate and Weather*, Oxford University Press, New York, vol. **2**. 1996, pp. 817–823.

Glossary Water and Wastewater Control Engineering, 3rd ed., PHA, ASCE, AWWA, WPCF, 1981, p. 259.

Goldstein et al., "Continuous Flow Water Pasteurizer," *J. Am. Water Works Assoc.* (February 1960): 247–254.

Gorham, P. R., "Toxic Algae as a Public Health Hazard," *J. Am. Water Works Assoc.*, **56**, 1487 (November 1964).

Gould, R. H., "Growing Data Disputes Algal Treatment Standards," *Water Wastes Eng.* (May 1978): 78–83.

Ground Water Basin Management, ASCE Manual No. 40, ASCE, New York, 1961; Large-Scale Ground-Water Development, United Nations Water Resources Centre, New York, 1960; Guidelines for Delineation of Wellhead Protection Areas, EPA Office of Ground-Water Protection, Washington, DC, June 1987.

Guidelines for Drinking-Water Quality, Vol. **1**, pp. 74–76, 85.

Guidelines for Drinking-Water Quality, Vol. **2**, p. 144.

Guidelines for Drinking-Water Quality, Vols. 1, 2, 3, WHO, Geneva, 1984, 1985; *Manual for Evaluating Public Drinking Water Supplies*, DHEW, Bureau of Water Hygiene, Cincinnati, OH, 1971.

Handbook—Ground Water, EPA/625/6-87/016, Center for Environmental Research Information, Cincinnati, OH, March 1987, p. 76.

Handbook—Ground Water, EPA/625/6-87/016, EPA, Center for Environmental Research Information, Cincinnati, OH, March 1987, pp. 136–145.

Harrington, J. Malcolm, et al., "An Investigation of the Use of Asbestos Cement Pipe for Public Water Supply and the Incidence of Gastrointestinal Cancer in Connecticut, 1935–1973," *Am. J. Epidemiol.*, **107**(2), 96–103 (1978).

Hauser, T. R., "Quality Assurance Update," *Environ. Sci. Technol.* (November 1979): 1356–1366.

Heath, R. C., *Basic Ground Water Hydrology*, USGS Paper 2220, U.S. Government Printing Office, Washington, DC, 1983.

Hoffer, R., The Delineation and Management of Wellhead Protection Areas, in N. Dee, W. F. McTernan, and E. Kaplan (Eds.), *Detection, Control, and Renovation of Contaminated Ground Water*, ASCE, New York, 1987, pp. 143–152.

"Home Filters to 'Purify' Water," *Changing Times* (February 1981): 44–47.

Hopkins, D. R., "Guinea Worm: The Next To Go?" *World Health* (April 1988): 27.

Hydrology Handbook, Manual 28, ASCE, New York, 1949.

Ingram, W. M., and C. G. Prescott, "Toxic Fresh-Water Algae," *Am. Midland Naturalist*, **52**(1), (July 1954): 75–87.

Jones, R. A., and G. Fred Lee, "Septic-Tank Disposal Systems as Phosphorus Sources for Surface Waters," *Tichardson Institute for Environmental Science*, Texas University at Dallas, Dallas, TX, November 1977.

Katadyn USA, Inc., Scottsdale, AZ.

Katz, W. E., and R. Eliassen, "Saline Water Conversion," *Water Quality & Treatment*, AWWA, McGraw-Hill, New York, 1971, p. 610.

Keller, R. A., "Seawater RO Desalting Moving into Big League," *World Water (London)*, March 1979, pp. 44–45.

Keswick, B. H., and C. P. Gerba, "Viruses in Groundwater," *Environ. Sci. Technol.* (November 1980): 1290–1297.

Kincannon, D. F., "Microbiology in Surface Water Sources," *OpFlow*, AWWA, Denver, CO, December 1978, p. 3.

Kosarek, L. J., "Radionuclides Removal from Water," *Environ. Sci. Technol.* (May 1979): 522–525.

Lee, R. G., W. C. Becker, and D. W. Collins, "Lead at the Tap: Sources and Control," *J. Am. Water Works Assoc.* (July 1989): 52–62.

Levin, A., "The Rural Water Survey," *J. Am. Water Works Assoc.* (August 1978): 446–452.

Maddaus, W. O., "The Effectiveness of Residential Water Conservation Measures," *J. Am. Water Works Assoc.* (March 1987): 52–58.

Manual of Individual Water Supply Systems, EPA-570/9-82-004, EPA Office of Drinking Water, U.S. Government Printing Office, Washington, DC, October 1982.

Manual of Water Well Construction Practices, EPA-570/9-75-c01, EPA Office of Water Supply, U.S. Government Printing Office, Washington, DC, 1976, pp. 104–116.

Mattson, M. E., "Membrane Desalting Gets Big Push," *Water & Wastes Eng.* (April 1975): 35–42.

McClelland, N. I., et al., "The Drinking-Water Additives Program," *Environ. Sci. Technol.* (January 1989): 14–18.

Means, E. G. III, and M. J. McGuire, "An Early Warning System for Taste and Odor Control," *J. Am. Water Works Assoc.* (March 1986): 77–83.

Meigs, J. Wister, et al., Asbestos Cement Pipe and Cancer in Connecticut, 1955–1974, *J. Environ. Health*. Microphor Inc., P.O. Box 1460, Willits, CA 95490.

Miller, E. F., "Demineralization of Brackish Municipal Water Supplies—Comparative Costs," *J. Am. Water Works Assoc.* (July 1977): 348–351.

M. Lee, "Analytical Note—Fluorescein Tracer Technique for Detection of Groundwater Contamination," *J. Am. Water Works Assoc.* (October 1979): 586–587.

Municipal Wastewater Reuse News, AWWA Research Foundation, Denver, CO, Aug. 1979, p. 7.

Municipal Wastewater Reuse News, AWWA Research Foundation, Denver, CO, November 1977, p. 4.

Murray, C. R., and E. B. Reeves, *Estimated Use of Water in the United States in 1975*, Geological Survey Circular 765, U.S. Department of the Interior, Washington, DC.

National Technical Advisory Committee Report, "Raw-Water Quality Criteria for Public Water Supplies," *J. Am. Water Works Assoc.*, **61**, 133 (March 1969).

New York Times, September 25, 1988, p. 34.

NIOSH Recommendations for Occupational Safety and Health Standards 1988, MMWR, Supplement, August 1988, p. 5.

Okun, D. A., "The Use of Polluted Sources of Water Supply," *APWA Reporter* (September 1976): 23–25.

Ongerth, J. E., "Giardia Cyst Concentrations in River Water," *J. Am. Water Works Assoc.* (September 1989): 71–80.

Palmer, C. M., *Algae in Water Supplies*, PHS Pub. No. 657, DHEW, Cincinnati, OH, 1959.

"PCBs in Old Submersible Well Pumps," *Dairy, Food and Environmental Sanitation* (January 1989): 35–36.

"Policy Statement on Use of the Ultraviolet Process for Disinfection of Water," HEW, PHS, Washington, DC, April 1, 1966.

"The Pollutants that Matter Most: Lead, Radon, Nitrate," *Consumer Reports* (January 1990): 31.

"Proposed by State of Washington Department of Ecology," *New York Times*, November 6, 1983.

Protection of Public Water Supplies from Ground-Water Contamination, Seminar Publication, EPA/625/4-85/016, Center for Environmental Research Information, Cincinnati, OH, September 1985.

R. B. Dole, "Use of Fluorescein in the Study of Underground Waters," U.S. Geol. Survey, Water-Supply Paper 160, 1906, pp. 73–85; S. Reznek, W. Hayden, and Recommended Standards for Water Works, A Report of the Committee of the Great Lakes-Upper Mississippi River Board of State Public Health and Environmental Managers, Health Education Service Division, Albany, NY, 1987, pp. 18–32.

Report on the Investigation of Travel of Pollution, Pub. No. 11, State Water Pollution Control Board, Sacramento, CA, 1954.

Residential Water Conservation Projects—Summary Report, HUD-PDR-903, HUD Washington, DC, June 1984.

Rozelle, L. T., "Point-of-Use and Point-of-Entry Drinking Water Treatment," *J. Am. Water Works Assoc.* (October 1987): 53–59.

Rural Wastewater Management, State of California Water Resources Board, Sacramento, CA, 1979, pp. 11–14.

Salvato, J. A., and A. Handley, *Rural Water Supply*, New York State Department of Health, Albany, 1972, pp. 47–50.

Schad, T. M., "Ground Water Classification: Goals and Basis," in P. Churchill and R. Patrick (Eds.), *Ground Water Contamination: Sources, Effects and Options to Deal with the Problem*, p. 335, Proceedings of the Third National Water Conference, Philadelphia, January 13–15, 1987, Academy of Natural Sciences, Philadelphia, PA, 1987.

Scheffer, S. L., "History of Desalting Operation, Maintenance, and Cost Experience," *J. Am. Water Works Assoc.* (November 1972): 726–734.

Schorr, P., and R. T. Dewling, "Reusing Water," *Civil Eng., ASCE* (August 1988): 69–71.

Schowengerdt, R. W., "An Overview of Ground Water Techniques." In N. Dee, W. F. McTernan, and E. Kaplan (Eds.), *Detection, Control, and Renovation of Contaminated Ground Water*, ASCE, New York, 1987, pp. 35–50.

Schroeder, H. A., "Environmental Metals: The Nature of the Problem," in W. D. McKee (Ed.), *Environmental Problems in Medicine*, Charles C. Thomas, Springfield, IL, 1974.

Schroeder, H. A., *New York State Environment*, New York State Department of Environmental Conservation, Albany, April 1976.

Sharpe, W. E., "Water and Energy Conservation with Bathing Shower Flow Controls," *J. Am. Water Works Assoc.* (February 1978): 93–97.

Sinclair, J. D., "More Efficient MSF Plants Are There to Be Specified," *World Water* (March 1979): 33–34.

Solley, W. B., E. B. Chase, and W. B. Mann IV, *Estimated Use of Water in the United States in 1980*, Geological Survey Circular 1001, U.S. Department of the Interior, Washington, DC, 1983.

Solt, G. S., "High Purity Water," *J. R. Soc. Health* (February 1984): 6–9.

Stiles, C. W., H. R. Crohurst, and G. E. Thomson, *Experimental Bacterial and Chemical Pollution of Wells via Ground Water, and the Factors Involved*, PHS Bull. 147, DHEW, Washington, DC, June 1927.

The Story of Drinking Water, Catalog No. 70001, AWWA, Denver, CO, p. 13.

Third Phase Update, Home Drinking Water Treatment Units Contract, EPA Criteria and Standards Division, Office of Drinking Water, Washington, DC, March 1982; F

"Update," *J. Am. Water Works Assoc.* (December 1978): 9.

"Update," *J. Am. Water Works Assoc.* (February 1989).

"Update," *J. Am. Water Works Assoc.* (July 1979): 8.

USEPA, *Needs Survey Report to Congress*, Environmental Protection Agency, Washington, DC, 1992.

U.S. EPA Statement, "Chlorinated and Brominated Compounds Are Not Equal," *J. Am. Water Works Assoc.*, October 1977, p. 12.

"Use of Reclaimed Waste Water as a Public Water Supply Source," *1978–79 Officers and Committee Directory Including Policy Statements and Official Documents*, AWWA, Denver, CO, 1979, p. 78.

Using Desalination Technologies for Water Treatment, Office of Technology Assessment, U.S. Government Printing Office, Washington, DC, March 1988, pp. 1–2.

Varma, M. M., S. G. Serdahely, and H. M. Katz, "Physiological Effects of Trace Elements and Chemicals in Water," *J. Environ. Health* (September/October 1976): 90–100.

Waste Disposal Practices and Their Effects on Ground Water, Report to Congress, Executive Summary, EPA, Washington, DC, January 1977.

Water Environ. Technol. (January 1991).

Water Quality Criteria 1972, NAS, National Academy of Engineering, Washington, DC, 1972, p. 87.

Water Quality Criteria 1972, National Academy of Sciences—National Academy of Engineering, EPA-R3-73-033, U.S. Government Printing Office, Washington, DC, March 1973, p. 177.

"Water Should Be Segregated by Use," *Water Sewage Works* (March 1979): 52–54; D. A. Okun, "Wastewater Reuse Dilemma," ESE Notes, University of North Carolina at Chapel Hill, November 1975.

Water Well Location by Fracture Trace Mapping, Technology Transfer, Office of Water Research and Technology, Department of the Interior, U.S. Government Printing Office, Washington, DC, 1978.

WHO Report, Global Water Supply and Sanitation Assessment 2000.

Wilen, B. O., "Options for Controlling Natural Organics," in R. B. Pojasek (Ed.), *Drinking Water Quality Enhancement through Source Protection*, Ann Arbor Science, Ann Arbor, MI; "Research Committee on Color Problems Report for 1966," *J. Am. Water Works Assoc.* (August 1967): 1023–1035.

Wolman, A., "Reflections, Perceptions, and Projections," *J. Water Pollut. Control Fed.* (December 1983): 1412–1416.

Wolman, A., "Water Supply and Environmental Health," *J. Am. Water Works Assoc.* (December 1970): 746–749.

CHAPTER 2

WATER TREATMENT

T. DAVID CHINN
Professional Engineer, Senior Vice President, HDR Engineering, Austin, Texas

TREATMENT OF WATER – DESIGN AND OPERATION CONTROL

Introduction

Safe, abundant, and affordable; these are the primary goals of water treatment professionals across the globe. No matter how poor the original source of water supply, the finished drinking water that emerges from the consumer's tap must be a high quality free from pathogenic (or disease-causing) microorganisims. It must also not contain concentrations of either natural or manmade contaminants in concentrations that could produce adverse health impacts. The aesthetics of drinking water are important since consumers will link the appearance, taste, and odor of the water to its safety. Although the amount of tap water that is actually consumed by humans or used for cooking and food preparation is less than 5 percent that enters the home, *all* of the water treated must meet these goals for safety and aesthetics. Other nonpotable uses such as washing, flushing wastes, irrigation, and so on must also meet acceptable standards so the water treatment plant must produce a sufficient quantity of water that meets all these needs. Achieving these objectives at a reasonable and affordable cost is perhaps the greatest challenge facing the water community today.

Water treatment in the twenty-first century accomplishes these goals using a range of principals and practices, some new, some not so modern. The quest for potable water dates as far back as the earliest recorded history. Water treatment is described in Sanskrit and Greek writings 6,000 years ago:

> Impure water should be purified by being boiled over a fire, or heated in the sun or by dipping a heated iron into it and then allowed to cool, or it may be purified through sand and coarse gravel.[1]

In 1854, Dr. John Snow's landmark epidemiological studies linked for the first time, a contaminated water well to an outbreak of cholera in London, although he didn't know exactly why. (The now infamous well located at No. 40 Broad Street, London is marked by a plaque commemorating this achievement.) This question was answered following Louis Pasteur's novel "germ theory" linking microorganisims to disease in the late 1880s. Soon thereafter, modern water treatment was born and at the turn of the last century, centralized water treatment facilities for communities became standard in the United States. By the 1920s, "state-of-the-art" water treatment consisted of sand filtration and chlorine disinfection and the threat of waterborne disease outbreaks such as typhoid and cholera were virtually eliminated.

Today, newer technologies such as low pressure polymetric membranes have improved the ability to filter impurities from raw water and will ultimately replace sand filters. Similarly, disinfection practices have improved to balance the chemical's use; a sufficient dose to destroy microorganisms but low enough not to produce harmful halenogated byproducts.

Surface Water

The quality of surface water depends on the watershed area drained, land use, location and sources of natural and manmade pollution, and natural agencies of purification, such as sedimentation, sunlight, aeration, nitrification, filtration, and dilution. Since these are variable, they cannot be depended on to continuously purify water effectively. However, large reservoirs providing extended storage permit natural purification to take place, but short-circuiting and direct contamination must be avoided. In addition, increasing urbanization, industrialization, and intensive farming have caused heavy organic and inorganic chemical discharges to streams, which are not readily removed by the usual water treatment. Treatment consisting of coagulation, flocculation, sedimentation, rapid sand filtration, and chlorination has little effect on some chemical contaminants noted. Because of these factors and to reduce risk, heavily polluted surface waters should be avoided as drinking water supplies, if possible, and upland protected water sources should be used and preserved consistent with multipurpose uses in the best public interest:

> The American Water Works Association (AWWA) is dedicated to securing drinking water from the highest quality sources available and protecting those sources to the maximum degree possible.[2]

The growing demand for use of reservoirs for recreational purposes requires that the public understand the need for strict controls to prevent waterborne diseases and watershed disturbance. Involved are added capital and maintenance and operating costs that may increase the charges for the water and use of the recreational facilities, if the multipurpose uses are permitted.

Treatment Required

The treatment required is dependent on the federal and state regulations and on the probable changing physical, chemical, and microbiological quality of the water source. This emphasizes the importance of adequate meteorological and hydrological information, the sanitary survey previously discussed, and its careful evaluation. The evaluation should take into consideration the existing land-use zoning and probable development. Water treatment plants should not have to bear the total burden and cost of elaborate treatment because of water pollution of its source water. The water purveyor should therefore take an active role in stream, lake, and land-use classifications and be aware of all existing and proposed industrial and municipal wastewater outlets and nonpoint pollution sources. The pollution from these sources should ideally be eliminated or minimized to the extent possible and adequately treated. Continual supervision and enforcement of watershed, land-use, and wellhead area protection rules and regulations must be assured.

The EPA rules, based on the 1986 Amendments to the Safe Drinking Water Act, require the following:[3]

1. Surface water complete filtration treatment if one of these two conditions applies:
   a. fecal coliforms exceed 20 per 100 ml
   b. total coliforms exceed 100 per 100 ml in more than 10 percent of the measurements for the previous 6 months, calculated each month
2. Minimum sampling frequency for fecal or total coliforms per week:
   a. 1 for systems serving fewer than 501 people
   b. 2 for systems serving 501 to 3,300
   c. 3 for systems serving 3,301 to 10,000
   d. 4 for systems serving 10,001 to 25,000
   e. 5 for systems serving 25,000 or more
3. Turbidity measurements every 4 hours; once a day for systems serving less than 501 people. Filtration treatment is required if turbidity level exceeds 5 NTU unless the state determines that the event is unusual.
4. Treatment to achieve at least 99.9 percent removal or inactivation of *Giardia lamblia* cysts (also *Cryptosporidium*) and 99.99 percent removal or inactivation of viruses, also *Legionella*.
5. Maintenance of disinfecting residuals in the distribution system—not less than 0.2 mg/l chlorine in at least 95 percent of the samples tested.
6. A watershed protection program; annual sanitary survey; absence of waterborne disease outbreaks; compliance with the total coliform and trihalomethane maximum contaminant levels (MCLs); turbidity of 0.5 NTU in 95 percent of monthly measurements; certified operators; and increased monitoring and reporting.

People expect the water to be safe to drink, attractive to the senses, soft, non-staining, and neither scale forming nor corrosive to the water system. The various treatment processes used to accomplish these results are briefly discussed under the appropriate headings below. In all cases, the water supply must meet the federal and state drinking water standards.[4] The untrained individual should not attempt to design a water treatment plant, for public health will be jeopardized. This is a job for a competent environmental engineer. Submission and approval of plans and specifications are usually required by the regulatory agency.[5] Computerized control of water treatment and distribution is considered essential to a greater or lesser degree, dependent on the operator skills and immediate availability of manufacturer assistance.

Disinfection

The more common chemicals used for the disinfection of drinking water are chlorine (gas and hypochlorite), chlorine–ammonia, chlorine dioxide, and ozone. Chlorine is discussed next; the others are discussed in relation to the removal or reduction of objectionable tastes and odors and trihalomethanes. Ozone and chlorine dioxide are receiving greater attention as primary disinfectants and chlorine–ammonia for maintenance of a residual in the distribution system. Other disinfectants that may be used under certain circumstances include UV radiation,* bromine, iodine, silver, and chlorinated lime.

The National Research Council–National Academy of Science, in a study of disinfectants, concluded that there had not been sufficient research under actual water treatment conditions for the reactions of disinfectants and their byproducts to be adequately understood and that the chemical side effects of disinfectants "should be examined in detail."[6]† There is need to identify the byproducts associated with the use of not only chlorine but also chloramines, ozone, and chlorine dioxide and their health significance.

Chlorination is the most common method of destroying the disease-producing organisms that might normally be found in water used for drinking in the United States. The water so treated should be relatively clear and clean with a pH of 8.0 or less and an average monthly MPN of coliform bacteria of not more than 50/100 ml.‡ Clean lake and stream waters and well, spring, and infiltration gallery

---

*UV disinfection is being used as a primary disinfectant more in Europe (approximately 2000), with free chlorine for residual maintenance. (R. L. Wolfe, "Ultraviolet Disinfection of Potable Water," *Environ. Sci. Technol.*, June 1990, pp. 768–772.)

†Greenburg points out that "the health effects of their (chlorine dioxide and ozone) reaction products, particularly the chlorite ion from chlorine dioxide and oxidized organic compounds from ozone are uncertain" and adds that "if unequivocal safety information becomes available, changes from chlorine to chlorine dioxide or ozone may be indicated but only if the manipulation of chlorination methods proves incapable of minimizing carcinogen hazard." (A. E. Greenburg, "Public Health Aspects of Alternative Water Disinfectants," *J. Am. Water Works Assoc.* (January 1981): 31–33.)

‡Suggested criteria include total coliform <100/100 ml, fecal coliform <20/100 ml, turbidity <1–5 NTU, color 15 units, chlorine demand $\not> 2$ mg/l, plus others.

supplies not subject to significant pollution can be made of safe sanitary quality by continuous and effective chlorination, but surface sources also usually require complete filtration treatment to protect against viruses, bacteria, protozoa, and helminths. The effectiveness of chlorine is dependent on the water pH, temperature, contact time, water clarity, and absence of interfering substances.

Operation of the chlorinator should be automatic, proportional to the flow of water, and adjusted to the temperature and chlorine demand of the water. A standby source of power and a spare machine including chlorine should be on the line. A complete set of spare parts for the equipment will make possible immediate repairs. The chlorinator should provide for the positive injection of chlorine and be selected with due regard to the pumping head and maximum and minimum water flow to be treated. The point of chlorine application should be selected to provide a contact time of 2 hours for surface water receiving free residual chlorination treatment and 3 hours with combined residual chlorination. A lesser time may be accepted for groundwater.[7] The chlorinator should have a capacity to provide at least 2 mg/l free chlorine residual after 30 minutes contact at maximum flow and chlorine demand.

Hypochlorinators are generally used to feed relatively small quantities of chlorine as 1 to 5 percent sodium or calcium hypochlorite solution. Positive feed machines are fairly reliable and simple to operate. Hypochlorite is corrosive and may produce severe burns. It should be stored in its original container in a cool, well-ventilated, dry place. Gas machines usually feed larger quantities of chlorine and require certain precautions as noted next. Chlorine addition, with either a hypochlorinator or gas machine, should be proportional to the flow, direct or through corrosion-resistant piping; iron or steel piping or fittings should not be used. Note that the addition of an acid such as ferric chloride to sodium hypochlorite will release chlorine gas.

Gas Chlorinator

When a dry feed gas chlorinator or a solution feed gas chlorinator is used, the chlorinator and liquid chlorine cylinders should be located in a separate gas-tight room that is mechanically ventilated to provide one air change per minute, with outside switch and the exhaust openings at floor level opposite the air inlets at ceiling level. Exhaust ducts must be separate from any other ventilating system of ducts and extend to a height and location that will not endanger the public, personnel, or property and ensure adequate dispersion. The door to the room should have a shatter-resistant glass inspection panel at least 12-inches square, and a chlorine gas mask, or preferably self-contained breathing apparatus, approved by the NIOSH, available just outside of the chlorinator and chlorine cylinder room. Vapor from a plastic squeeze bottle containing aqua ammonia will produce a white cloud at a chlorine leak.* The chlorine canister-type of mask is only suitable for

---

*In an emergency, do not try to neutralize chlorine; leave this to the professionals. Call CHEMTREC at 800-424-9300 or the nearest supplier or producer. The permissible 8 hours concentration exposure is 1 ppm, 3 ppm for 15 min.

low concentrations of chlorine in air and only for a brief period. It does not supply oxygen. The self-contained breathing apparatus* with full-face piece (pressure demand) with at least 30-minute capacity meeting NIOSH standards is usually required. It can be used during repairs and for high concentrations of chlorine. A factory-built chlorinator housing, completely equipped, is available.

The temperature around the chlorine cylinders should be cooler than the temperature of the chlorinator room to prevent condensation of chlorine in the line conducting chlorine or in the chlorinator. Cylinders must be stored at a temperature below 140°F.[8][†] A platform scale is needed for the weighing of chlorine cylinders in use to determine the pounds of chlorine used each day and anticipate when a new cylinder will be needed. Cylinders should be in a safety bracket or chained to prevent being tipped. They should be connected to a manifold to allow chlorine to be drawn from several cylinders at a time and to facilitate cylinder replacement without interrupting chlorination. It is advisable to not draw more than 35 to 40 pounds of chlorine per day at a continuous rate from a 100- or 150-pound cylinder to prevent clogging by chlorine ice. Liquid chlorine comes in 100- and 150-pound cylinders, in 1-ton containers, and in 16- to 90-ton rail-tank cars. Smaller cylinders, as little as 1 pound, are available. The major factors affecting withdrawal rates are ambient air temperature and size and type cylinder. The normal operating temperature is 70°F (21°C).

A relatively clear source of water of adequate volume and pressure is necessary to prevent clogging of injectors and strainers and ensure proper chlorination at all times. The water pressure to operate a gas chlorinator should be at least 15 psi and about three times the back pressure (water pressure at point of application plus friction loss in the chlorine solution hose and a difference in elevation between the point of application and the chlorinator) against which the chlorine is injected. About 40 to 50 gpd of water is needed per pound of chlorine to be added. Residual chlorine recorders and alarms and chlorine feed recorders provide additional protection and automatic residual chlorine control.

Testing for Residual Chlorine

The recommended field tests for measuring residual chlorine in water are the N,N-diethyl-p-phenylenediamine (DPD) colorimetric and the stabilized neutral orthotolidine (SNORT) methods.[9] The DPD and amperometric titration methods are approved by the EPA. In any case, all tests should be made in accordance with accepted procedures such as in *Standard Methods for the Examination of Water and Wastewater*.[10]

The DPD test procedure for residual chlorine measurement and the Free Available Chlorine Test, syringaldazine (FACTS) test procedure are reported to be

---

*At least two units are recommended, including worker protective clothing.
†The fusible plugs are designed to soften or melt at a temperature between 158 and 165°F (70 and 74°C). The chlorinator should have automatic shutoff if water pressure is lost or if chlorine piping leaks or breaks. See *The Chlorine Manual*, 5th ed., Chlorine Institute, Washington, DC, 1986.

equivalent. The FACTS and amperometric procedures are also equivalent.[11] A comprehensive evaluation of residual chlorine, chlorine dioxide, and ozone measurement methods is available.[12] The use of dry reagents is recommended for the DPD test as the liquid form is unstable. High concentrations of iron and manganese and dirty glassware cause interference with residual chlorine readings. The evaluation should be read immediately to also minimize interference from chloramines.

Chlorine Treatment for Operation and Microbiological Control

To ensure that only properly treated water is distributed, it is important to have a competent and trustworthy person in charge of the chlorination plant. He or she should keep daily records showing the gallons of water treated, the pounds of chlorine or quarts of chlorine solution used and its strength, the gross weight of chlorine cylinders if used, the setting of the chlorinator, the time residual chlorine tests made, the results of such tests, and any repairs or maintenance, power failures, modifications, or unusual occurrences dealing with the treatment plant or water system. Where large amounts of chlorine are needed, the use of ton containers can effect a saving in cost, as well as in labor, and possibly reduce chlorine gas leakage, although if a chlorine leak does occur, it can be of major consequence.

The required chlorine dosage should take into consideration the appearance as well as the quality of a water. Pollution of the source of water, the type of micro-organisms likely to be present, the pH of the water, contact time, interfering substances, temperature, and degree of treatment a water receives are all very important. Disinfection effectiveness is also dependent on the absence of turbidity, less than 1 NTU.

The chlorine residual that will give effective disinfection of a relatively demand-free *clear* water has been studied by Butterfield[13] and others. The germicidal efficiency of chlorine is primarily dependent on the percent-free chlorine that is in the form of hypochlorous acid (HOCl), which in turn is dependent on the pH, contact time, and temperature of the water, as can be seen in Table 2.1. Hypochlorous acid is about 80 to 150 times more effective than the hypochlorite ion, 150 times more effective than monochloramine, and 80 times more effective than dichloramine. The percentage of hypochlorous acid is the major factor determining destruction or inactivation of enteric bacteria and amebic cysts.[14] *Giardia* cysts are almost always present in raw sewage.

In a review of the literature, Greenberg and Kupka concluded that a chlorine dose of at least 20 mg/l with a contact time of 2 hours is needed to adequately disinfect a biologically treated sewage effluent containing tubercle bacilli.[15]

Laboratory studies by Kelly and Sanderson indicated that, depending on pH level and temperature, residual chlorine values of greater than 4 ppm, with 5-minutes contact, or contact periods of at least 4 hours with a residual chlorine value of 0.5 ppm, are necessary to inactivate viruses, and that the recommended standard for disinfection of sewage by chlorine (0.5 ppm residual after 15-minutes contact) does not destroy viruses.[16]

TABLE 2.1 Chlorine Residual for Effective Disinfection of Demand-Free Water (mg/l)

pH	Approximate Percent at 32–68°F[b]		Bactericidal Treatment[a]		Cysticidal Treatment Free Available Chlorine after 30 min		
	HOCl	OCl⁻	Free Available Chlorine after 10 min, 32–78°F	Combined Available Chlorine after 60 min, 32–78°F	36–41°F[c]	60°F[c]	78°F[b]
5.0	—	—	—			2.3	1.9[d]
6.0	98–97	2–3	0.2	1.0	7.2		2.5[d]
7.0	83–75	17–25	0.2	1.5	10.0	3.1	2.6[d]
7.2	74–62	26–38	—				2.8[d]
7.3	68–57	32–43	—				3.0[d]
7.4	64–52	36–48	—				3.2[d]
7.5	58–47	42–53	—		14.0[d]	4.7	3.5[d]
7.6	53–42	47–58	—				3.8[d]
7.7	46–37	53–64	—		16.0[d]	6.0	4.2[d]
7.8	40–32	60–68	—				5.0[d]
8.0	32–23	68–77	0.4	1.8	22.0	9.9	
9.0	5–3	95–97	0.8	Reduce pH of water to below 9.0		78.0	20.0[d]
10.0	0	100	0.8			761	170[d]

[a] Ethylene Glycol Intoxication Due to Contamination of Water Systems," MMWR (September 18, 1987): 611–614.

[b] Water Treatment Plant Design, American Water Works Association, New York, 1969, pp. 153, 165; E. W. Moore, "Fundamentals of Chlorination of Sewage and Waste," Water Sewage Works, 130–136 (March 1951).

[c] S. L. Chang, "Studies on Endamoeba histolytica," War Med., 5, 46 (1944); see also W. Brewster Snow, "Recommended Chlorine Residuals for Military Water Supplies," J. Am. Water Works Assoc., 48, 1510 (December 1956).

[d] Approximations. All residual chlorine results reported as milligrams per liter. One milligram per liter hypochlorous acid gives 1.35 mg/l free available chlorine as HOCl and OCl⁻ distributed as noted above. The HOCl component is the markedly superior disinfectant, about 80–150 times more effective than the hypochlorite ion (OCl⁻).

Note: Free chlorine = HOCl. Free available chlorine = HOCl + OCl⁻. Combined available chlorine = chlorine bound to nitrogenous matter as chloramine. Only free available chlorine or combined available chlorine is measured by present testing methods; therefore, to determine actual free chlorine (HOCl), correct reading by percent shown above. "Chlorine residual," as the term is generally used, is the combined available chlorine and free available chlorine, or total residual chlorine. When the chlorine–ammonia ratio reaches 15:1 or 20:1 and pH < 4.4, nitrogen trichloride is formed; it is acrid and highly explosive. Ventilate.

Viricidal treatment requires a free available chlorine of 0.53 mg/l at pH 7 and 5 mg/l at pH 8.5 in 32°F demand-free water. For water at a temperature of 77–82.4°F and pH 7–9, a free available chlorine of 0.3 mg/l is adequate. (See Manual for Evaluating Public Drinking Water Supplies, PHS Pub. 1820, Environmental Control Administration, Cincinnati, OH, 1969.) At a pH 7 and temperature of 77°F at least 9 mg/l combined available chlorine is needed with 30 minutes contact time. Turbidity should be less than one turbidity unit. The above results are based on studies made under laboratory conditions using water free of suspended matter and chlorine demand.

Another study showed that inactivation of partially purified poliomyelitis virus in water required a free-residual chlorine after 10 minutes of 0.05 mg/l at a pH of 6.85 to 7.4. A residual chloramine value of 0.50 to 0.75 mg/l usually inactivated the virus in 2 hours.[17] Weidenkopf reported on Polio 1 inactivation by free chlorine as 0.1 mg/l at pH 6.0 and 0.53 mg/l at pH 8.5.[18] Destruction of coxsackievirus required 7 to 46 times as much free chlorine as for *E. coli*.[19] Infectious hepatitis virus was not inactivated by 1.0 mg/l total chlorine after 30 minutes or by coagulation, settling, and filtration (diatomite), but coagulation, settling, filtration, and chlorination to 1.1 mg/l total and 0.4 mg/l free chlorine was effective.[20] Bush and Isherwood suggest the following:

> The use of activated sludge with abnormally high sludge volume index followed by sand filtration may produce the kind of control necessary to stop virus. Chlorination with five tenths parts per million chlorine residual for an eight hour contact period seems adequate to inactivate Coxsackie virus.[21]

Malina[22] summarized the effectiveness of water and wastewater treatment processes on the removal of viruses. The virus concentration in untreated municipal wastewater was found to range from about 200 plaque-forming units per liter (PFU/l) in cold weather to about 7,000 in warm months in the United States, with 4,000 to 7,000 PFU/l common. In contrast, the virus concentration in South Africa was found to be greater than 100,000 PFU/l. Virus removal in wastewater is related in part to particulate removal. Possible virus removal values by various wastewater treatment systems are as follows:

Primary sedimentation	0–55%
Activated sludge	64–99%
Contact stabilization	74–95%
Trickling filters	19–94%
Stabilization ponds	92–100%
Coagulation–flocculation	86–100%
Chlorine (as final treatment)	99–100%
Iodine	100%
Ozone	100%
Anaerobic digestion	62–99%

Chemical coagulation, with adequate concentrations of aluminum sulfate or ferric chloride, of surface water used as a source of drinking water or of wastewater that has received biological treatment can remove 99 percent of the viruses. Hepatitis A virus, rotavirus, and poliovirus removal of 98.4 to 99.7 percent was also achieved in a pilot plant by softening during Ca^{2+} and Mg^{2+} hardness reduction.[23] A high pH of 10.8 to 11.5, such as softening with excess lime, can achieve better than 99 percent virus removal, but pH adjustment is then necessary.

Filtration using sand and/or anthracite following coagulation, flocculation, and settling can remove 99 percent or more of the viruses, but some viruses penetrate the media with floc breakthrough and turbidity at low alum feed.[24] Diatomaceous earth filtration can remove better than 98 percent of the viruses, particularly if the water is pretreated. Activated-carbon adsorption is not suitable for virus removal. The infectivity of hepatitis A virus is destroyed by 2.0 to 2.5 mg/l free residual chlorine. Reverse osmosis and ultra filtration, when followed by disinfection, can produce a virus-free water. However, it has been found that both enteroviruses and rotaviruses could be isolated from water that received complete treatment containing more than 0.2 mg/l free chlorine, less than one coliform bacteria per 100 ml, and turbidity of less than 1 NTU.[25] The WHO states that a contaminated source water may be considered adequately treated for viruses infectious to humans if it has a turbidity of 1 NTU or less and is disinfected to provide a free residual chlorine of at least 0.5 mg/l after a contact period of at least 30 minutes at a pH below 8.0.[26]

A conventional municipal biological wastewater treatment plant can produce an effluent with less than 10 PFU/l. When followed by conventional water treatment incorporating filtration and chlorination, a virus-free water can be obtained.[27] The product of the contact time (t) in minutes and free residual chlorine, or other approved disinfectant, in milligrams per liter (C) produces a value that is a measure of the adequacy of disinfection. The Ct value for a particular organism will vary with the water pH, temperature, degree of mixing, turbidity, and presence of interfering substances, in addition to disinfectant concentration and contact time. For example, a turbidity less than 1 to 5 NTUs and a free chlorine as HOCl (that penetrates the cell wall of microorganisms and destroys their nucleic acid) are necessary. A smaller Ct value is effective at lower pH and at higher temperature.[28] The Ct value effective to inactivate 99.9 percent of the Giardia cyst will also inactivate 99.99 percent or greater of the bacteria and viruses at a given pH and temperature.

For protozoa (*Giardia lamblia*) a Ct value of 150 to 200 at pH 8.0 or less is required with water at 50°F (10°C). Experimental results based on animal infectivity data show that a 99.99 percent cyst inactivation can be obtained at Ct values of 113 to 263, at pH 6, temperature 33°F (0.5°C), and chlorine concentration of 0.56 to 3.93 mg/l for 39 to 300 minutes. At pH 8, temperature 33°F (0.5°C), chlorine concentration of 0.49 to 3.25 mg/l, and contact time of 132 to 593 minutes, the Ct values varied from 159 to 526. If a large enough Ct value can be maintained to ensure adequate *Giardia* cyst disinfection to EPA satisfaction, then filtration may not be required.[29]

Chlorine dioxide can achieve 99.9 percent *Giardia lamblia* inactivation at Ct values of 63, in water at 34°F (1°C) or less, to 11 at 77°F (25°C) or greater. Inactivation using ozone is achieved at Ct values of 2.9, in water at 34°F (1°C) or less, to 0.48 at 77°F (25°C) or greater. These Ct values also achieve greater than 99.99 percent inactivation of enteric viruses. See state regulatory agency for required Ct values to inactivate *Giardia lamblia* and enteric viruses using chlorine, chloramine, chlorine dioxide, and ozone.

Naegleria fowleri cyst is a pathogenic flagellated protozoan. It causes primary amebic meningoencephalitis, a rare disease generally fatal to humans. The organism is free living, nonparasitic, found in soil and water. *Naegleria gruberi*, a nonpathogenic strain, was used in experimental inactivation studies. At pH 5.0 the *N. gruberi* cyst was inactivated in 15.8 to 2.78 minutes by 0.45 to 2.64 mg/l free chlorine residual at 25°C (77°F). At pH 7.0 it was inactivated in 21.5 to 2.9 minutes by 0.64 to 3.42 mg/l; and at pH 9.0 in 11.5 to 2.36 minutes by 15.4 to 87.9 mg/l residual chlorine. Also, it was reported that *Acanthamoeba* sp. 4A cysts (pathogenic) were inactivated after 24 hours by an initial chlorine dose of 8.0 mg/l but ending with a chlorine residual of 6.0 mg/l. *Naegleria fowleri* was reported to be inactivated by 4 mg/l chlorine residual in 10 minutes at a temperature of 77°F (25°C) and a pH of 7.2 to 7.3.[30] *Giardia lamblia* cysts are inactivated in 60 minutes by 2.0 mg/l free chlorine residual at pH 6.0 and 41°F (5°C); in 60 minutes by 2.5 mg/l free chlorine at pH 6.0, 7.0, and 8.0 and 60°F (15°C); in 10 minutes by 1.5 mg/l free chlorine at pH 6.0, 7.0, and 8.0 and 77°F (25°C); and in 30 minutes by 6.2 mg/l total chlorine at pH 7.9 and 37°F (3°C).[31]

Entamoeba histolytica cysts are inactivated by 2 mg/l free chlorine in 15 minutes at a temperature of 68°F (20°C) and pH 7.0,[32] by 2.5 mg/l free chlorine in 10 minutes at a temperature of 86°F (30°C) and pH 7.0, by 5.0 mg/l free chlorine in 15 minutes at a temperature of 50°F (10°C) and pH 7.0, and by 7.0 mg/l free chlorine in 10 to 15 minutes at a temperature of 86°F (30°C) and pH 9.0.[33]

The removal of nematodes requires prechlorination to produce 0.4 to 0.5 mg/l residual after a 6-hour retention period followed by settling. The pathogenic fungus *Histoplasma capsulatum* can be expected in surface-water supplies, treated water stored in open reservoirs, and improperly protected well-water supplies. Fungicidal action is obtained at a pH of 7.4 and at a water temperature of 78.8°F (26°C) with 0.35 mg/l free chlorine after 4 hour contact and with 1.8 mg/l free chlorine after 35 minutes contact. Complete rapid sand filter treatment completely removed all viable spores even before chlorination.[34]

Cysts of *E. histolytica* and *Giardia lamblia* (also worms and their eggs) are removed by conventional water treatment, including coagulation, flocculation, sedimentation, and filtration (2–6 gpm/ft^2). Direct and high-rate filtration, diatomaceous earth filtration with good precoat (1 kg/m^2), and special cartridge filters ($<7 \times 8$-μm pore size) can also be effective. The slow sand filter is also considered effective. Pressure sand filtration is not reliable. The inactivation of *Giardia lamblia* by free chlorine is similar to that for *E. histolytica*.[35]

Coliform bacteria can be continually found in a chlorinated surface-water supply (turbidity 3.8 to 84 units, iron particles, and microscopic counts up to 2,000 units) containing between 0.1 and 0.5 mg/l of free residual chlorine and between 0.7 and 1.0 mg/l total residual chlorine after more than 30 minutes contact time.[36]

It is evident from available information that the coliform index may give a false sense of security when applied to waters subject to intermittent doses of pollution. The effectiveness of proper disinfection, including inactivation of viruses, other conditions being the same, is largely dependent on the freedom from suspended material and organic matter in the water being treated. Treated water having a turbidity of less than 5 NTU (ideally less than 0.1), a pH less than 8, and an HOCl residual of 1 mg/l after 30 minutes contact provides an acceptable level of protection.[37]

Free residual chlorination is the addition of sufficient chlorine to yield a free chlorine residual in the water supply in an amount equal to more than 85 percent of the total chlorine present. When the ratio of chlorine to ammonia is 5:1 (by weight), the chlorine residual is all monochloramine; when the ratio reaches 10:1, dichloramine is also formed; when the ratio reaches 15:1 or 20:1, nitrogen trichloride is formed, reaching a maximum at pH less than 4.5 and at a higher pH in polluted waters. Nitrogen trichloride as low as 0.05 mg/l causes an offensive and acrid odor that can be removed by carbon, aeration (natural or forced draft), exposure to sunlight, or forced ventilation indoors.[38] It can titrate partly as free chlorine and is also highly explosive. The reaction of chlorine in water is shown in Figure 2.1.

The minimum free chlorine residual at distant points in the distribution system should be 0.2 to 0.5 mg/l. Combined chlorine residual, if use is approved, should be 1.0 to 2.0 mg/l at distant points in the distribution system.[39]

In the presence of ammonia, organic matter, and other chlorine-consuming materials, the required chlorine dosage to produce a free residual will be high. The water is then said to have a high chlorine demand. With free residual chlorination, water is bleached, and iron, manganese, and organic matter are oxidized by chlorine and precipitated, particularly when the water is stored in a reservoir or basin for at least 2 hours. Most taste- and odor-producing compounds are destroyed; the reduction of sulfates to taste- and odor-producing sulfides is prevented; and objectionable growths and organisms in the mains are controlled or eliminated, provided a free chlorine residual is maintained in the water. An indication of accidental pollution of water in the mains is also obtained if the free chlorine residual is lost, provided chlorination is not interrupted.

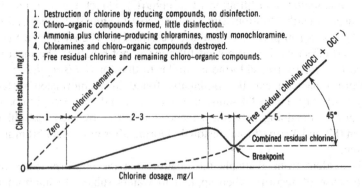

FIGURE 2.1 Reaction of chlorine in water. (Adapted from *Manual of Instruction for Water Plant Chlorinator Operators*, New York State Department of Health, Albany.)

The formation of trihalomethanes and other chloro-organics, their prevention, control, and removal, and the use of other disinfectants are discussed later in this chapter.

Distribution System Contamination

Once a water supply distribution system is contaminated with untreated water, the presence of coliform organisms may persist for an extended period of time. A surface-water supply or an inadequately filtered water supply may admit into a distribution system organic matter, minerals, and sediment, including fungi, algae, macroscopic organisms, and microscopic organisms. These flow through or settle in the mains or become attached and grow inside the mains when chlorination is marginal or inadequate to destroy them. Suspended matter and iron deposits will intermingle with and harbor the growths. Hence, the admission of contaminated water into a distribution system, even for a short time, will have the effect of inoculating the growth media existing inside the mains with coliform and other organisms. Elimination of the coliform organisms will therefore involve removal of the growth media and harborage material, which is not always readily possible, and disinfectant penetration. Bacteriological control of the water supply is lost until the biofilm and incrustation harboring coliform and other organisms are removed, unless a free chlorine residual of at least 0.2 to 0.4 mg/l is maintained in active parts of a distribution system. Even this may be inadequate if unfiltered water is admitted or if contaminants or particulates are released in the distribution system, such as after fire flows.

If a positive temporary program of continuous heavy chlorination at the rate of 5 to 10 mg/l, coupled with routine flushing of the main, is maintained, it is possible in most cases to eliminate the coliform on the inside surface of the pipes[40] and hence the effects of accidental contamination in 2 to 3 weeks or less. If a weak program of chlorination is followed, with chlorine dosage of less than 5 mg/l, the contamination may persist for an extended period of time. The rapidity with which a contaminated distribution system is cleared will depend on many factors: admission of only low-turbidity filtered water; uninterruption of chlorination even momentarily; the chlorine dosage and residual maintained in the entire distribution system; the growths in the mains and degree of pipe incrustation and their removal; conscientiousness in flushing the distribution system; the social, economic, and political deterrents; and, mostly, the competency of the responsible individual. A solution to main bacterial growths might be main cleaning and relining. The deterioration of water quality in a distribution system may be due to biological, physical, and chemical factors. The causes are usually complex and require laboratory participation and evaluation of data to identify possible causes and action measures. Species identification may be helpful in determining the significance of coliform-positive samples collected from a water system.[41] Physical analyses may include temperature, suspended and attached solids, chemical analyses including iron, dissolved oxygen, pH, alkalinity, nitrate and nitrite ions, ammonium, microbiological analyses including heterotrophic and direct microscopic counts, and residual chlorine over the long term.[42]

Plain Sedimentation

Plain sedimentation is the quiescent settling or storage of water, such as would take place in a reservoir, lake, or basin, without the aid of chemicals, preferably for a month or longer, particularly if the source water is a sewage-polluted river water. This natural treatment results in the settling out of suspended solids; reduction of hardness, ammonia, lead, cadmium, and other heavy metals; breakdown of organic chemicals and fecal coliform; removal of color (due to the action of sunlight); and die-off of pathogenic microorganisms principally because of the unfavorable temperature, lack of suitable food, and sterilizing effect of sunlight. Certain microscopic organisms, such as protozoa, consume bacteria, thereby aiding in purification of the water. Experiments conducted by Sir Alexander Houston showed that polluted water stored for periods of 5 weeks at 32°F (0°C), 4 weeks at 41°F (5°C), 3 weeks at 50°F (10°C), or 2 weeks at 64.4°F (18°C) effected the elimination of practically all bacteria.[43] A bacteria and virus removal of 80 to 90 percent can be expected after 10 to 30 days storage.[44] Plain sedimentation, however, has some disadvantages that must be taken into consideration and controlled. The growth of microscopic organisms causing unpleasant tastes and odors is encouraged, and pollution by watershed surface wash, fertilizers, pesticides, recreational uses, birds and animals, sewage, and industrial wastes may occur unless steps are taken to prevent or reduce these possibilities. Although subsidence permits bacteria, including pathogens, to die off, it also permits bacteria to accumulate and grow in reservoir bottom mud under favorable conditions. In addition, iron and manganese may go into solution, carbon dioxide may increase, and hydrogen sulfide may be produced.

Presettling basins or upflow roughing filters are sometimes used to eliminate heavy turbidity or pollution and thus better prepare the water for treatment by coagulation, flocculation, settling, and filtration. Ordinarily, at least two basins are provided to permit one to be cleaned while the other is in use. A capacity sufficient to give a retention period of at least two or three days is desirable. When heavily polluted water is to be conditioned, provision can be made for preliminary coagulation at the point of entrance of the water into the basins followed by chlorination or other disinfection at the exit. Consideration must be given to the possible formation of trihalomethanes and their prevention.

Microstraining

Microstraining is a process designed to reduce the suspended solids, insects, and nuisance organisms, including plankton, in water. The filtering surface may consist of very finely woven fabrics of stainless steel, nylon, bronze, or other resistant material on a revolving drum. Water flows into the drum, which is closed at the other end, and out through the filtering surface. Applications to water supplies are primarily the clarification of relatively clean surface waters low in true color and colloidal turbidity, in which microstraining and disinfection constitute the pretreatment, before water coagulation and clarification, ahead of slow or rapid sand filters and diatomite filters. Removal of the more common types of algae have been as high as 95 percent. Washwater consumption by the outside cleaning jets may run from 1 to 3 percent of the flow through the unit. The wastewater is collected and carried off by a trough in the upper part of the drum for proper disposal. Blinding of the fabric rarely occurs but may be due to inadequate washwater pressure or the presence of bacterial slimes. Cleansing is readily accomplished with commercial sodium hypochlorite.[45] Small head losses and low maintenance costs may make the microstrainer attractive for small installations.

Unit sizes start at about 2.5 feet in diameter by 2 feet wide. These have a capacity varying between 50,000 and 250,000 gpd, depending on the type and amount of solids in the water and the fabric used. Larger units have capacities in excess of 10 mgd.

Coagulation, Flocculation, and Settling

Adding a coagulant such as alum (aluminum sulfate) to water permits particles to come together and results in the formation of a flocculent mass, or floc, which enmeshes and agglomerates microorganisms, suspended particles, and colloidal matter, removing and attracting these materials in settling out. Removal of 90 to 99 percent of the bacteria and viruses and more than 90 percent of the protozoa and phosphate can be expected.[46] Total organic carbon and THM precursors and around 80 percent of the color and turbidity are also removed. The common coagulants used, in addition to alum, are copperas (ferrous sulfate), ferric sulfate, ferric chloride, sodium aluminate, pulverized limestone, bentonite, and clays. Sodium silicate and polyelectrolytes, including polymers, are also used at times as coagulant aids to improve coagulation and floc strength, usually resulting in less sludge and lower chemical dosages. The use of ozone as a microflocculant has also led to the need for less alum.

Proper respiratory protection should be provided for water plant operators handling water treatment plant chemicals, including chlorine and fluoride. Safety professionals, safety equipment suppliers, and chemical manufacturers can be of assistance. All chemicals must meet EPA purity standards.

To adjust the chemical reaction (alkalinity and pH) for improved coagulation, it is sometimes necessary to first add soda ash, hydrated lime, quicklime, or sulfuric acid. Color is best removed at a pH of 6.0 to 6.5. The mixing of the coagulant is usually done in two steps. The first step is rapid or flash mix and the second is slow mix, during which flocculation takes place. Rapid mix is a violent agitation for a few seconds, not more than 30 seconds, and may be accomplished by a mechanical agitator, pump impeller and pipe fittings, baffles, hydraulic jump, Parshall flume, in-line mixer, or other means. Slow mix and flocculation are accomplished by means of baffles or a mechanical paddle mixer to promote formation of a floc and provide a detention of at least 30 minutes with a flow-through velocity of 0.5 to 1.5 fpm. The flocculated water then flows to the settling basin designed to provide a retention of 4 to 6 hours, an overflow rate of about 500 gpd/ft^2 of area, or 20,000 gpd/ft of weir length. The velocity through the basin should not exceed 0.5 fpm. Cold water has a higher viscosity than warm water, hence the rate of particle or floc settling is much less in cold water; this must be taken into consideration in the design of a sedimentation basin. It is always recommended that mixing tanks and settling basins be at least two in number to permit cleaning and repairs without completely interrupting the water treatment, even though mechanical cleaning equipment is installed for sludge removal.

For the control of coagulation, jar tests are made in the laboratory to determine the approximate dosage of the chemicals (not laboratory grade) used, at the actual water temperature, that appear to produce the best results.[47] The best pH and coagulant and coagulant aid are also determined. Then, with this as a guide, the chemical dosing equipment, dry feed, or solution feed is adjusted to add the desired quantity of chemical proportional to the flow of water treated to give the best results. See Figure 2.5 later in the chapter. The dosages may be further adjusted or refined based on actual operating conditions. Aluminum breakthrough is minimized with coagulation pH control at the prevailing water temperature. It should be remembered that the algal level in a surface-water source will affect the dissolved-oxygen, carbon dioxide, and pH levels in the raw water and produce changes between night and day.

Zeta potential is also used to control coagulation. It involves determination of the speed at which particles move a given distance through an electric field caused by a direct current passing through the raw water. Best flocculation takes place when the charge approaches zero, giving best precipitation when a coagulant such as aluminum sulfate, assisted by a polyelectrolyte (polymer) if necessary, is added. Polymers may contain hazardous impurities. Quality control specifications should be met.

The addition of a polymer, silicate, and special clays may assist coagulation and clarification of certain waters as previously noted. A faster settling and more filterable floc is reported, which is less affected by temperature change or excessive flows. Less plugging of filters, longer filter runs, more consistent effluent turbidity, less backwash water, less sludge volume, and easier dewatering of sludge are claimed for polymer, clay–alum treatment.[48] High-rate filtration however may require surface wash in order to adequately clean the filter during the backwash cycle.

Another device for coagulating and settling water consists of a unit in which the water, to which a coagulant has been added, is introduced near the bottom, mixes with recirculated sludge, and flows upward through a blanket of settled floc. The clarified water flows off at the top. Sludge is drawn off at the bottom. These basins are referred to as upflow suspended-solids contact clarifiers. The detention period used in treating surface water is 4 hours, but may be as little as 1.5 to 2 hours, depending on the quality of the raw water. The normal upflow rate is 1,440 gpd/ft^2 of clarifier surface area and the overflow rate is 14,400 gpd per foot of weir length. A major advantage claimed, where applicable, is a reduction of the detention period and hence savings in space. Disadvantages include possible loss of sludge blanket with changing water temperature and variable water quality.

Tube settlers are shallow tubes, usually inclined at an angle of approximately 60 degrees from horizontal. Flow is up through the tubes that extend from about middepth to a short distance below the water surface and are inclined in the direction of water flow. Solids settle in the tube bottom, should slide down against the flow, and accumulate on the bottom of the basin. Effective operation requires laminar flow, adequate retention, nonscouring velocities, and floc particle settling with allowance for sludge accumulation and desludging at maximum flow rates.[49] Pilot plant studies are advisable prior to actual design and construction. Algal growths may clog tubes. Inclined plate settlers are similar to the tube settlers except that 45 to 60 degree inclined plates are used instead of tubes; the settled sludge slides down the smooth plates (plastic) opposite the direction of flow; the water enters through the sides and flows upward. If the depth is adequate, tube and inclined plate settlers can be used in existing settling basins to increase their capacity and improve their efficiency.

Filtration

Filters are of the slow sand, rapid sand, or other granular media (including multimedia) and pressure (or vacuum) type. Each has application under various conditions. The primary purpose of filters is to remove suspended materials, although microbiological organisms and color are also reduced. Of the filters mentioned, the slow sand filter is recommended for use at small communities, in developing areas, and in rural places, where adaptable. A rapid sand filter is not recommended because of the rather complicated control required to obtain satisfactory results unless competent supervision and operation can be ensured. This precaution also applies to package plants. The pressure filter, including the diatomaceous earth type, is commonly used for the filtration of industrial water supplies and swimming pool water; it is not generally recommended for the treatment of drinking water, except where considered suitable under the conditions of the proposed use. Variations of the conventional rapid sand filter, which may have application where raw water characteristics permit, are direct filtration, deep-bed filtration (4 to 8 feet media depth and 1.0 to 2.0 mm media size), high-rate filtration (up to 10 gpm/ft^2), declining flow rate filtration, and granular activated-carbon filters. In all cases, pilot plant studies at the site should first be conducted to demonstrate their feasibility and effectiveness.

Slow Sand Filter

A slow sand filter consists of a watertight basin, usually covered, built of concrete, and equipped with a rate controller and loss-of-head gauge. The basin holds a special sand 30 to 48 inches deep that is supported on a 12- to 18-inch layer of graded gravel placed over an underdrain system that may consist of open-joint, porous, or perforated 2- to 4-inch-diameter pipe or conduit spaced no greater than 3 feet. The velocity in the underdrain system should not exceed 0.75 fps. The sand should have an effective size of 0.25 to 0.35 mm and a uniformity coefficient not greater than 2.5. Operation of the filter is controlled so that filtration will take place at a rate of 1 to 4 million gallons per acre per day, with 2.5 million gallons as an average rate. This would correspond to a filter rate of 23 to 92 gal/ft^2 of sand area per day or an average rate of 57 gal. A rate up to 10 million gallons may be permitted by the approving authority if justified.

From a practical standpoint, the water that is to be filtered should have low color, less than 30 units, and low coliform concentration (less than 1,000 per 100 ml) and be low in suspended matter and algae, with a low turbidity not exceeding an occasional 50 units; otherwise, the filter will clog quickly. A plain sedimentation basin, roughing filter, or other pretreatment ahead of the filter can be used to reduce the suspended matter, turbidity, and coliform concentration of the water if necessary. It could also serve as a balancing tank. A loss-of-head gauge should be provided on the filter to show the resistance the sand bed offers to the flow of water through it and to show when the filter needs cleaning, usually 30 to 60 days, more or less, depending on raw water quality and filter rate. This is done by draining the water down to 6 inches below the surface of the sand bed and scraping about 1 inch of sand with adhering particles and *schmutzdecke* off the top of the bed. The sand is washed and replaced when the depth of sand is reduced to about 24 inches. A scraper or flat shovel is practical for removing the top layer of clogged sand with the aid of a motorized cart. The sand surface can also be washed in place by a special washer traveling over the sand bed. Filtered water is readmitted to a depth of several inches above the sand to prevent scour when placed in operation. Slow sand filters should be constructed in pairs as a minimum. These filters, *operated without interruption*, are easily controlled and, when followed by disinfection, produce a consistently satisfactory water. The filtration process is primarily biological rather than chemical/physical.

A well-operated plant will remove 98 to 99.5 percent of the coliform bacteria, protozoa, and viruses, as well as some organic and inorganic chemicals in the raw water (after a biological film has formed on the surface and within the sand bed). Effluent turbidity of less than 0.3 NTU and coliform of 1 per 100 ml or less can (and *must*) be regularly obtained. Chlorination of the filtered water is necessary to destroy microorganisms passing through the filter and growing or entering the storage basin and water system. This type plant will also remove about 25 to 40 percent of the color in the untreated water. Chlorination of the sand filter itself is desirable either continuously or periodically to destroy microorganisms growing within the sand bed, supporting gravel, or underdrain system. Continuous prechlorination at a rate that produces 0.3 to 0.5 mg/l in the water on top of the filter will not harm the filter film; it will increase the length of the filter run. A high chlorine residual (5 mg/l) is detrimental. Sand filters should also have the capability to discharge to the waste-stream to allow the filter to ripen. Filtration to waste may be required for 6 hours to 2 weeks after cleaning, depending on the age of the filter, particulate matter in the raw water, and filtrate quality.[50] In any case, continuous postchlorination should be provided.

A slow sand filter suitable for a small rural water supply is shown in Figure 2.2. Details relating to design are given in Table 2.2. The rate of filtration in this filter is controlled by selecting an orifice and filter area that will deliver not more than 50 gal/ft^2 of filter area per day, thus preventing excessive rates of filtration that could endanger the quality of the treated

water. Where competent and trained personnel are available, the rate of flow can be controlled by manipulating a gate or butterfly valve on the effluent line from each filter, provided a venturi, orifice, triangular weir, or other suitable meter, with indicating and preferably recording instruments, is installed on the outlet to measure the rate of flow. The valve can then be adjusted to give the desired rate of filtration until the filter needs cleaning. Another practical method of controlling the rate of filtration is by installing a float valve on the filter effluent line, as shown in Figure 2.3. The valve is actuated by the water level in a float chamber, which is constructed to maintain a reasonably constant head over an orifice in the float chamber that would yield the desired filter rate of flow. A hydraulically operated float can be connected to a control valve by tubing and located at some distance from the valve. A solenoid valve can accomplish the same type of control. A modulating float valve is more sensitive to water level control than the ordinary float valve. Riddick describes a remote float-controlled weighted butterfly valve, with spring-loaded packing glands and stainless steel shafts.[51] A special rate control valve can also be used if it is accurate within the limits of flow desired. The level of water over the orifice or filter outlet must be *above* the top of the sand to prevent the development of a negative head. If a negative head is permitted to develop, the mat on the surface of the sand may be broken and dissolved air in the water may be released in the sand bed, causing the bed to become air bound. At least 3 inches of water over the sand or a flexible influent hose will minimize any possible disturbance of the sand when water from the influent line falls into the filter. Postchlorination should be provided. A cartridge-type tablet or stack chlorinator may meet the needs in a rural situation. The filter should operate without interruption to produce uniform results, and daily operation reports should be kept.

Rapid Sand (Granular Media) Filter

$$\text{Rate of filtration} = \frac{7.48}{\text{minutes for water in filter to fall 1 ft}}$$

Fill filter with water, shut off influent, and open drain.

Backwash time = 10 to 15 minutes minimum, until water entering through is clear

Normal washwater usage = 2 to 2.5% or less of water filtered

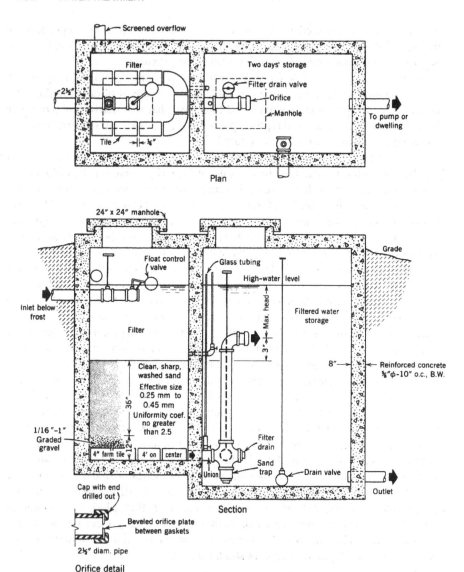

FIGURE 2.2 Slow sand filter for a small water supply. Minimum of two units. The difference in water level between the two glass tubes represents the frictional resistance to the flow of water through the filter. When this difference approaches the maximum head and the flow is inadequate, the filter needs cleaning. To clean, scrape the top 1 in. of sand bed off with a mason's trowel, wash in a pan or barrel, and replace clean sand on bed. Float control valve may be omitted where water on filter can be kept at a desirable level by gravity flow or an overflow valve. Add a meter, venturi, or other flow-measuring device on the inlet to the filter. Rate of flow can also be controlled by maintaining a constant head with a weighted float valve over an orifice or weir. See 1.18. Filtered water should be disinfected before use. Allow sufficient head room for cleaning the filter. Separation of the filter box from the filtered water storage clear well is usually required by health officials for public water supplies. Thoroughly ventilate filter box and water storage tank before entering and during occupancy.

TABLE 2.2 Orifice Flow for Selected Orifice Diameters

Maximum Head, (ft of water)	Maximum Flow by Orifice Diameter[a] (gpd)[b]														
	$\frac{1}{16}$	$\frac{3}{32}$	$\frac{1}{8}$	$\frac{3}{16}$	$\frac{1}{4}$	$\frac{5}{16}$	$\frac{3}{8}$	$\frac{7}{16}$	$\frac{1}{2}$	$\frac{9}{16}$	$\frac{5}{8}$	$\frac{11}{16}$	$\frac{3}{4}$	$\frac{7}{8}$	1
1	67	149	266	597	1,060	1,660	2,390	3,240	4,240	5,370	6,640	8,010	9,550	13,030	17,000
$1\frac{1}{2}$	82	183	326	732	1,305	2,040	2,930	3,990	5,220	6,580	8,130	9,850	11,700	15,950	20,800
2	96	213	380	852	1,520	2,370	3,410	4,650	6,060	7,680	9,480	11,480	14,300	18,600	24,200
$2\frac{1}{2}$	107	236	421	945	1,680	2,620	3,780	5,150	6,720	8,500	10,500	12,680	15,100	20,600	26,800
3	116	259	462	1,036	1,840	2,880	4,140	5,640	7,380	9,130	11,520	13,920	16,550	22,600	29,500
$3\frac{1}{2}$	126	279	498	1,120	1,990	3,100	4,470	6,060	7,950	10,050	12,420	14,900	17,900	24,300	30,700
4	135	300	534	1,200	2,125	3,330	4,790	6,530	8,520	10,800	13,300	16,100	19,150	26,100	34,000
$4\frac{1}{4}$	145	322	574	1,290	2,290	3,580	5,160	7,020	9,150	11,600	14,350	17,300	20,600	28,000	36,600
5	150	333	594	1,332	2,370	3,700	5,340	7,260	9,480	12,000	14,800	17,950	21,400	29,000	38,700
$5\frac{1}{2}$	157	350	624	1,400	2,490	3,890	5,600	7,620	9,960	12,600	15,550	18,850	22,400	30,500	39,300
6	164	366	650	1,460	2,600	4,070	5,850	7,980	10,400	13,150	16,250	19,900	23,400	31,800	41,500
$6\frac{1}{2}$	169	376	672	1,510	2,680	4,180	6,030	8,220	10,700	13,580	16,750	20,200	24,200	32,800	42,800

[a] In inches.

[b] No loss head through sand and gravel or pipe is assumed; flow is based on $Q = C_d VA$, where $V = \sqrt{2gh}$ and $C_d = 0.6$, with free discharge. (Design filter for twice the desired flow to ensure an adequate delivery of water as the frictional resistance in the filter to the flow builds up. *Use two or more units in parallel.*)

Note: The loss of head through a clean filter is about 3 in.; hence add 3 in. to the "maximum head" in the table and sketch to obtain the indicated flow in practice. A minimum 3–4 ft of water over the sand is advised. Example: To find the size of a filter that will deliver a maximum of 500 gpd: From above, a filter with a $\frac{1}{8}$-in. orifice and a head of water of 3 ft 9 in. will meet the requirements. Filtering at the rate of 50 gpd/ft² of filter area, the required filter area is 500 gpd/[50 gal/(ft²)(day)] = 10 ft². Provide at least 2 days storage capacity.

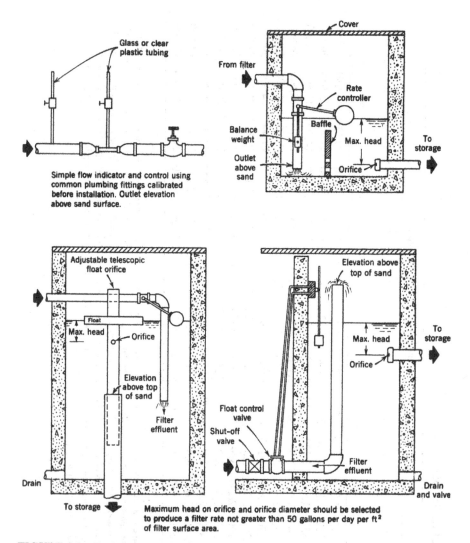

FIGURE 2.3 Typical devices for the control of the rate of flow or filtration. Plant capacity: 50 to 100 percent greater than average daily demand, with clear well.

Sand expansion = 40 to 50% = 33.6 to 36 in. for 24-in. sand bed
= 25 to 35% for dual media, anthracite, and sand

$$\text{Rate of backwash} = \frac{7.48}{\text{minutes for water in filter to rise 1 ft}}$$

Lower water level to sand, slowly open backwash valve, 20 to 25 gpm/ft^2 minimum (32 to 40 in. rise/min).

Orifice area $= 0.25$ to 0.30% of filter area

Lateral area $= 2 \times$ Orifice area

Manifold area $= 1.5$ to $2 \times$ Total area of laterals

The rate controller is omitted in a declining-rate filter. A gate valve or orifice is used in its place. An acceptable rate of flow (filtration) is set at the start of a run but decreases as the head loss increases. The use of this concept is recommended in developing countries and other areas where skilled operators are not generally available.[52] See also "Slow Sand Filter" (filtration rate control) in this chapter and Table 2.2.

In a combination anthracite over sand bed filter (dual media), use is made of the known specific gravity of crushed anthracite of 1.35 to 1.75 and the specific gravity of sand of 2.5 to 2.65. The relative weight of sand in water is three times that of anthracite. The anthracite effective size is 0.9 to 1.1 and uniformity coefficient 1.6 to 1.8. The sand effective size is 0.45 to 0.55 and uniformity coefficient 1.5 to 1.7 (see Figure 2.4). Anthracite grains can be twice as large as sand grains; after backwashing, the sand will settle in place before the anthracite in two separate layers.[53] High-rate sand–anthracite filters require careful operating attention and usually use of a filter conditioner to prevent floc passing through while at the same time obtaining a more uniform distribution of suspended solids throughout the media depth. Longer filter runs, such as two to three times the conventional filter, at a rate of 4 to 6 gpm/ft^2 and up to 8 or 10 gpm/ft^2 and less washwater are reported. A mixed or multimedia may consist of anthracite on top, effective size 0.95 to 1.0 mm and uniformity coefficient 1.55 to 1.75; silica sand in the middle, effective size 0.45 to 0.55 and uniformity coefficient 1.5 to 1.65; and garnet sand* on the bottom, effective size 0.2 to 0.35 mm and uniformity coefficient 1.6 to 2.0. Total media depth is 24 to 48 in. Under the same conditions, filter runs will increase with dual media and increase further with mixed media over single-media sand. The turbidity of the effluent can be expected to be less than 0.5 NTU in a well-operated plant. Granular activated-carbon media (GAC) in place of sand and anthracite can function both as an adsorbent of organics and as a filter medium, but backwash rates will have to be reduced and carefully controlled. If the organics are synthetic chemicals, frequent regeneration will be required. If the organics are taste- and odor-producing compounds, activated carbon may remain effective for several years. The GAC will dechlorinate water that has been chlorinated.

Treatment of the raw water by coagulation, flocculation, and settling to remove as much as possible of the pollution is usually necessary and an important preliminary step in the rapid sand filtration of water. Without coagulation, virus removal is very low. The settled water, in passing to the filter, carries with it some flocculated suspended solids, color, and microorganisms. This material forms a mat on top of the sand that aids greatly, together with adsorption on the

---

*Garnet sand has a specific gravity of 4.0–4.2.

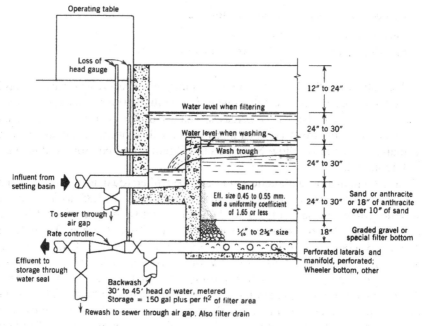

FIGURE 2.4   Essential parts of a rapid sand filter. The minimum total depth is 8.5 ft, 12 ft preferred.

bed granular material, in the straining and removal of other suspended matter, color, and microorganisms, but this also causes rapid clogging of the sand. Special arrangement is therefore made in the design for washing the filter (every 48 to 72 hours, depending on the head loss) by forcing water backward up through the filter at a rate that will provide a sand expansion of 40 to 50 percent based on the backwash rate, water temperature, and sand effective size. For example, with a 0.4-mm-effective-size sand, a 40 percent sand expansion requires a washwater rate rise of 21 in./min with 32°F (0°C) water and a rise of 32.5 inches with water at 70°F (21°C).[54] The dirty water is carried off to waste by troughs built in above the sand bed 5 to 6 feet apart. Separate air laterals in the underdrain system can increase the backwash efficiency. A system of water jets or rakes or a 1.5- to 2-in. pressure line at 45 to 75 psi with a hose connection (including vacuum breaker) should be provided to scour the surface of the sand to assist in loosening and removing the material adhering to it, in the pores of the sand on the surface, and in the filter depth. Air scour or wash is also very effective for cleaning the entire bed depth, especially in beds 4 to 6 feet deep. A backwash rate of 10 gpm/ft^2 may be acceptable with an anthracite or granular activated-carbon bed. Effective washing of the sand is essential. Backwash should start when the turbidity begins to rise above 0.1 to 0.2 NTU, not after the turbidity reaches 1.0 NTU. Before being placed back into service the filter effluent should be sent to waste for 5 to 20 minutes to reestablish filtration efficiency. Aluminum in the filtered water should be less than 0.15 mg/l. Filters that have been out of service should be backwashed before being returned to service. The filter sand should be inspected periodically. Its condition can be observed by lowering the water level below the sand and looking for mounds and craters, debris on the surface, or cracks on the surface or along the walls. Depth samples of the sand may also be taken for laboratory observation. During the washing operation the samples should be inspected for uneven turbulence and the presence of small lumps known as mud balls. Any of these conditions requires further investigation and correction. A conventional rapid sand filter plant flow diagram and unit processes are shown in Figure 2.5.

When properly operated, a filtration plant, including coagulation, flocculation, and settling, can be expected to remove 90 to 99.9 percent of the bacteria, protozoa, and viruses,[55] a great deal of the odor and color, and practically all of the suspended solids. Adequate pretreatment is essential. Nevertheless, chlorination must be used to ensure that the water leaving the plant is safe to drink. Construction of a rapid sand filter should not be attempted unless it is designed and supervised by a competent environmental engineer. Pilot plant studies, including preliminary treatment for heavily polluted water, may be required to ensure the proposed treatment will produce a water meeting drinking water standards at all times. Adequate coagulation, flocculation, and settling, in addition to granular media filtration and disinfection, are necessary to ensure the removal of bacteria, protozoa (*Giardia* and *E. histolytica* cyst), and viruses. Improper pH control can result in weak floc and passage of dissolved coagulant.

A flow diagram of a typical treatment plant is shown in Figure 2.6.

Direct Filtration

Direct filtration of waters with low suspended matter and turbidity, color, coliform organisms, and plankton, and free of paper fiber, has been attractive because of the lower cost in producing a good quality water, if substantiated by prior pilot plant studies reflecting seasonal variations in raw water quality. Direct filtration removals of bacteria, viruses, and turbidity tend to be more erratic than with conventional treatment.[56] In direct filtration, the sedimentation basin is omitted. The unit processes prior to filtration (dual or mixed media) may consist of only rapid mix, rapid mix and flocculation, or rapid mix and contact basin (1-hour detention) without sludge collector. A flocculation or contact basin is recommended for better water quality control. Rapid sand filtration with coagulation and flocculation is reported to remove 90 to 99 percent of the viruses, bacteria, and protozoa.[57] A polymer is normally used in addition to a coagulant. Direct filtration is a good possibility[58] if either of three conditions holds:

1. The raw water turbidity and color are each less than 25 units.
2. The color is low and the maximum turbidity does not exceed 200 NTU.
3. The turbidity is low and the maximum color does not exceed 100 units.

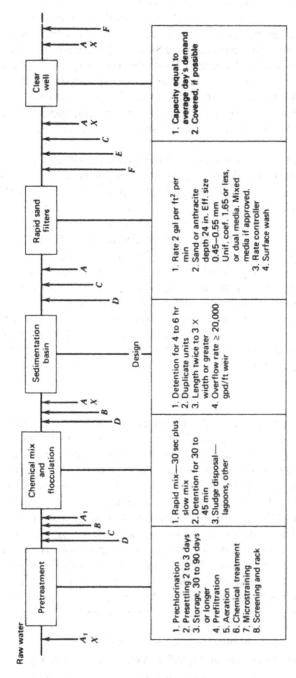

FIGURE 2.5 Conventional rapid sand filter plant flow diagram. Possible chemical combinations: *A*: Chlorine. *A₁* Eliminate if THMs formed. *B*: Coagulant; aluminum sulfate (pH 5.5–8.0), 10 to 50 mg/l; ferric sulfate (pH 5.0–11.0), 10 to 50 mg/l; ferrous sulfate (pH 8.5–11.0), 5 to 25 mg/l; ferric chloride (pH 5.0–11.0); sodium aluminate, 5 to 20 mg/l; activated silica, organic chemicals (polyelectrolytes). *C*: Alkalinity adjustment; lime, soda ash, or polyphosphate. *D*: Activated carbon, potassium permanganate. *E*: Dechlorination; sulfur dioxide, sodium sulfite, sodium bisulfite, activated carbon. *X*: Chlorine dioxide, ozone, chlorine-ammonia. Note that the chlorinator should be selected to postchlorinate at 3 mg/l. Provide for a dose of 3 mg/l plus chlorine demand for groundwater. Additional treatment processes may include softening (ion exchange, lime-soda, excess lime and recarbonation), iron and manganese removal (ion exchange, chemical oxidation and filtration, ozone oxidation, sequestering), organics removal (activated carbon, superchlorination, ozone oxidation), and demineralization (distillation, electrodialysis, reverse osmosis, chemical oxidation and filtration, freezing).

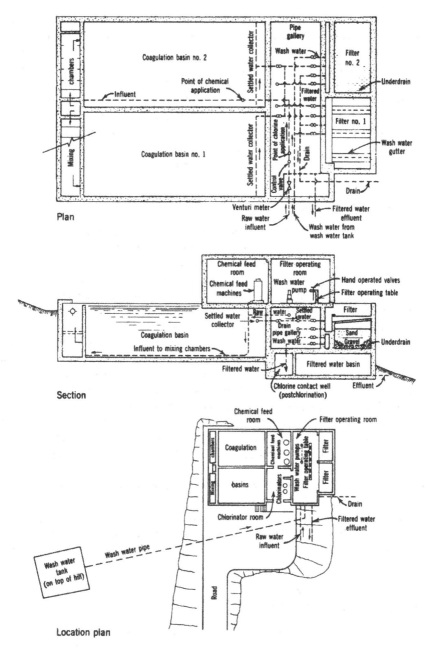

FIGURE 2.6 Flow diagram of typical treatment plant. This plant is compactly arranged and adaptable within a capacity range of 0.25 to 1.0 mgd. Operation is simple as the emphasis is on manual operation with only the essentials in mechanical equipment provided. Design data are described in the text. (*Source: Water Treatment Plant Design*, American Water Works Association, New York, 1969. Copyright 1969 by the American Water Works Association. Reprinted with permission.)

The presence of paper fiber or of diatoms in excess of 1,000 areal standard units per milliliter (asu/ml) requires that settling (or microscreening) be included in the treatment process chain. Diatom levels in excess of 200 asu/ml may require the use of special coarse coal on top of the bed in order to extend filter runs. Coliform MPNs should be low. Filter rates of 4 to 5 gpm/ft^2 and as high as 15 gpm/ft^2 may produce satisfactory results with some waters, but caution is advised. Decreased chemical dosage, and hence sludge production but increased filter washwater, will usually result in reduced net cost as compared to conventional treatment.[59] Surface wash, subsurface wash, or air scour is required. Good pretreatment and operation control are essential.

Pressure Sand Filter

A pressure filter is similar in principle to the rapid sand gravity filter except that it is completely enclosed in a vertical or horizontal cylindrical steel tank through which water under pressure is filtered. The normal filtration rate is 2 gpm/ft^2 of sand. Higher rates are used. Pressure filters are most frequently used in swimming pool and industrial plant installations and for precipitated iron and manganese removal. It is possible to use only one pump to take water from the source or out of the pool (and force it through the filter and directly into the plant water system or back into the pool), which is the main advantage of a pressure filter. This is offset by difficulty in introducing chemicals under pressure, inadequate coagulation facilities, and lack of adequate settling. The appearance of the water being filtered and the condition of the sand cannot be seen; the effectiveness of backwashing cannot be observed; the safe rate of filtration may be exceeded; and it is difficult to look inside the filter for the purpose of determining loss of sand or anthracite, need for cleaning, replacing of the filter media, and inspection of the washwater pipes, influent, and effluent arrangements. Because of these disadvantages and weaknesses, a pressure filter is not considered dependable for the treatment of contaminated water to be used for drinking purposes. It may, however, have limited application for small, slightly contaminated water supplies and for turbidity removal if approved. In such cases, the water should be coagulated and flocculated in an open basin before being pumped through a pressure filter. This will require double pumping. Pressure filters are not used to filter surface water or other polluted water or following lime-soda softening.

Diatomaceous Earth Filter

The pressure-filter type consists of a closed steel cylinder, inside of which are suspended septa, the filter elements. In the vacuum type, the septa are in an open tank under water that is recirculated with a vacuum inside the septa. Normal rates of filtration are 1 to 1.5 gpm/ft^2 of element surface. To prepare the filter for use, a slurry or filter aid (precoat) of diatomaceous earth is introduced with the water to be treated at a rate of about 1.5 oz/ft^2 of filter septum area, which results in about 1/16 inch depth of media being placed evenly on the septa, and the water is recirculated for at least 3 minutes before discharge. Then more filter aid (body feed) is continuously added with the water to maintain the permeability of the filter media. Use of a cationic polymer enhances the removal of bacteria and viruses. The rate of feed is roughly 2 to 3 mg/l per unit of turbidity in the water. Filter aid comes in different particle sizes. It forms a coating or mat around the outside of each filter element. Because of smaller media pore size, it is more efficient than sand in removing from the water suspended matter and organisms such as protozoal cysts (which cause amebiasis and giardiasis), cercariae (which cause schistosomiasis), flukes (which cause paragonimiasis and clonorchiasis), and worms (which cause dracunculiasis, ascariasis, and trichuriasis). Expected bacteria removal is 90 to 99 percent, viral 95 percent, and protozoan 99 percent. These organisms, except for *Giardia* cysts, are not common in the United States. Effluent turbidity of 0.10 NTU or less is normal with proper operation.

The diatomite filter has found its greatest practical application in swimming pools, iron removal for groundwaters, and industrial and military installations. It has a special advantage in the removal of oil from condensate water, since the diatomaceous earth is wasted. It should not be used to treat a public water supply unless pilot plant study results on the water to be treated meet the regulatory agency requirements. The filter should not be used to treat raw water with greater than 2400 MPN per 100 ml, 30 turbidity units, or 3,000 areal standard microscopic units per 100 ml. It does not remove color or taste- and odor-producing substances. In any case, disinfection is considered a necessary adjunct to filtration.

A major weakness in the diatomite filter is that failure to add diatomaceous earth to build up the filtering mat, either through ignorance or negligence, will render the filter entirely ineffective and give a false sense of security. In addition, the septa will become clogged and require replacement or removal and chemical cleaning. During filtration, the head loss through the filter should not exceed 30 lb/in.^{2}, thereby requiring a pump and motor with a wide range in the head characteristics.[60] The cost of pumping water against this higher head is therefore increased. Diatomite filters cannot be used where pump operation is intermittent, as with a pressure tank installation, for the filter cake will slough off when the pump stops, unless sufficient continuous recirculation of 0.1 gpm/ft^2 of filter area is provided by a separate pump. A reciprocating pump should not be used. Head loss should not exceed a vacuum of 15 in. of mercury in a vacuum system.[61]

The filter is backwashed by reversing the flow of the filtered water back through the septa, thereby forcing all the diatomite to fall to the bottom of the filter shell, where it is flushed to waste. Only about 0.5 percent of the water filtered is used for backwash when the filter run length equals the theoretical or design length. As with other filters, the diatomite filter must be carefully operated by trained personnel in order to obtain dependable results, where its use is approved.

Package Water Treatment Plant

These plants are usually predesigned gravity rapid sand filter plants. They are compact and include chemical feeders, coagulation, flocculation, and/or settling, filtration, and water conditioning if needed. Filter design rates are usually 3 to 5 gal/ft^2. Provision must be made for adequate sludge storage and removal and for chlorine contact time. Because of variable raw water quality, it is necessary to first demonstrate that a water of satisfactory sanitary quality can be produced under all conditions. Since these plants include automated equipment, it is essential that a qualified operator be available to make treatment adjustments as needed. Approval of the regulatory agency is usually required.

Water Treatment Plant Wastewater and Sludge

Water treatment plant sludge from plain sedimentation and coagulation–flocculation settling basins and backwash wastewater from filters are required by the Clean Water Act (PL 72–500) to be adequately treated prior to discharge to a surface-water course. Included are softening treatment sludge, brines, iron and manganese sludge, and diatomaceous earth filter wastes. The wastes are characteristic of substances in the raw water and chemicals added in water treatment; they contain suspended and settleable solids, including organic and inorganic chemicals as well as trace metals, coagulants (usually aluminum hydroxide), polymers, clay, lime, powdered activated carbon, and other materials. The aluminum could interfere with fish survival and growth.

The common waste treatment and disposal processes include sand sludge drying beds where suitable, lagooning where land is available, natural or artificial freezing and thawing, chemical conditioning of sludge using inorganic chemicals and polymers to facilitate dewatering, and mechanical dewatering by centrifugation, vacuum filtration, and pressure filtration.[62]

Sludge dewatering increases sludge solids to about 15 to 20 percent. The use of a filter press involves a sludge thickener, polymer, sludge decant, lime, retention basin, addition of a precoat, and mechanical dewatering by pressure filtration. The filter cake solids concentration is increased to 40 percent. The use of a polymer with alum for coagulation could reduce the amount of alum used to less than one-fifth, the cost of coagulant chemicals by one-third, and the sludge produced by over 50 percent. Lime softening results in large amounts of sludge, increasing with water hardness. Recovery and recycling of lime may be economical at large plants. Sludge may be disposed of by lagooning, discharge to a wastewater treatment plant, or mechanical dewatering and landfilling, depending on feasibility and regulations.[63] Brine wastes may be discharged at a controlled rate to a stream if adequate dilution is available or to a sanitary sewer if permitted.

The ultimate disposal of sludge can be a problem in urban areas and land disposal where the runoff or leachate might be hazardous to surface or underground waters. Sludge analyses may be required for sludge disposal approval.

Causes of Tastes and Odors

Tastes and odors in water supplies are caused by oils, minerals, gases, organic matter, and other compounds and elements in the water. Some of the common causes are oils and products of decomposition exuded by algae and some other microorganisms; wastes from gas plants, coke ovens, paper mills, chemical plants, canneries, tanneries, oil refineries, and dairies; high concentrations of iron, manganese, sulfates, and hydrogen sulfide in the water; decaying vegetation such as leaves, algae, weeds, grasses, brush, and moss in the water; and chlorine compounds and high concentrations of chlorine. The control of taste and odor-producing substances is best accomplished by eliminating or controlling the source when possible. When this is not possible or practical, study of the origin and type of the tastes and odors should form the basis for the necessary treatment.

Control of Microorganisms

For the most part, microorganisms that cause tastes and odors are harmless. They are visible under a microscope and include plankton, fungi, bacteria, viruses, and others. Plankton are aquatic organisms; they include algae, protozoa, rotifers, copepods, and certain larvae. Phytoplankton are plant plankton. Zooplankton are animal plankton that feed on bacteria and small algae. Crenothrix, gallionella, and leptothrix, also known as iron bacteria, can also be included. *Thiobacillus thiooxidans* and sulfur bacteria have been implicated in the corrosion of iron. Phytoplankton, including algae, contain chlorophyll; utilize carbon, nitrogen, phosphorus, and carbon dioxide in water; produce oxygen; are eaten by zooplankton; and serve as a basic food for fish. All water is potentially a culture medium for one or more kinds of algae. Bacteria and algae are dormant in water below a temperature of about 48°F (10°C). Heavy algal growths cause a rise in pH and a decrease in water hardness during the day and the opposite at night.

Crenothrix and leptothrix are reddish-brown, gelatinous, stringy masses that grow in the dark inside distribution systems or wells carrying water devoid of oxygen but containing iron in solution. Control, therefore, may be effected by the removal of iron from the water before it enters the distribution system, maintenance of pH at 8.0 to 8.5, increase in the concentration of dissolved oxygen in the water above about 2 mg/l, or continuous addition of chlorine to provide a free residual chlorine concentration of about 0.3 mg/l, or 0.5 to 1.0 mg/l total chlorine. Chemical treatment of the water will destroy and dislodge growths in the mains, with resulting temporary intensification of objectionable tastes and odors, until all the organisms are flushed out of the water mains. A consumer information program should precede the treatment. Iron bacteria may grow in ditches draining to reservoirs. Copper sulfate dosage of 3 mg/l provides effective control.[64] The slime bacteria known as actinomycetes are also controlled by this treatment.

High water temperatures, optimum pH values and alkalinities, adequate food such as mineral matter (particularly nitrates, phosphorus, potassium, and carbon dioxide), low turbidities, large surface area, shallow depths, and sunlight favor the growth of plankton. Exceptions are diatoms, such as asterionella, which grow also in cold water at considerable depth without the aid of light. Fungi can also grow in the absence of sunlight. Extensive growths of anabaena, oscillaria, and microcystis resembling pea green soup are encouraged by calcium and nitrogen. Protozoa such as synura are similar to algae, but they do not need carbon dioxide; they grow in the dark and in cold water. The blue-green algae do not require direct light for their growth, but green algae do. They are found in higher concentrations within about 5 feet of the water surface.

Sawyer[65] has indicated that any lake having, at the time of the spring overturn, inorganic phosphorus greater than 0.01 mg/l and inorganic nitrogen greater than 0.3 mg/l can be expected to have major algal blooms. Reduction of nutrients therefore should be a major objective, where possible. In a reservoir, this can be accomplished by minimizing the entrance of nutrients such as farm and forest drainage, by watershed control, by removal of aquatic weeds before the fall die-off, and by draining the hypolimnion (the zone of stagnation) during periods of stratification since this water stratum has the highest concentration of dissolved minerals and nutrients. See also "Aquatic Weed Control" and "Reservoir Management, Intake Control, and Stratification," this chapter.

Inasmuch as the products of decomposition and the oils given off by algae cause disagreeable tastes and odors, preventing their growth will remove the cause of difficulty. Where it is practical to cover storage reservoirs to exclude light, this is the easiest way to prevent the growth of those organisms that require light and cause difficulty. Where this is not possible, copper sulfate, potassium permanganate, or chlorine can be applied to prevent the growth of the organisms. A combination of chlorine, ammonia, and copper sulfate has also been used with good results. However, in order that the proper chemical dosage required may be determined, it is advisable to make microscopic examinations of samples collected at various depths and locations to determine the type, number, and distribution of organisms and the chemical crystal size for maximum contact. This may be supplemented by laboratory tests using the water to be treated and the proposed chemical dose before actual treatment. In New England, diatoms usually appear in the spring; blue-green algae appear between the diatoms and green algae. Shallow areas usually have higher concentrations of algae.

In general, the application of about 2.5 pounds of copper sulfate per million gallons of water treated at intervals of two to four weeks between April and October in the temperate zone will prevent difficulties from most microorganisms. A chelating agent such as citric acid improves the performance of the copper sulfate in a high-alkalinity water. Follow-up treatment with potassium permanganate will also kill and oxidize the algae. More exact dosages for specific microorganisms are given in Table 2.3. The required copper sulfate dose can be based on the volume of water in the upper 10 feet of a lake or reservoir, as most plankton are found within this depth. Bartsch[66] suggests an arbitrary dosage related to the alkalinity of the water being treated. A copper sulfate dosage of 2.75 pounds per million gallons of water in the reservoir is recommended when the methyl orange alkalinity is less than 50 mg/l. When the alkalinity is greater than 40 mg/l, a dosage of 5.4 lb/acre of reservoir surface area is recommended.[67] Higher doses are required for the more resistant organisms. The dose needed should be based on the type of algae making their appearance in the affected areas, as determined

TABLE 2.3   Dosage of Copper Sulfate to Destroy Microorganisms

Organism	Taste, Odor, Other	Dosage (lb/10^6 gal)
Diatomaceae (Algae)	(Usually brown)	
Asterionella	Aromatic, geranium, fishy	1.0–1.7
Cyclotella	Faintly aromatic	Use chlorine
Diatoma	Faintly aromatic	—
Fragilaria	Geranium, musty	2.1
Meridon	Aromatic	—
Melosira	Geranium, musty	1.7–2.8
Navicula	—	0.6
Nitzchia	—	4.2
Stephanodiscus	Geranium, fishy	2.8
Synedra	Earthy, vegetable	3.0–4.2
Tabellaria	Aromatic, geranium, fishy	1.0–4.2
Chlorophyceae (Algae)	(Green algae)	
Cladophora	Septic	4.2
Closterium	Grassy	1.4
Coelastrum	—	0.4–2.8
Conferva	—	2.1
Desmidium	—	16.6
Dictyosphaerium	Grassy, nasturtium, fishy	Use chlorine
Draparnaldia	—	2.8
Entomophora	—	4.2
Eudorina	Faintly fishy	16.6–83.0
Gloeocystis	Offensive	—
Hydrodictyon	Very offensive	0.8
Miscrospora	—	3.3
Palmella	—	16.6
Pandorina	Faintly fishy	16.6–83.0
Protococcus	—	Use chlorine
Raphidium	—	8.3
Scenedesmus	Vegetable, aromatic	8.3
Spirogyra	Grassy	1.0
Staurastrum	Grassy	12.5
Tetrastrum	—	Use chlorine
Ulothrix	Grassy	1.7
Volvox	Fishy	2.1
Zygnema		4.2
Cyanophyceae (Algae)	(Blue-green algae)	
Anabaene	Moldy, grassy, vile	1.0
Aphanizomenon	Moldy, grassy, vile	1.0–4.2
Clathrocystis	Sweet, grassy, vile	1.0–2.1
Coelosphaerium	Sweet, grassy	1.7–2.8
Cylindrosphermum	Grassy	1.0
Gloeocopsa	(Red)	2.0
Microcystis	Grassy, septic	1.7
Oscillaria	Grassy, musty	1.7–4.2
Rivularia	Moldy, grassy	—

(continues)

TABLE 2.3 (*continued*)

Organism	Taste, Odor, Other	Dosage (lb/10⁶ gal)
Protozoa		
Bursaria	Irish moss, salt marsh, fishy	—
Ceratium	Fishy, vile (red-brown)	2.8
Chlamydomonas	—	4.2–8.3
Cryptomonas	Candied violets	4.2
Dinobryon	Aromatic, violets, fishy	1.5
Entamoeba histolytica		Use chlorine
(cyst)	—	5–25 mg/l
Euglena	—	4.2
Glenodinium	Fishy	4.2
Mallomonas	Aromatic, violets, fishy	4.2
Peridinium	Fishy, like clam shells, bitter taste	4.2–16.6
Synura	Cucumber, musk melon, fishy	0.25
Uroglena	Fishy, oily, cod liver oil	0.4–1.6
Crustacea		
Cyclops	—	16.6
Daphnia	—	16.6
Schizomycetes		
Beggiatoa	Very offensive, decayed	41.5
Cladothrix	—	1.7
Crenothrix	Very offensive, decayed	2.8–4.2
Leptothrix	Medicinal with chlorine	—
Sphaerotilis natans	Very offensive, decayed	3.3
Thiothrix (sulfur bacteria)		Use chlorine
Fungi		
Achlya	—	—
Leptomitus	—	3.3
Saprolegnia	—	1.5
Miscellaneous		
Blood worm	—	Use chlorine
Chara	—	0.8–4.2
Nitella flexilis	Objectionable	0.8–1.5
Phaetophyceae	(Brown algae)	—
Potamogeton	—	2.5–6.7
Rhodophyceae	(Red algae)	—
Xantophyceae	(Green algae)	—

Note: Chlorine residual 0.5 to 1.0 mg/l will also control most growths, except melosira, cysts of *Entamoeba histolytica*, Crustacea, and Synura (2.9 mg/l free).

by periodic microscopic examinations. An inadequate dosage is of very little value and is wasteful. Higher dosages than necessary have caused wholesale fish destruction. For greater accuracy, the copper sulfate dose should be increased by 2.5 percent for each degree of temperature above 59°F (15°C) and 2 percent for each 10 mg/l organic matter. Consideration must also be given to the dosage

TABLE 2.4   Dosage of Copper Sulfate and Residual Chlorine That If Exceeded May Cause Fish Kill

Fish	Copper Sulfate lb/10⁶ gal	Copper Sulfate mg/l	Free Chlorine (mg/l)	Chloramine (mg/l)
Trout	1.2	0.14	0.10 to 0.15	0.4
Carp	2.8	0.33	0.15 to 0.2	0.76 to 1.2
Suckers	2.8	0.33		
Catfish	3.5	0.40		
Pickerel	3.5	0.40		
Goldfish	4.2	0.50	—	0.25
Perch	5.5	0.67		
Sunfish	11.1	1.36	—	0.4
Black bass	16.6	2.0		
Minnows	—	—	0.4	0.76–1.2
Bullheads	—	—	—	0.4
Trout fry	—	—	—	0.05–0.06
Gambusia	—	—	—	0.5–1.0

applied to prevent the killing of fish. If copper sulfate is evenly distributed, in the proper concentration, and in accordance with Table 2.4, there should be very little destruction of fish. Fish can withstand higher concentrations of copper sulfate in hard water. If a heavy algal crop has formed and then copper sulfate applied, the decay of algae killed may clog the gills of fish and reduce the supply of oxygen to the point that fish will die of asphyxiation, especially at times of high water temperatures. Tastes and odors are of course also intensified. Certain blue-green algae* may produce a toxin that is lethal to fish and animals. Other conditions may also be responsible for the destruction of fish. For example, a lower dissolved oxygen, a pH value below 4 to 5 or above 9 to 10, a free ammonia or equivalent of 1.2 to 3 mg/l, an unfavorable water temperature, a carbon dioxide concentration of 100 to 200 mg/l or even less, free chlorine of 0.15 to 0.3 mg/l, chloramine of 0.4 to 0.76 mg/l, 0.5 to 1.0 mg/l hydrogen sulfide and other sulfides, cyanogen, phosphine, sulfur dioxide, and other waste products are all toxic to fish.[68†] Even a chlorine residual of greater than 0.1 mg/l may be excessive.[69] Lack of food, overproduction, and species survival also result in mass "fish kills." The total chlorine residual to protect fish in the "full-channel mixing zone" should not exceed 0.005 mg/l.

Copper sulfate may be applied in several ways. The method used usually depends on such things as the size of the reservoir, equipment available, proximity of the microorganisms to the surface, reservoir inlet and outlet arrangement, and

---

*Some belonging to the genera *Microcystis* and *Anabaena*. *Prymnesium parvum* is incriminated in fish mortality in brackish water. Marine dinoflagellates *Gymnodinium* and *Gonyaulax* toxins cause death of fish and other aquatic life.

†Trout are usually more sensitive.

time of year. One of the simplest methods of applying copper sulfate is the burlap-bag method. A weighed quantity of crystals (blue-stone) is placed in a bag and towed at the desired depth behind a rowboat or, preferably, motor-driven boat. The copper sulfate is then drawn through the water in accordance with a planned pattern, first in one direction in parallel lanes about 25 feet apart and then at right angles to it so as to thoroughly treat the entire body of water, including shallow areas. The rapidity with which the chemical goes into solution may be controlled by regulating the fabric of the bag used, varying the velocity of the boat, using crystals of large or small size, or combinations of these variables. In another method, a long wedge-shaped box (12 × 6 in.) is attached vertically to a boat. Two bottom sides have double 24-mesh copper screen openings 1 foot high; one has a sliding cover. Copper sulfate is added to a hopper at the top. The rate of solution of copper sulfate is controlled by raising or lowering the sliding cover over the screen, by the boat speed, and by the size of copper sulfate crystals used. Where spraying equipment is available, copper sulfate may be dissolved in a barrel or tank carried in the boat and sprayed on the surface of the water as a 0.5 or 1 percent solution. Pulverized copper sulfate may be distributed over large reservoirs or lakes by means of a mechanical blower carried on a motor-driven boat. Larger crystals are more effective against algae at lower depths. Where water flows into a reservoir, it is possible to add copper sulfate continuously and proportional to the flow, provided fish life is not important. This may be accomplished by means of a commercial chemical feeder, an improvised solution drip feeder, or a perforated box feeder wherein lumps of copper sulfate are placed in the box and the depth of submergence in the water is controlled to give the desired rate of solution. In the winter months, when reservoirs are frozen over, copper sulfate may be applied if needed by cutting holes in the ice 20 to 50 feet apart and lowering and raising a bag of copper sulfate through the water several times. If an outboard motor is lowered and rotated for mixing, holes may be 1,000 feet apart. Scattering crystals on the ice is also effective in providing a spring dosage when this is practical. Dosage should limit copper to less than 1 mg/l in the water treatment plant effluent. Potassium permanganage crystals may be used where copper sulfate is ineffective.

It is possible to control microorganisms in a small reservoir, where chlorine is used for disinfection and water is pumped to a reservoir, by maintaining a free residual chlorine concentration of about 0.3 mg/l in the water. However, chlorine will combine with organic matter and be used up or dissipated by the action of sunlight unless the reservoir is covered and there is a sufficiently rapid turnover of the reservoir water. Where a contact time of 2 hours or more can be provided between the water and disinfectant, the chlorine–ammonia process may be used to advantage. Chlorine may also be added as chloride of lime or in liquid form by methods similar to those used for the application of copper sulfate.

Mackenthum[70] cautions that the control of one nuisance may well stimulate the occurrence of another under suitable conditions and necessitate additional control actions. For example, the control of algae may lead to the growth of weeds. Removal of aquatic weeds may promote the growth of phytoplankton or bottom algae such as chara. The penetration of sunlight is thereby facilitated, but nutrients are released by growth and then decay of chara. The primary emphasis should be elimination of nutrients.

Gnat flies sometimes lay their eggs in reservoirs. The eggs develop into larvae, causing consumer complaints of worms in the water. The best control measure is covering the reservoir or using fine screening to prevent the entrance of gnats.

Zebra Mussel and Its Control

The zebra mussel is 1 to 2 in. (2.5–5 cm) long. An adult female can release 30,000 to 40,000 eggs per year. The eggs hatch into larvae (veligers) in several days, drift with the currents, hatch within 3 weeks, and attach themselves to any hard surface such as water intakes, boat and ship hulls, rock reefs, and canals, where they grow. Waters having less than 15 mg/l calcium, pH less than 6.5, temperatures below 45°F (7°C) or above 90°F (32°C), and salinity greater than 600 mg/l chloride ion limit growth. The mussels mature in about two years and have a life span of three to five years, depending on the environment. A major concern is the accumulation of the zebra mussels inside industrial plant, power plant, and drinking water intakes, causing restriction in flows and eventual clogging. Flows of 3.3 to 5.0 ft/sec (1.0–1.5 m/s) prevent attachment of the mussel, but infestation has extended to intake screens, raw water wells, settling tanks, condensers, and cooling towers. The mussel imparts a very disagreeable taste to drinking water when it dies.

The zebra mussel is believed to have been introduced into the United States via the St. Lawrence River and the Great Lakes through international ship freshwater ballast discharges. This practice is prohibited but is not entirely effective. The mussels are spreading to inland waters by attachment to recreation boats and by waterfowl carrying the veligers.

Suggested control measures include location of water intakes under sand, cleanable screens, mechanical scraping or pigging of intake pipes, electrical currents, certain sound frequencies, flushing intakes with hot water [113–131°F (45–55°C)] for not less than 10 min, oxygen deprivation, chlorine if trihalomethane production is not a problem, and ozone injection. Dual cleanable intake lines may be needed.

Although the zebra mussel improves water clarity (it filters about 1 liter of water per day and consumes phytoplankton and organic material), it deprives fish of algae and other food. The problems created, prevention or treatment, and costs involved remain to be fully resolved.[71]

Aquatic Weed Control

The growth of aquatic plants (and animals) is accelerated in clear water by nutrients in the water and bottom sediment and when the temperature of the surface water is about 59°F (15°C). Vegetation that grows and remains below the water surface does not generally cause difficulty. Decaying submergent, emergent, and floating aquatic vegetation as well as decaying leaves, brush, weeds, grasses, and debris in the water can cause tastes and odors in water supplies. The discharge of organic wastes from wastewater treatment plants, overflowing septic tank systems, storm sewers, and drainage from lawns, pastures, and fertilized fields contains nitrogen and phosphorus, which promote algal and weed growth. The contribution of phosphorus from sewage treatment plants and septic tank systems can be relatively small compared to that from surface runoff. Unfortunately, little can be done to permanently prevent the entrance of all wastes and drainage or destroy growths of rooted plants, but certain chemical, mechanical, and biological methods can provide temporary control.

Reasonably good temporary control of rooted aquatic plants may be obtained by physically removing growths by dredging; wire or chain drags, rakes, and hand pulling; and mechanical cutting. Winter drawdown and deepening of reservoirs, lakes, and ponds edges to a depth of 2 feet or more will prevent or reduce plant growths. Weeds that float to the surface should be removed before they decay. Sandy, gravelly, rocky, or clayey bottoms inhibit plant growths.

Where it is possible, the water level should be drained or lowered 3 to 6 feet to expose the affected areas of the reservoir for about 1 month during the freezing winter months, followed by drying the weeds and roots and clearing and removal. Drying out the roots and burning and removing the ash are effective for a number of years. Flooding 3 feet or more above normal is also effective where possible.

Biological control with plant-eating fish, such as white amur or grass carp, is illegal in many states. They eat aquatic insects and other invertebrates and are detrimental to other fish and water quality.

As a last resort, aquatic weeds may be controlled by chemical means. Tastes and odors may result if the water is used for drinking purposes; the chemical may kill fish and persist in the bottom mud, and it may be hazardous to the applicator. The treatment must be repeated annually or more often, and heavy algal blooms may be stimulated, particularly if the plant destroyed is allowed to remain in the water and return its nutrients to the water. Chemical use should be restricted and permitted only after careful review of the toxicity to humans and fish, the hazards involved, and the purpose to be served. Copper sulfate should not be used for the control of aquatic weeds, except for algae, since the concentration required to destroy the vegetation will assuredly kill any fish present in the water and probably exceed permissible levels in drinking water. Diquat and endothal have been approved by the EPA, if applied according to directions. Diquat use requires a 10-day waiting period. Endothal use requires a waiting period of 14 days with the amine salt formulation and 7 days with the potassium or sodium salts formulation.[72] The health and conservation departments should be consulted prior to any work. A permit is usually required.

Other Causes of Tastes and Odors

In new reservoirs, clearance and drainage reduce algal blooms by removing organic material beneficial to their growth. Organic material, which can cause anaerobic decomposition, odors, tastes, color, and acid conditions in the water, is also removed. If topsoil is valuable, its removal may be worthwhile.

Some materials in water cause unpleasant tastes and odors when present in excessive concentrations, although this is not a common source of difficulty. Iron and manganese, for example, may give water a bitter, astringent taste. In some cases, sufficient natural salt is present, or salt water enters to cause a brackish taste in well water. It is not possible to remove the salt in the well water without going to great expense. Elimination of the cause by sealing off the source of the salt water, groundwater recharge with fresh water, or controlling pump drawdown is sometimes possible.

Other causes of tastes and odors are sewage and industrial or trade wastes and spills. Sewage would have to be present in very large concentrations to be noticeable in a water supply. If this were the case, the dissolved oxygen in the water receiving the sewage would most probably be used up, with resultant nuisance conditions. On the other hand, the billions of microorganisms introduced, many of which would cause illness or death if not removed or destroyed before consumption, are the greatest danger in sewage pollution. Trade or industrial wastes introduce in water suspended or colloidal matter, dissolved minerals and organic chemicals, vegetable and animal organic matters, harmful bacteria, and other materials that are toxic and produce tastes and odors. Of these, the wastes from steel mills, paper plants, and coal distillation (coke) plants have proved to be the most troublesome in drinking water, particularly in combination with chlorine. Tastes produced have been described as "medicinal," "phenolic," "iodine," "carbolic acid," and "creosote." Concentrations of 1 part phenol to 500 million parts of water will cause very disagreeable tastes even after the water has traveled 70 miles.[73] The control of these tastes and odors lies in the prevention and reduction of stream pollution through improved plant operation and waste treatment. Chlorine dioxide has been found effective in treating a water supply not too heavily polluted with phenols. The control of stream pollution is a function and responsibility of federal and state agencies, municipalities, and industry. Treatment of water supplies to eliminate or reduce objectionable tastes and odors is discussed separately.

Sometimes high uncontrolled doses of chlorine produce chlorinous tastes and chlorine odors in water. This may be due to the use of constant feed equipment rather than a chlorinator, which will vary the chlorine dosage proportional to the quantity of water to be treated. In some installations, chlorine is added at a point that is too close to the consumers, and in others, the dosage of chlorine is marginal or too high or chlorination treatment is used where coagulation, filtration, and chlorination should be used instead. Where superchlorination is used and high concentrations of chlorine remain in the water, dechlorination with sodium sulfite, sodium bisulfite, sodium thiosulfate, sulfur dioxide, or activated carbon is indicated. Sulfur dioxide is most commonly used in manner similar to that used for liquid chlorine and with the same precautions; dosage must be carefully controlled to avoid lowering the pH and dissolved oxygen, as reaeration may then be necessary.

Methods to Remove or Reduce Objectionable Tastes and Odors

Some of the common methods used to remove or reduce objectionable tastes and odors in drinking water supplies, not in order of their effectiveness, are as follows:

- Free residual chlorination or superchlorination
- Chlorine–ammonia treatment
- Aeration or forced-draft degasifier
- Application of activated carbon
- Filtration through granular activated-carbon or charcoal filters
- Coagulation and filtration of water (also using an excess of coagulant)
- Control of reservoir intake level
- Elimination or control of source of trouble
- Chlorine dioxide treatment
- Ozone treatment
- Potassium permanganate treatment
- Hydrogen sulfide removal
- Removal of gasoline, fuel oil, and other organics

Bench and pilot studies over a representative period of time are advised, including laboratory studies, possible treatment or combinations of treatment, and source control.

Free Residual Chlorination   Free residual chlorination will destroy, by oxidation, many taste- and odor-producing substances and inhibit growths inside water mains. Biochemical corrosion is also prevented in the interior of water mains by destroying the organisms associated with the production of organic acids. The reduction of sulfates to objectionable sulfides is also prevented. However, chloro-organics (THMs), which are suspected of being carcinogenic, may be formed, depending on the precursors in the water treated, pH, temperature, contact time, and point of chlorination.

Nitrogen trichloride is formed in water high in organic nitrogen when a very high free chlorine residual is maintained. It is an explosive, volatile, oily liquid that is removed by aeration or carbon. Nitrogen trichloramine exists below pH 4.5 and at higher pH in polluted waters. See "Chlorine Treatment for Operation and Microbiological Control," this chapter.

Chlorine–Ammonia Treatment   In practice, chlorine–ammonia treatment is the addition of about three or four parts chlorine to one part ammonia. Ammonia is available as a liquid and as a gas in 50-, 100-, and 150-pound cylinders. The ammonia is added a few feet ahead of the chlorine. Nitrogenous organic compounds reduce the effectiveness of inorganic chloramines and give misleading residual readings. Chloramines are weak disinfectants. Chloramine (monochloramine) concentration should be increased by 25 and contact time 100 times to obtain the same effectiveness as free chlorine disinfection. Because of this, chlorine–ammonia treatment is not recommended as the primary disinfectant. The disinfection efficiency of chloramines decreases with increases in pH. Chloramines prevent chlorinous tastes due to the reaction of chlorine with taste-producing substances in water. However, other taste and odor compounds may develop. An excess of ammonia can cause bacterial growth in the extremities of distribution systems. Chloramines may stimulate the growth of algae and bacteria in open reservoirs and will interfere with the maintenance of a free chlorine residual if insufficient chlorine is used. Chloramines are toxic to fish, including tropical fish, if not removed prior to use. Hospitals should also be informed that chloramines may adversely affect dialysis patients. Soft-drink manufacture may also be affected.

Chloroform is not formed as in free residual chlorination; but other chloro-organic compounds that may cause adverse health effects are formed.[74] Free residual chlorination followed by dechlorination and then chlorination of the water distributed is sometimes practiced with good bacteriological control.[75] Chlorine–ammonia treatment of the filtered water to maintain a chloramine residual in the distribution system, instead of a free residual, is sometimes used to minimize trihalomethane formation, but microbiological quality must not be compromised.

Aeration Aeration is a natural or mechanical process of increasing the contact between water and air for the purpose of releasing entrained gases, adding oxygen, and improving the chemical and physical characteristics of water. Some waters, such as water from deep lakes and reservoirs in the late summer and winter seasons, cistern water, water from deep wells, and distilled water, may have an unpleasant or flat taste due to a deficient dissolved-oxygen content. Aeration will add oxygen to such waters and improve their taste. In some instances, the additional oxygen is enough to make the water corrosive. Adjustment may be needed. Free carbon dioxide and hydrogen sulfide will be removed or reduced, but tastes and odors due to volatile oils exuded by algae are not effectively removed.[76] Aeration is advantageous in the treatment of water containing dissolved iron and manganese in that oxygen will change or oxidize the dissolved iron and manganese to insoluble ferric and manganic forms that can be removed by settling, contact, and filtration. It is also useful to remove carbon dioxide before lime–soda ash softening.

Aeration is accomplished by allowing the water to flow in thin sheets over a series of steps, weirs, splash plates, riffles, or waterfalls; by water sprays in fine droplets; by allowing water to drip out of trays, pipes, or pans that have been slotted or perforated with 3/16- to 1/2-inch holes spaced 1 to 3 inches on centers to maintain a 6-inch head; by causing the water to drop through a series of trays containing 6 to 9 inch of coke or broken stones; by means of spray nozzles; by using air-lift pumps; by introducing finely divided air in the water; by permitting water to trickle over 1 × 3-inch cypress wood slats with 1/2- to

3/4-inch separations in a tank through which air is blown up from the bottom; by forced or induced draft aeration; and by similar means. Water is applied at the rate of 1 to 5 gpm/ft^2 of total tray area.[77] Coke will become coated, and hence useless if the water is not clear, if the coke is not replaced. Slat trays are usually 12 inches apart.

Louvered enclosures are necessary for protection from wind and freezing. Many of these methods are adaptable to small rural water supplies, but care should be taken to protect the water from insects and accidental or willful contamination. Screening of the aerator is necessary to prevent the development of worms. Aerated water must be chlorinated before distribution for potable purposes. Corrosion control may also be necessary.

Activated Carbon — Powdered and Granular   The sources of raw material for activated carbon include bituminous coal, lignite, peat, wood, bone, petroleum-based residues, and nutshells. The carbon is activated in an atmosphere of oxidizing gases such as CO_2, CO, O_2, steam, and air at a temperature of between 572° and 1,832°F (300°–1,000°C), usually followed by sudden cooling in air or water. The micropores formed in the carbonized particles contribute greatly to the adsorption capacity of the activated carbon. Granular carbon can be reactivated by heat treatment as, for example, in a multihearth furnace at a temperature of 1508° to 1,706°F (820°–930°C) in a controlled low-oxygen steam atmosphere, where dissolved organics in the carbon pores are volatilized and released in gaseous form. The regenerated carbon is cooled by water quenching.[78] In any case, the spent carbon, whether disposed of in a landfill or incinerator or regenerated, must be handled so as not to pollute the environment.

Granular activated-carbon (GAC) filters (pressure type) are used for treating water for soft drinks and bottled drinking water. The GAC filter beds are used at water treatment plants to remove taste- and odor-producing compounds, as well as color and synthetic organic chemicals suspected of being carcinogenic. Colloids interfere with adsorption if not removed prior to filtration. The GAC filters or columns normally follow conventional rapid sand filters but can be used alone if a clear, clean water is being treated.

Granular activated carbon is of limited effectiveness in the removal of trihalomethane precursor compounds. It is effective for only a few weeks. In contrast, GAC beds for taste and odor control need regeneration every three to six years.[79] When the GAC bed becomes saturated with the contaminant being removed, the contaminant appears in the effluent (an event known as breakthrough) if the GAC is not replaced or regenerated.

Activated carbon in the powdered form is used quite generally and removes by adsorption, if a sufficient amount is used, practically all tastes and odors found in water. The powdered carbon may be applied directly to a reservoir as a suspension with the aid of a barrel and boat (as described for copper sulfate) or released slowly from the bag in water near the propeller, but the reservoir should be taken out of service for one to two days, unless the area around the intake can be isolated. The application of copper sulfate within this time will improve the settling of the carbon.

Doses vary from 1 to 60 pounds or more of carbon to 1 million gallons of water, with 25 pounds as an average. In unusual circumstances, as much as 1,000 pounds of carbon per million gallons of water treated may be needed, but cost may make this impractical. Where a filtration plant is provided, carbon is fed by means of a standard chemical dry-feed machine or as a suspension to the raw water, coagulation basin, or filters. However, carbon can also be manually applied directly to each filter bed after each wash operation. Ten to 15 minutes contact time between the carbon and water being treated and good mixing will permit efficient adsorption of the taste and color compounds. Activated carbon is also used in reservoirs and settling basins to exclude sunlight causing the growth of algae. This is referred to as "blackout" treatment. The dosage of carbon required can be determined by trial and error and tasting the water, or by a special test known as the "threshold odor test," which is explained in *Standard Methods*.[80] If the water is pretreated with chlorine, after 15 to 20 minutes, the activated carbon will remove up to about 10 percent of its own weight of chlorine; hence, they should *not* be applied together. Careful operation control can make possible prompt detection of taste- and odor-producing compounds reaching the plant and the immediate application of corrective measures.

The GAC filters are usually 2.5 to 3 feet deep and operate at rates of 2 to 5 gpm/ft^2. They are supported on a few inches of sand. Pressure filters containing sand and activated carbon are used on small water supplies. The GAC columns are up to 10 feet deep. The water, if not clear, must be pretreated by conventional filtration, including coagulation and clarification.

Charcoal Filters   Charcoal filters, either of the open-gravity or closed-pressure type, are also used to remove substances causing tastes and odors in water. The water so treated must be clear, and the filters must be cleaned, reactivated, or replaced when they are no longer effective in removing tastes and odors. Rates of filtration vary from 2 to 4 gpm/ft^2 of filter area, although rates as high as 10 gpm/ft^2 are sometimes used. Trays about 4 ft^2 containing 12 inch of coke are also used. The trays are stacked about 8 inches apart, and the quantity is determined by the results desired.

Coagulation   Coagulation of turbidity, color, bacteria, organic matter, and other material in water followed by flocculation, settling, and then filtration will also result in the removal of taste- and odor-producing compounds, particularly when activated carbon is included. The use of an excess of coagulants will sometimes result in the production of a better tasting water. In any case, a surface-water supply should be treated to produce a very clear water so as to remove the colloids, which together with volatile odors account for the taste and odors of most finished waters.[81]

Reservoir Management, Intake Control, and Stratification   The quality of reservoir and lake water varies with the depth, season of the year or temperature, wave action, organisms and food present, condition of the bottom, clarity of the water, and other factors. Stratification is more likely to be pronounced in deep-water bodies. Lakes are classified as eutrophic (productive) or oligotrophic (unproductive).

Temperature is important in temperate zones. At a temperature of 39.2°F (4°C), water is heaviest, with a specific gravity of 1.0. Therefore, in the fall of the year, the cool air will cause the surface temperature of the water to drop, and when it reaches 39.2°F, this water, with the aid of wind action, will move to the bottom and set up convection currents, thereby forcing the bottom water up. Then in the winter the water may freeze, and conditions will remain static until the spring, when the ice melts and the water surface is warmed. A condition is reached when the entire body of water is at a temperature of about 39.2°F, but a slight variation from this temperature, aided by wind action, causes an imbalance, with the bottom colored, turbid water deficient in oxygen (usually also acidic and high in iron, manganese, and nutrient matter) rising and mixing with the upper water. The warm air will cause the temperature of the surface water to rise, and a temporary equilibrium is established, which is upset again with the coming of cold weather. This phenomenon is known as reservoir turnover. In a shallow reservoir or lake, less than about 25 feet deep, wind action rather than water density induces water mixing.

In areas where the temperature does not fall below 39.2°F and during warm months of the year, the water in a deep reservoir or lake will be stratified in three layers: the top mixed zone (epilimnion), which does not have a permanent temperature stratification and which is high in oxygen and algae; the middle transition zone (metalimnion or thermocline), in which the drop in temperature equals or exceeds 1.8°F (1°C) per meter and oxygen decreases; and the bottom zone of stagnation (hypolimnion), about one-half or more of the depth, which is generally removed from surface influence. The hypolimnion is cold, below 54°F (12°C), and often deficient in oxygen. The metalimnion is usually the source of the best quality water. The euphotic zone, in the epilimnion, extends to the depth at which photosynthesis fails to occur because of inadequate light penetration. The reservoir or upper lake layer or region, in which organic production from mineral substances takes place because of light penetration, is called the trophogenic region. The layer in the hypolimnion in which the light is deficient and in which nutrients are released by dissimilation (the opposite of assimilation) is called the tropholytic region. Hydrogen sulfide, manganese, iron, and ammonia may occur at the bottom, making for poor-quality, raw water.

A better quality water can usually be obtained by drawing from different depth levels, except during reservoir turnover. To take advantage of this, provision should be made in deep reservoirs for an intake tower with inlets at different elevations so that the water can be drawn from the most desirable level. Where an artificial reservoir is created by the construction of a dam, it is usually better to waste surplus water through a bottom blowoff rather than over a spillway. Then stagnant hypolimnion bottom water, usually the colder water except in the winter, containing decaying organic matter (hydrogen sulfide), phosphorus, color, manganese, iron, and silt, can be flushed out.

Chlorine Dioxide Treatment   Chlorine dioxide treatment was originally developed to destroy tastes produced by phenols. However, it is also a strong disinfectant over a broad pH range and effective against other taste-producing compounds such as from algae, decaying vegetation, and industrial wastes. It also oxidizes iron and manganese and aids in their removal.

Chlorine dioxide is manufactured at the water plant where it is to be used. Sodium chlorite solution and chlorine water are usually pumped into a glass cylinder where chlorine dioxide is formed and from which it is added to the water being treated, together with the chlorine water previously acidified with hydrochloric acid. A gas chlorinator is needed to form chlorine water, and for a complete reaction with full production of chlorine dioxide, the pH of the solution in the glass reaction cylinder must be less than 4.0—usually 2.0 to 3.0. Where hypochlorinators are used, the chlorine dioxide can be manufactured by adding hypochlorite solution, a dilute solution of hydrochloric acid, and a solution of sodium chlorite in the glass reaction cylinder so as to maintain a pH of less than 4.0. Three solution feeders are then needed. Cox gives the theoretical ratio of chlorine to sodium chlorite as 1.0 to 2.57 with chlorine water or hypochlorite solution and sodium chlorite to chlorine dioxide produced as 1.0 to 0.74.[82] A chlorine dioxide dosage of 0.2 to 0.3 mg/l will destroy most phenolic taste-producing compounds. Chlorine dioxide does not react with nitrogenous compounds or other organic materials in solution having a chlorine demand. Chloramine and trihalomethane formation is prevented or reduced, provided free chlorine is not present. It is an effective disinfectant, about equivalent to hypochlorous acid. In contrast to free chlorine, chlorine dioxide efficiency increases as the pH increases: Chlorine dioxide is a more efficient bactericide and virucide at pH 8.5 to 9.0 than at pH 7.0.[83] Chlorine dioxide may have to be supplemented by chlorine to maintain an effective residual in the distribution system. The EPA requires that chlorine dioxide residual oxidants be controlled so that the sum of chlorine dioxide, chlorite ion, and chlorate ion does not exceed 0.5 to 1.0 mg/l. The chlorite ion is said to be very toxic. Because of this and other uncertainties, caution is urged. Chlorine dioxide is a very irritating gas and is more toxic than chlorine. It explodes on heating. The permissible 8-hour exposure concentration is 0.1 ppm and 0.3 ppm for 15 minutes.[84]

Ozone Treatment   Ozone in concentrations of 1.0 to 1.5 mg/l has been used for many years as a disinfectant and as an agent to remove color, taste, and odors from drinking water.[85] It effectively eliminates or controls color, taste, and odor problems not amenable to other treatment methods; controls disinfection byproducts formation; and improves flocculation of surface waters in low concentrations. Doses of 5 mg/l are reported to interfere with flocculation and support bacterial growths. Ozone also oxidizes and permits iron and manganese removal by settling and filtration and aids in turbidity removal. About 2 mg/l or less is required.[86] Like chlorine, ozone is a toxic gas. Source water quality affects ozonation effectiveness. Pilot plant studies are indicated.

In contrast to chlorine, ozone is a powerful oxidizing agent over a wide pH and temperature range. It is an excellent virucide, is effective against amoebic and *Giardia* cysts, and destroys bacteria, humic acid, and phenols. The potential for the formation of chlorinated organics such as THMs is greatly reduced with preozonation; the removal of soluble organics in coagulation is also reported to be improved.[87] Ozone is reported to be 3,100 times faster and more effective than chlorine in disinfection.[88] Ozone attacks the protein covering protecting a microorganism; it inactivates the nucleic acid, which leads to its destruction. Ozonation provides no lasting residual in treated water but increases the dissolved oxygen; it has a half-life of about 20 minutes in 70°F (21°C) distilled water. Ozone is more expensive as a disinfectant than either chlorine or chlorine dioxide. The disadvantage of no lasting residual can be offset by adding chlorine or chlorine–ammonia to maintain a chlorine residual in the distribution system.

New products can also be formed during the ozonation of wastewaters; not all low-molecular-weight organic compounds are oxidized completely to CO_2 and H_2O. Careful consideration must be given to the possibility of the formation of compounds with mammalian toxicity during ozonation of drinking water.[89]

However, at least one study concludes that the probability of potentially toxic substances being formed is small.[90] Ozone disinfection before chloramination yielded less than 1 $\mu g/1$ total trihalomethane.[91]

Ozone must be generated at the point of use; it cannot be stored as a compressed gas. Although ozone can be produced by electrolysis of perchloric acid and by UV lamps, the practical method for water treatment is by passage of dry, clean air between two high-voltage electrodes. Pure oxygen can be added in a positive-pressure injection system. High-purity oxygen will produce about twice the amount of ozone from the same ozonator at the same electrical input.[92] The ozonized air is injected in a special mixing and contact chamber (30 min) with the water to be treated. The space above the chamber must be carefully vented, after its concentration is reduced, using an ozone-destruct device to avoid human exposure, as ozone is very corrosive and toxic.

The vented ozone may contribute to air pollution. It should not exceed 0.12 ppm in the ambient air. As with chlorine, special precautions must be taken in the storage, handling, piping, respiratory protection, and housing of ozone. Exhaust air and plant air must be continuously monitored. The permissible 8-hours exposures are 0.1 and 0.3 ppm for 15 minutes.[93] Exposure to 0.05 ppm 24 hr/day, 7 days/week, is reported to be detrimental. Greater than 0.1 ppm calls for investigation. A concentration of 10,000 ppm is lethal in 1 minute; 500 ppm is lethal after 16 hours.

The generation of ozone results in the production of heat, which may be utilized for heating.

Hydrogen Sulfide, Sources and Removal Hydrogen sulfide is undesirable in drinking water for aesthetic and economic reasons. Its characteristic "rotten-egg" odor is well known, but the fact that it tends to make water corrosive to iron, steel, stainless steel, copper, and brass is often overlooked. The permissible 8-hours occupational exposure to hydrogen sulfide is 20 ppm, but only 10 minutes for 50 ppm exposure.[94] Death is said to result at 300 ppm. As little as 0.2 mg/l in water causes bad taste and odor and staining of photographic film.

The sources of hydrogen sulfide are both chemical and biological. Water derived from wells near oil fields or from wells that penetrate shale or sandstone frequently contain hydrogen sulfide. Calcium sulfate, sulfites, and sulfur in water containing little or no oxygen will be reduced to sulfides by anaerobic sulfur bacteria or biochemical action, resulting in liberation of hydrogen sulfide. This is more likely to occur in water at a pH of 5.5 to 8.5, particularly in water permitted to stand in mains or in water obtained from close to the bottom of deep reservoirs. Organic matter often contains sulfur that, when attacked by sulfur bacteria in the absence of oxygen, will release hydrogen sulfide. Another source of hydrogen sulfide is the decomposition of iron pyrites or iron sulfide.

The addition of 2.0 to 4 mg/l copper sulfate or the maintenance of at least 0.3 mg/l free residual chlorine in water containing sulfate will inhibit biochemical activity and also prevent the formation of sulfides. The removal of H_2S already formed is more difficult, for most complete removal is obtained at a pH of around 4.5. Aeration removes hydrogen sulfide, but this method is not entirely effective; carbon dioxide is also removed, thereby causing an increase in the pH of the water, which reduces the efficiency of removal. Therefore, aeration must be supplemented. Aeration followed by settling and filtration is an effective combination. Chlorination alone can be used without precipitation of sulfur, but large amounts, theoretically 8.4 mg/l chlorine to each milligram per liter of hydrogen sulfide, would be needed. The alkalinity (as $CaCO_3$) of the water is lowered by 1.22 parts for each part of chlorine added. Chlorine in limited amounts, theoretically 2.1 mg/l chlorine for each milligram per liter of hydrogen sulfide, will result in formation of flowers of sulfur, which is a fine colloidal precipitate requiring coagulation and filtration for removal. If the pH of the water is reduced to 6.5 or less by adding an acid to the water or a sufficient amount of carbon dioxide as flue gas, for example, good hydrogen sulfide removal should be obtained. But pH adjustment to reduce the aggressiveness of the water would be necessary. Another removal combination is aeration, chlorination, and filtration through an activated-carbon pressure filter. Pilot plant studies are indicated.

Pressure tank aerators, that is, the addition of compressed air to hydropneumatic tanks, can reduce the entrained hydrogen sulfide in well water from 35 to 85 percent, depending on such factors as the operating pressures and dissolved oxygen in the hydropneumatic tank effluent.[95] The solubility of air in water increases in direct proportion to the absolute pressure. Carbon dioxide is not removed by this treatment. Air in the amount of 0.005 to 0.16 ft³/gal water and about 15 minutes detention is recommended, with the higher amount preferred. The air may be introduced through perforated pipe or porous media in the tank bottom or with the influent water. Unoxidized hydrogen sulfide and excess air in the tank must be bled off. Air-relief valves or continuous air bleeders can be used for this purpose. It is believed that oxidation of the hydrogen sulfide through the sulfur stage to alkaline sulfates takes place, since observations show no precipitated sulfur in the tank. Objections to pressure tank aerators are milky water caused by dissolved air and corrosion. The milky water would cause air binding or upset beds in filters if not removed.

A synthetic resin has been developed that has the property of removing hydrogen sulfide with pH control. It can be combined with a resin to remove hardness so that a low-hardness water can be softened and deodorized. The resin is manufactured by Rohm and Haas Company (Philadelphia).

Removal of Gasoline, Fuel Oil, and Other Organics in an Aquifer   Leaking storage tanks and piping (50 percent of old tanks leak after 20 years), overflow from air vent or in filling, or accidental spillage near a well may cause gasoline or fuel oil to seep into an aquifer. Correction requires on-site studies and identification, location, and elimination of the cause. Removal of the source is a first and immediate step, followed possibly by pumping out and recovery of the contaminant, bioremediation, or other measures, based on the site geology, soils, and plume.

Sometimes lowering the pump or drop pipe intake in a well may help, if this is possible. However, the contaminant is likely to coat or fill soil pore spaces and persist in the contaminated zone a long time. Gasoline and oil tend to adhere to soil particles by surface tension and attraction, particularly in the unsaturated soil zone until dissipated through adsorption, dispersion, diffusion, and ion exchange[96] or flushed out or broken down by soil microorganisms. The gasoline, benzene, or fuel oil will gradually collect on the groundwater surface and in the well and can be skimmed off over a period of time. Denser liquids, such as chlorinated solvents, tend to move down through the groundwater.

A well that is contaminated by gasoline or fuel oil might be rehabilitated by extended pumping after removal of the source. The objective would be to create a cone of depression so that the zone of influence due to pumping encompasses the underground contaminated area or plume. This is not very effective if a large area is involved. Many withdrawal wells would be needed. One pump would lower the water table around the well to create a cone of depression; another with the intake close to the water surface in the well serves as an oil or gasoline skimmer. Of course, the contaminated water pumped must be treated before disposal. Continual leaching of gasoline and oil or other contaminant in pore spaces, on soil particles in both the saturated and unsaturated zones, and in rock fractures and faults can be expected for some time. The process may be expedited, under professional direction, by purging and the use of nontoxic, biodegradable detergents; however, it may not be completely effective in restoring the aquifer.

An aquifer was cleaned by treatment of the water from extraction wells using a combination of high-temperature air stripping, biological treatment, and granular activated-carbon filtration. Volatile organics, nonvolatile organics, and trace metals were removed. Soluble organics and some nonvolatile organics were destroyed.[97] Different options should be compared.

Soils contaminated with gasoline from underground storage tanks have been successfully treated by excavating and stockpiling when mixed with bacteria specifically cultivated for petroleum decomposition. Nutrients were applied to enhance growth and soil piles turned intermittently to ensure adequate concentrations of oxygen and soil/bacteria mixing. Spraying of a ditch contaminated by diesel oil with a solution of bacteria and nutrients was also successful.[98]

An activated-carbon filter will remove small amounts of oil or gasoline from a contaminated water supply. It may become expensive if large quantities must be removed and the activated carbon must be replaced frequently. Air stripping is effective for the removal of petroleum products but is more suitable for large water systems. Air stripping may also be used to decontaminate an unsaturated sandy soil. Preheated air is injected through injection wells and released through venting wells. Vacuum wells may also be used to extract volatile contaminants in unsaturated loose soil.

Ozone has also been found effective in removing volatile organic chemicals from drinking water, including 1,1-dichloroethene, 1,1,2-tetrachloroethene, trichloroethene, vinyl chloride, chlordane, polychlorinated biphenyls, and toxaphene. The treatment may include UV–ozone, UV–hydrogen peroxide, and ozone–hydrogen peroxide.[99]

Treatment methods for the removal of synthetic organic chemicals are also discussed later in this chapter.

Containment of a chemical contaminant in the aquifer may also be possible. This may be accomplished by the use of barriers such as a bentonite slurry trench, grout curtain, sheet piling, or freshwater barrier and by the provision of an impermeable cap over the offending source if it cannot be removed. The method used would depend on the problem and the hydrogeological conditions. However, there are many uncertainties in any method used, and no barrier can be expected to be perfect or maintain its integrity forever.

Bioremediation and Aquifer Restoration   In situ aerobic microbial degradation, also referred to as bioremediation, biorestoration, or bioreclamation, has also been used to treat soil contaminated with biodegradable, nonhalogenated organics. The process is complex in view of the many variables and unknowns. In general, as much as possible of the contaminated water and surface soil is removed; oxygen, nutrients such as nitrogen and phosphorus, enzymes, and/or bacteria are added, then reinjected through injection wells and recirculated. Favorable soil conditions include neutral pH, high soil permeability, and 50 to 75 percent moisture content.[100]

Progress is also being made in the use of genetically engineered microorganisms to break down (metabolize) pentachlorophenol (PCP) compounds into water carbon dioxide and cell protoplasm. Microorganisms are also being engineered to break down other toxic chemicals.[101] Naturally occurring soil bacteria that use hydrocarbons for food may be present at shallow depths and possibly in soil at greater depths.[102]

The decontamination procedure should be tailored to the soils and pollutants present and their characteristics. Competent prior background information—including interviews; hydrogeological, land-use, and soil investigations; microbiological and chemical analyses; laboratory bench-scale studies; and well monitoring—are usually necessary to characterize the problem, determine the best remediation treatment, and evaluate effectiveness. Aerial photographs, topographic maps, and nearby well logs are valuable aids. It is also important to know the hydraulic conductivity of the affected formation(s), depth to groundwater, direction of groundwater flow, and types and extent of contamination. Adequate hydraulic conductivities are required to obtain in situ bioreclamation. The feasibility of biodegradation is determined in the laboratory using soil samples from several locations on the site.[103]

Iron and Manganese Occurrence and Removal

Iron in excess of 0.3 to 0.5 mg/l will stain laundry and plumbing fixtures and cause water to appear rusty. When manganese is predominant, the stains will be brown or black. Neither iron nor manganese is harmful in the concentrations found in water. Iron may be present as soluble ferrous bicarbonate in alkaline well or spring waters; as soluble ferrous sulfate in acid drainage waters or waters containing sulfur; as soluble organic iron in colored swamp waters; as suspended insoluble ferric hydroxide formed from iron-bearing well waters, which are subsequently exposed to air; and as a product of pipe corrosion producing red water.

Most soils, including gravel, shale, and sandstone rock, and most vegetation contain iron and manganese in addition to other minerals. Decomposing organic matter in water, such as in the lower levels of reservoirs, removes the dissolved oxygen usually present in water. This anaerobic activity and acidic condition dissolves mineral oxides, changing them to soluble compounds. Water containing carbon dioxide or carbonic acid, chlorine, or other oxidizing agent will have the same effect. In the presence of air or dissolved oxygen in water, soluble ferrous bicarbonate and manganous bicarbonate will change to insoluble ferric iron and manganic manganese, which will settle out in the absence of interfering substances. Ferrous iron and manganous manganese may be found in the lower levels of deep reservoirs, flooding soils, or rock containing iron and manganese or their compounds; hence, it is best to draw water from a higher reservoir level but below the upper portion, which supports microscopic growths like algae. This requires the construction and use of multiple-gate intakes, as previously mentioned. Consideration must be given to vertical circulation, such as in the spring and fall when the ferrous iron and manganous manganese are brought into contact with dissolved oxygen and air and convert to the insoluble state and settle out, if not drawn out in the intake.

The presence of as little as 0.1 mg/l iron in a water will encourage the growth of such bacteria as leptothrix and crenothrix. Carbon dioxide also favors their growth. These organisms grow in distribution systems and cause taste, odor, and color complaints. Mains, service lines, meters, and pumps may become plugged by the crenothrix growths. Gallionella bacteria can grow in wells and reduce capacity. Complaints reporting small gray or brownish flakes or masses of stringy or fluffy growths in water would indicate the presence of iron bacteria. The control of iron bacteria in well water is also discussed under "Control of Microorganisms," and "Corrosion Cause and Control," this chapter.

Corrosive waters that are relatively free of iron and manganese may attack iron pipe and house plumbing, particularly hot-water systems, causing discoloration and other difficulties. Such corrosion will cause red water, the control of which is discussed separately.

Iron and manganese can be removed by aeration or oxidation with chlorine, chlorine dioxide, ozone, potassium permanganate, or lime and lime–soda softening followed by filtration. A detention of at least 20 minutes may be required following aeration if the raw water is high in manganese or iron. Manganese can also be removed by filtration through manganese green sand capped with at least 6 inches of anthracite, but potassium permanganate is added ahead of the filter. Iron and manganese can be kept in solution (sequestered) if present in combination or individually at a concentration of 1 mg/l or less. Sodium silicates may be used to sequester up to 2 mg/l iron and/or manganese in well water prior to air contact.[104] Each method has limitations and requirements that should be determined by on-site pilot plant studies. A summary of processes used to remove iron and manganese is given in Table 2.5.

Most of the carbon dioxide in water is removed by aeration; then the iron is oxidized and the insoluble iron is removed by settling or filtration. If organic matter and manganese are also present, the addition of lime or chlorine will assist in changing the iron to an insoluble form and hence simplify its removal.

The open coke-tray aerator is a common method to oxidize and remove iron and manganese. Two or more perforated wooden trays containing about 9 inches of coke are placed in tiers. A 20- to 40-minute detention basin is provided beneath the stack of trays; there the heavy precipitate settles out. The lighter precipitate is pumped out with the water to a pressure filter, where it is removed. Carbon dioxide and hydrogen sulfide are liberated in the coke-tray aerator, and when high concentrations of carbon dioxide are present, it may be necessary to supplement the treatment by the addition of soda ash, caustic soda, or lime to neutralize the excess carbon dioxide to prevent corrosion of pipelines.

Open slat-tray aerators operate similarly to the coke-tray type but are not as efficient; however, they are easier to clean than the coke tray, and there is no coke to replace. When the trays are enclosed and air under pressure is blown up through the downward falling spray, a compact unit is developed in which the amount of air can be proportioned to the amount of iron to be removed. Theoretically, 0.14 mg/l oxygen is required to precipitate 1 mg/l iron. The unit may be placed indoors or outdoors.

Another method for iron removal utilizes a pressure tank with a perforated air distributor near the bottom. Raw water admitted at the bottom of the pressure tank mixes with the compressed air from the distributor and oxidizes the iron present. The water passes to the top of a pressure tank, at which point air is released and automatically bled off. The amount of air injected is proportioned to the iron content by a manually adjusted needle valve ahead of a solenoid valve on the air line.[105]

At a pH of 7.0, 0.6 parts of chlorine removes 1 part iron and 0.9 parts alkalinity. At a pH of 10.0, 1.3 parts of chlorine removes 1 part of manganese and 3.4 parts alkalinity.[106]

TABLE 2.5 Processes of Iron and Manganese Removal

Treatment Process	Oxidation Required	Character of Water	Equipment Required	pH Range Required	Chemicals Required	Remarks
1. Aeration, sedimentation, filtration	Yes	Iron alone in absence of appreciable concentrations of organic matter	Aeration, settling basin, sand filter	>7	None	Easily operated, no chemical control required
2. Aeration, contact oxidation, sedimentation, sand filtration	Yes	Iron and manganese loosely bound to organic matter but no excessive carbon dioxide or organic acid content	Contact aerator of coke, gravel, or crushed pyrolusite; settling basin; sand filter	>7 for iron removal, 7.5–10 for manganese	None	Double pumping required; easily controlled
3. Aeration, contact filtration	Yes	Iron and manganese bound to organic matter but no excessive organic acid content	Aerator and filter bed of manganese-coated sand, Birm, crushed pyrolusite ore, or manganese zeolite	>7 for iron removal, 7.5–10 for manganese	Lime for manganese removal	Double pumping required unless air compressor or "sniffler" valve used to force air into water; limited air supply adequate; easily controlled

4. Contact filtration	Yes, but not by aeration	Iron and manganese bound to organic matter but no excessive carbon dioxide or organic acid content	Filter bed of manganese-coated sand, Birm, crushed pyrolusite ore, or manganese zeolite	>7 for iron removal, >8.5 for manganese	Filter bed reactivated or oxidized with chlorine at intervals or with potassium permanganate applied continuously	Single pumping; aeration not required
5. Catalytic action, aeration, sedimentation, filtration	Yes	Manganese in combination with organic matter	Closed pyrolusite bed, aerator, second open-contact bed, sand filter	>7	None	Manganese changed to manganous hydroxide by catalytic action in absence of air, then oxidized
6. Aeration, chlorination, sedimentation, sand filtration	Yes	Iron and manganese loosely bound to organic matter	Aerator and chlorinator or chlorinator alone, settling basin, sand filter	7–8	Chlorine or potassium permanganate	Required chlorine dose reduced by previous aeration, but chlorination alone permits single pumping pH control required
7. Aeration, lime treatment, sedimentation, sand filtration	Yes	Iron and manganese in combination with organic matter; organic acids	Effective aerator, lime feeder mixing basin, settling basin, sand filter	8.5–10	Lime	pH control required

(*continues*)

TABLE 2.5 (continued)

Treatment Process	Oxidation Required	Character of Water	Equipment Required	pH Range Required	Chemicals Required	Remarks
8. Aeration, coagulation and lime treatment, sedimentation, sand filtration	Yes	Colored turbid surface water containing iron and manganese combined with organic matter	Conventional rapid sand filtration plant	8.5–9.6	Lime and ferric chloride or ferric sulfate, or chlorinated copperas, or lime and copperas	Complete laboratory control required
9. Zeolite softening	No	Well water devoid of oxygen, containing < 0.5 ppm iron and manganese for each 17.0 ppm hardness removed	Conventional sodium zeolite unit, with manganese zeolite unit (or equivalent) for treatment of bypassed water	> 6.5	None added continuously but bed is regenerated at intervals with salt solution	Only soluble ferrous and manganese bicarbonate can be removed by base exchange, so aeration or double pumping not required
10. Lime treatment, sedimentation, sand filtration	No	Soft well water devoid of oxygen, containing iron as ferrous bicarbonate	Lime feeder, enclosed mixing and settling tanks, pressure filter	8.1–8.5	Lime	Iron precipitated as ferrous carbonate in absence of oxygen; minimizes or prevents corrosion; double pumping not required

Source: C. R. Cox, *Operation and Control of Water Treatment Processes*, WHO Monograph Series, No. 49, World Health Organization, Geneva, 1964, pp. 212–213. Reproduced with permission.

Corrosion Cause and Control

Internal Pipe Corrosion   Internal pipe corrosion usually occurs in unlined metal distribution system piping and building plumbing in contact with soft water of low hardness, pH, and alkalinity containing carbon dioxide and oxygen. In serious cases, water heaters are damaged, the flow of water is reduced, the water is red or rusty where unprotected iron pipe is used, and the inside surface of pipe and fittings is dissolved, with consequent release of trace amounts of possibly harmful chemicals and weakening or pitting of pipe. Dissolved iron may be redeposited as tubercules with a reduction of pipe diameter and water flow. Biochemical changes take place in pipe where iron bacteria such as crenothrix and leptothrix use iron in their growth. High water velocities, carbon dioxide, dissolved solids, and high water temperatures [(140°–150°F) (60°–66°C)] all accelerate corrosion. Free chlorine residual less than 2 mg/l in water at pH 7 to 8 results in minimal corrosion. However, significant metal leaching (copper, cadmium, zinc, and lead) can occur in home water systems served with private wells when the water has high pH and hardness.[107]

Although much remains to be learned concerning the mechanism of corrosion, a simple explanation as related to iron may aid in its understanding. Water in contact with iron permits the formation of soluble ferrous oxide and hydrogen gas. Gaseous hydrogen is attracted to the pipe and forms a protective film if allowed to remain. But gaseous hydrogen combines with oxygen usually present in "aggressive" water, thereby removing the protective hydrogen film and exposing the metal to corrosion. High water velocities also remove the hydrogen film. In addition, ferrous oxide combines with the water and part of the oxygen usually present to form ferric hydroxide when the carbonate concentration is low, which redeposits in other sections of pipe or is carried through with the water. When the carbonate concentration is high, ferrous carbonate is formed. Another role is played by carbon dioxide. It has the effect of lowering the pH of the water since more hydrogen ions are formed, which is favorable to corrosion.

Pipe Materials and Corrosion   Lined steel and ductile iron pipe, asbestos-cement, wood-stave, plastic, vitrified clay, and concrete pressure pipe are corrosion resistant.* Plastic pipe may be polybutylene (PB), polyethylene (PE), or polyvinyl chloride (PVC). The PVC pipe comes in diameters of 0.5 to 30 inches, 10 to 20 foot lengths, and 100 to 235 psi working pressures. It is very resistant to corrosion. Fiberglass-reinforced plastic pipe is available in diameters up to 144 inches and lengths up to 60 feet. Polyethylene pipe comes in 18 to 120 inches in diameter and 20 foot lengths. Fiber–epoxy pipe comes in 20-foot lengths and 2 to 12 inches in diameter and is easily installed. It combines light weight with high tensile and compressive strength. The pipe withstands pressures of 300 psi, electrolytic attack, as well as embrittlement associated with

---

*See the AWWA/ANSI C104/A 21.4–85 Standard for Cement-Mortar Lining for Ductile-Iron Pipe and Fittings for water. Thermoplastic pipe should have the National Sanitation Foundation seal of approval. The AWWA standards C900-89 and C905 apply to 4- to 12-in. PVC pipe.

cold temperatures and aging. Ductile iron pipe comes in diameters of 4 to 54 inches and for pressures of 250 to 350 psi. Concrete pressure pipe withstands pressures of 400 psi and is available in diameters from 16 to 60 inches. With soft waters, calcium carbonate tends to be removed from new concrete, cement-lined, and asbestos-cement pipe for the first few years. Salt used in deicing can seep through the ground and greatly weaken reinforced concrete pipe and corrode the steel. Wood-stave, vitrified clay, and concrete pipes have limited applications. Iron and steel pipes are usually lined or coated with cement, tar, paint, epoxy, or enamel, which resist corrosion provided the coating is unbroken. Occasionally, coatings spall off or are imperfect, and isolated corrosion takes place. It should be remembered that even though the distribution system is corrosion resistant, corrosive water should be treated to protect household plumbing systems.

Polycyclic aromatic carbons, some of which are known to be carcinogenic, are picked up from bituminous lining of the water distribution system, not from oil-derived tarry linings. On general principles, bituminous linings are being discontinued in England by the Department of the Environment.[108] The WHO recommends that polynuclear aromatic hydrocarbon (PAH) levels in drinking water not exceed 10 to 50 $\mu g/1$, the levels found in unpolluted groundwater. Since some PAHs are carcinogenic in laboratory animals and may be carcinogenic in humans, the WHO also recommends that the use of coal-tar based and similar materials for pipe linings and coatings on water storage tanks *should be discontinued*. This recommendation was made with the knowledge that food contributes almost 99 percent of the total exposure to PAHs and that drinking water contributes probably less than 1 percent. Tetrachloroethylene can leach from vinyl-toluene-lined asbestos–cement pipe at dead-end or low-flow sections. The health risk is considered negligible.[109] Petroleum distillates, such as gasoline, can pass through PB and PE pipe and impart taste and odor to drinking water, but PVC pipe is penetrated to a lesser extent by gasoline. The PE, PB, PVC, asbestos–cement, and plastic joining materials may permit permeation by lower molecular weight organic solvents or petroleum products. The manufacturer should be consulted as to whether pipe may pass through contaminated soil.[110] See also "Drinking Water Additives," in Chapter 1.

Corrosion Control   The control of corrosion involves the removal of dissolved gases, treatment of the water to make it noncorrosive, building up of a protective coating inside pipe, use of resistant pipe materials or coating, cathodic protection, the insulation of dissimilar metals, prevention of electric grounding on water pipe, and control of growths in the mains. Therefore, if the conditions that are responsible for corrosion are recognized and eliminated or controlled, the severity of the problem will be greatly minimized. The particular cause(s) of corrosion should be determined by proper chemical analyses of the water as well as field inspections and physical tests. The applicable control measures should then be employed.

The gases frequently found in water and that encourage corrosion are oxygen and carbon dioxide. Where practical, as in the treatment of boiler water or hot water for a building, the oxygen and carbon dioxide can be removed by heating or by subjecting the water, in droplets, to a partial vacuum. Some of the oxygen is restored if the water is stored in an open reservoir or storage tank.

Dissolved oxygen can also be removed by passing the water through a tank containing iron chips or filings. Iron is dissolved under such conditions, but it can be removed by filtration. The small amount of oxygen remaining can be treated and removed with sodium sulfite. Ferrous sulfate is also used to remove dissolved oxygen.

All carbon dioxide except 3 to 5 mg/l can be removed by aeration, but aeration also increases the dissolved-oxygen concentration, which in itself is detrimental. Sprays, cascades, coke trays, diffused air, and zeolite are used to remove most of the carbon dioxide. A filter rate of 25 gpm/ft^2 in coke trays 6 inches thick may reduce the carbon dioxide concentration from 100 to 10 mg/l and increase pH from about 6.0 to 7.0.[111] The carbon dioxide remaining, however, is sufficient to cause serious corrosion in water having an alkalinity caused by calcium carbonate of less than about 100 mg/l. It can be removed where necessary by adding sodium carbonate (soda ash), lime, or sodium hydroxide (caustic soda). With soft waters having an alkalinity greater than 30 mg/l, it is easier to add soda ash or caustic soda in a small water system to eliminate the carbon dioxide and increase the pH and alkalinity of the water. The same effect can be accomplished by filtering the water through broken limestone or marble chips. Well water that has a high concentration of carbon dioxide but no dissolved oxygen can be made noncorrosive by adding an alkali such as sodium carbonate, with pH adjusted to 8.1 to 8.4. Soft waters that also have a low carbon dioxide content (3 to 5 mg/l) and alkalinity (20 mg/l) may need a mixture of lime and soda ash to provide both calcium and carbonate for the deposition of a calcium carbonate film.*

Sodium and calcium hexametaphosphate, tetrasodium pyrophosphate, zinc phosphates, sodium silicate (water glass), lime, caustic soda, and soda ash are used to build up an artificial coating inside of pipe. Health department approval of chemical use and pilot plant studies are usually required. The sodium concentration in drinking water is increased when sodium salts are added.

Sodium hexametaphosphate dissolves readily and can be added alone or in conjunction with sodium hypochlorite by means of a solution feeder. Concentrated solutions of metaphosphate are corrosive. A dosage of 5 to 10 mg/l is normally used for 4 to 8 weeks until the entire distribution system is coated, after which the dosage is maintained at 1 to 2 mg/l with pH maintained at 7.2 to 7.4. The initial dosage may cause precipitated iron to go into solution with resultant temporary complaints, but flushing of the distribution system will minimize this problem. Calcium metaphosphate is a similar material, except that it dissolves slowly and can be used to advantage where this property is desirable.

---

*One grain per gallon (17.1 mg/l) of lime, caustic soda, and soda ash remove, respectively, 9.65, 9.55, and 7.20 mg/l free CO^2; the alkalinity of the treated water is increased by 23.1, 21.4, and 16.0 mg/l, respectively. One milligram per liter of chlorine decreases alkalinity (as $CaCO^3$) 0.7 to 1.4 mg/l and 1 mg/l alum decreases natural alkalinity 0.5 mg/l.

Inexpensive and simple pot-type feeders that are particularly suitable for small water supplies are available. Sodium pyrophosphate is similar to sodium hexametaphosphate. All these compounds are reported to coat the interior of the pipe with a film that protects the metal, prevents lime scale and red water trouble, and resists the corrosive action of water. However, heating of water above 140° to 150°F (60°–66°C) will nullify any beneficial effect. The phosphate in these compounds may stimulate biological growths in mains. In any case, the corrosion control method used should be monitored to determine its effectiveness.

Sodium silicate in solution is not corrosive to metals and can easily be added to a water supply with any type of chemical feeder to form calcium silicate, provided the water contains calcium. Doses vary between 25 and 240 lb/million gal, 70 lb/million gal being about average. The recommendations of the manufacturer should be followed in determining the treatment to be used for a particular water.

Adjustment of the pH and alkalinity of a water so that a thin coating is maintained on the inside of piping will prevent its corrosion. Any carbon dioxide in the water must be removed before this can be done, as previously explained. Lime* is added to water to increase the alkalinity and pH so as to come within the limits shown in Figure 2.7. The approximate dosage may be determined by the marble test, but the Langelier saturation index, Ryznar index, and Enslow stability indicator are more accurate methods. Under these conditions, calcium carbonate is precipitated from the water and deposited on the pipe to form a protective coating, provided a velocity of 1.5 to 3.0 fps is maintained to prevent heavy precipitation near the point of treatment and none at the ends of the distribution

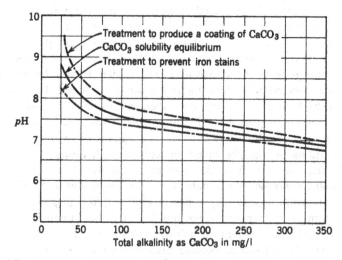

FIGURE 2.7 Solubility of $CaCO_3$ at 71°F (Baylis curve). (*Source*: C. R. Cox, *Water Supply Control*, Bulletin No. 22, New York State Department of Health, Albany, 1952, p. 185.)

*At a pH above 8.3, calcium carbonate is soluble to 13–15 mg/l.

system. The addition of 0.5 to 1.5 mg/l metaphosphate will help obtain a more uniform calcite coating throughout the distribution system. The addition of lime must be carefully controlled so as not to exceed a pH of 8.0 to 8.5 to maintain chlorination disinfection effectiveness. Calcium carbonate is less soluble in hot water than in cold water. It should be remembered that the disinfecting capacity of chlorine (HOCl) decreases as the pH increases; hence, the free available chlorine concentration maintained in the water should be increased with the higher pH. See Table 2.1. Also note that soft corrosive water with a high pH will increase corrosion of copper and zinc; old, yellow brass plumbing can be dezincified and galvanizing can be removed from iron pipe.[112]

The Langelier saturation index (the difference between the measured pH and the calculated pH) can be used to determine the point of calcium carbonate stability for corrosion control with waters having an alkalinity greater than 35 to 50 mg/l. A positive Langelier index is indicative primarily of calcium carbonate (scale) deposition; a negative index number is indicative of increasing water corrosivity with −2.0 considered high. Slightly positive is the goal. The point of calcium carbonate stability is also indicated by the Ryznar index. A Ryznar index number of less than about 6.0 is indicative primarily of the start of calcium carbonate (scale) deposition; an index number greater than 6.0 to 7.0 is indicative of increasing water corrosivity. Other measures are the Enslow stability indicator and the aggressiveness index. The Caldwell–Lawrence diagram[113] is useful for solving water-conditioning problems, but raw water concentration of calcium, magnesium, total alkalinity, pH, and TDS values must be known. See *Standard Methods*[114] for procedures. Do not rely on only one method.

The AWWA recognizes the coupon test to measure the effects of physical factors and substances in water on small sections of stainless steel and galvanized iron inserted in a water line for 90 days. Measurement of the weight loss due to corrosion or weight gained due to scale formation can thus be determined under the actual use conditions. The gain on stainless steel should not exceed 0.05 mg/cm^2; the loss from the galvanized iron should not exceed 5.0 mg/cm^2. Temperature, pH, velocity, dissolved oxygen, and water quality affect corrosion rates. Coupons should preferably remain in the pipe for 1 year or longer. The test does not show the inside condition of the pipe.[115]

The danger of lead or zinc poisoning and off-flavors due to copper plumbing can be greatly reduced when corrosive water is conducted through these pipes by simply running the water to waste in the morning. This will flush out most of the metal that has had an opportunity to go into solution while standing during the night. Maintenance of a proper balance between pH, calcium carbonate level, and alkalinity as calcium carbonate is necessary to reduce and control lead corrosion by soft aggressive water. Formation and then *maintenance* of a carbonate film are necessary. See Figure 2.7. In a soft, corrosive water, sodium hydroxide can be used for pH adjustment and sodium bicarbonate for carbonate addition. Lead pipe should not be used to conduct drinking water. Low lead solders and use of plastic pipe and glass-lined water heaters will minimize the problems associated with corrosion in the home. See "Lead Removal," this chapter.

Biochemical actions such as the decomposition of organic matter in the absence of oxygen in the dead end of mains, the reduction of sulfates, the biochemical action within tubercles, and the growth of crenothrix and leptothrix, all of which encourage corrosion in mains, can be controlled by the maintenance of at least 0.3 mg/l free residual chlorine in the distribution system.

External Pipe Corrosion External corrosion of underground pipe may be caused by stray direct electric currents; buried defective electric, telephone, and TV cables and grounding connections to water mains; grounding of household systems, appliances, and equipment; direct current welding equipment; acidic soils; abrasions and breaks in external coating; anaerobic bacteria; and dissimilar metals in contact. Stray currents from electric trolleys and subways also contribute to the problem. Soil around pipe serves as the electrolyte and the pipe serves as the conductor.

Corrosion caused by electrolysis or stray direct electric currents can be prevented by making a survey of the piping and removing grounded electrical connections and defective electric cables. Moist soils will permit electric currents to travel long distances. A section of nonconducting pipe in dry soil may confine the current. In the vicinity of power plants, this problem is very serious and requires the assistance of the power company involved.

Where dissimilar metals are to be joined, a plastic, hard rubber, or porcelain fitting can be used to separate them. It must be long enough to prevent the electric charge from jumping the gap. A polyethylene tube or encasement around cast-iron and ductile iron pipe and mastic, coal-tar enamel, epoxy, or similar coating protected with a wrapping will protect pipe from corrosive soil.

Corrosion of water storage tanks and metal pipelines can be controlled by providing "cathodic protection," in which a direct current is imposed to make the metal (cathode) more electronegative than an installed anode. But it is necessary to repaint the metal above the water line in a water storage tank with an approved coating. Consult with the provider of cathodic protection equipment. A number of galvanic anodes, which are higher in the galvanic electromotive series,* such as magnesium or zinc, may be used adjacent to pipelines. The higher metal in the electromotive series will be the anode and will corrode; the lower metal (the pipe) is the cathode. The current flows from the anode to the cathode. The moist soil serves as the electrolyte. Eventually, the anode will have to be replaced.

Well Clogging and Cleaning A common problem with wells in anaerobic zones is the reduction in production capacity, usually due to clogging of the formation or incrustation of the well screen openings. This may be due to mineral scale precipitation formed around the screen and on the screen; to bacteria that oxidize iron such as crenothrix, leptothrix, and gallionella; and plugging when silt, fine sand, and clay build-up in the formation or gravel pack around the well

---

*Galvanic series from most active to least active: sodium, magnesium, zinc, aluminum, steel, iron, lead, tin, brass, copper, nickel, silver, gold, platinum.

screen. Anaerobic waters may contain sulfides or iron and possibly manganese. If sulfates predominate, the water will contain sulfides. Iron bacteria are found where dissolved oxygen and dissolved iron are present. The source may also be surface water contamination.

As much information as possible should be obtained concerning the well-water characteristics to suggest a possible cause of reduced well capacity before any unclogging work is done. Chemical analysis of a representative water sample and a marble test or calculation of the Langelier saturation index or Ryznar index can show if calcium carbonate could precipitate out on the well screen or if iron or manganese and incrustations are present. Comparison with analyses made when the well was new may provide useful information. High bacterial plate counts may indicate organic growths in or on the well screen. Microscopic examination can show if iron bacteria or other objectionable growths are present.[116]

Treatment methods include the use of acids to dissolve mineral scale and bacterial iron precipitate, but care is necessary to minimize corrosion of the well, screen, and pump. Chlorination (sodium hypochlorite) to disinfect a well will also remove and retard growth of the iron bacteria. A 1-mg/l copper sulfate solution or a quaternary ammonium compound might also be effective. Sodium polyphosphates have been found effective in unplugging wells caused by clay and silt particles.[117] Repeat treatment is usually needed, including well surging to purge the well screen and adjacent aquifer.

Water Softening   Water softening is the removal of minerals causing hardness from water. For comparative purposes, one grain per gallon of hardness is equal to 17.1 mg/l. Water hardness is caused primarily by the presence of calcium bicarbonate, magnesium bicarbonate (carbonate hardness), calcium sulfate (gypsum), magnesium sulfate (epsom salts), calcium chloride, and magnesium chloride (noncarbonate hardness) in solution. In the concentrations usually present these constituents are not harmful in drinking water. The presence of hardness is demonstrated by the use of large quantities of soap in order to make a lather*; the presence of a gritty or hard curd in laundry or in a basin; the formation of hard chalk deposits on the bottom of pots and inside of piping causing a reduced water flow; and the lowered efficiency of heat transfer in boilers caused by the formation of an insulating scale. Hard water is not suitable for use in boilers, laundries, textile plants, and certain other industrial operations where a zero hardness of water is needed.

In softening water the lime or lime–soda ash process, zeolite process, and organic resin process are normally used. In the lime–soda ash method, the soluble bicarbonates and sulfates are removed by conversion to relatively insoluble forms. In the zeolite process, the calcium and magnesium are replaced with sodium,

---

*With a water hardness of 45 mg/l the annual per-capita soap consumption was estimated at 29.23 lb; with 70 mg/l hardness, soap consumption was 32.13 lb; with 298 mg/l hardness, soap consumption was 39.89 lb; and with 555 mg/l, soap consumption was 45.78 lb. (M. L. Riehl, *Hoover's Water Supply and Treatment*, National Lime Association, Washington, DC, April 1957.)

forming sodium compounds in the water that do not cause hardness but add to the sodium content. With synthetic organic resins, dissolved salts can be almost completely removed. Table 2.6 gives ion exchange values. Caustic soda can also remove both carbonate and noncarbonate hardness, but it is more costly.

Lime–soda ash softening requires the use of lime to convert the soluble bicarbonates of calcium and magnesium (carbonate hardness) to insoluble calcium carbonate and magnesium hydroxide, which are precipitated. Prior aeration of excess lime is needed to remove carbon dioxide if it is present in the raw water. The soluble calcium and magnesium sulfate and chlorides (noncarbonate hardness) are converted to insoluble calcium carbonate and magnesium carbonate by the addition of soda ash and lime and precipitated. Lime softening will also remove 90 to 99 percent of the bacteria and viruses but does not remove the need for disinfection. The sodium chloride and sodium sulfate formed remain in the water. Excess lime is needed to achieve a pH of about 9.5 to precipitate calcium carbonate, and a pH of about 11 is needed to precipitate magnesium hydroxide when it is greater than about 40 mg/l. Then pH adjustment is needed to control calcium carbonate precipitation on filters and in the distribution system. Carbon dioxide gas is usually added (recarbonation) to change the calcium hydroxide to calcium carbonate to improve precipitation and to adjust the pH to 8.6 or less in the finished water. Carbon dioxide is usually produced by the burning of gas, oil, coke, or coal. A coagulant such as aluminum sulfate (filter alum), ferrous sulfate (copperas), ferric sulfate, or sodium aluminate is usually used to coagulate and settle the compounds formed, followed by filtration to remove turbidity and color. Large volumes of sludge with high water content are produced. Disposal may present a problem. Options include reclamation and land disposal. The lime–soda ash method is not suitable for softening small quantities of water because special equipment and technical control are necessary. The process is more economical for softening moderately hard water. As water hardness increases, the lime requirement increases, which makes the zeolite process more attractive. The lime–soda ash process is usually controlled to reduce hardness to about 50 to 80 mg/l.

The zeolite and synthetic resin softening methods are relatively simple ion exchange processes that require little control. Only a portion of the hard water need be passed through a zeolite softener since a water of zero hardness is produced by the zeolite filter. The softener effluent can be mixed with part of the untreated water to produce a water of about 50 to 80 mg/l hardness. The calcium and magnesium in water to be treated replace the sodium in the zeolite filter media, and the sodium passes through with the treated water. This continues until the sodium is used up, after which the zeolite is regenerated by bringing a 5 to 10 percent solution of common salt in contact with the filter media. Units are available to treat the water supply of a private home or a community. Water having a turbidity of more than 5 units will coat the zeolite grains and reduce the efficiency of a zeolite softener. Iron in the ferric form and organic substances are also detrimental. Iron or manganese or iron plus manganese should not exceed 0.3 mg/l. Pretreatment to remove turbidity, organic matter, and iron would be

TABLE 2.6 Ion Exchange Materials and Their Characteristics

Exchange Material	Exchange Capacity (grains/ft^3)	Effluent Content	Regeneration Material	Remarks
Natural zeolites	3,000–5,000	Sodium bicarbonate, chloride, sulfate	0.37–0.45 lb salt per 1,000 grains hardness removed	Ferrous bicarbonate and manganous bicarbonate also removed from well water devoid of oxygen, pH of water must be 6.0–8.5, moderate turbidity acceptable. Use 5–10% brine solution. Saturated brine is about 25%.
Artificial zeolites	9,000–12,000	Sodium bicarbonate, chloride,	0.37–0.45 lb salt per 1,000 grains hardness removed	
Carbonaceous zeolites	9,000–12,000	Carbon dioxide and acids, sodium chloride and sulfate	0.37–0.45 lb salt per 1,000 grains hardness removed	Acid waters may be filtered. CO_2 in effluent removed by aeration, acid by neutralization with bypassed hard water or addition of caustic soda.
Synthetic organic resins	10,000–30,000	Carbon dioxide and acids	0.2–0.3 lb salt per 1,000 grains hardness removed	Dissolved salts are removed by resins. To remove CO_2 and acids, add soda ash or caustic soda, or CO_2 by aeration, and acids by synthetic resin filtration.

Note: One gallon of saturated brine weighs 10 lb and contains 2.5 lb of salt. Hardness is caused by calcium bicarbonate, magnesium bicarbonate, calcium sulfate, magnesium sulfate, and calcium chloride. Natural zeolite is more resistant to waters of low pH than artificial zeolite. Natural zeolite is also known as greensand.

indicated. The filters are not less than 3 feet deep. Downward-flow filters generally operate at rates between 3 and 5 gpm/ft^2; upward-flow filters operate at 4 to 6 gpm/ft^2. The maximum rate should not exceed 7 gpm/ft^2.

Synthetic resins for the removal of salts by ion exchange are discussed under "Desalination" in Chapter 1. Consideration must be given to the disposal of the brine waste from the ion exchange process.

Small quantities of water can be softened in batches for laundry purposes by the addition of borax, washing soda, ammonia, or trisodium phosphate. Frequently, insufficient contact time is allowed for the chemical reaction to be completed, with resultant unsatisfactory softening.

Lime softening removes arsenic, barium, cadmium, chromium, fluoride, lead, mercury, selenium, radioactive contaminants, copper, iron, manganese, and zinc.

The extent to which drinking water is softened should be evaluated in the light of the relationship of soft water to cardiovascular diseases. In view of the accumulating evidence, the wisdom of constructing municipal softening plants is being questioned. There is evidence associating the ingestion of sodium with cardiovascular diseases, kidney disease, and cirrhosis of the liver. See "Hardness and Sodium," in Chapter 1.

Fluoridation   Since about 1943 fluorides have been added to public water supplies in controlled amounts to aid in the reduction of tooth decay. The compounds commonly used are sodium fluoride (NaF), sodium silicofluoride (Na_2SiF_6), and hydrofluosilicic acid (H_2SiF_6), also called fluosilicic acid. They are preferred because of cost, safety, and ease of handling. Ammonium silicofluoride may be used in conjunction with chlorine where it is desired to maintain a chloramine residual in the distribution system, if permitted by the regulatory agency. Calcium fluoride (fluospar) does not dissolve readily. Hydrofluoric acid is hazardous; unsealed storage containers should be vented to the atmosphere. Backflow devices are required on all fluoride and water feed lines.

Solution and gravimetric or volumetric dry feeders are used to add the fluoride, usually after filtration treatment and before entry into the distribution system. Fluoride solutions for small water systems are usually added by means of a small positive-feed displacement pump. Corrosion-resistant piping must be used. Calcium hypochlorite and fluoride should not be added together, as a calcium fluoride precipitate would be formed. Fluoride compounds should not be added before lime–soda ash or ion exchange softening. Personnel handling fluorides are required to wear protective clothing. Proper dust control measures, including exhaust fans, must be included in the design where dry feeders are used. Dosage must be carefully controlled.

The average annual per-capita cost of fluoridation of a public water supply is small. Softened water should be used to prepare a sodium fluoride solution whenever the hardness, as calcium carbonate, of the water used to prepare the solution is greater than 75 mg/l, or even less. This is necessary to prevent calcium and magnesium precipitation, which clogs the feeder. Small quantities of water can be softened by ion exchange, or polyphosphates may be used.[118] See "Fluorides" in Chapter 1.

Removal of Inorganic Chemicals

The sources, health effects, permissible concentrations, and control measures related to certain inorganic chemicals are also discussed under the appropriate headings earlier in Chapter 1 under "Chemical Examinations". Fundamental to the control of toxic inorganic chemicals in drinking water is a sanitary survey and identification of the sources, types, and amounts of pollutants followed by their phased elimination as indicated, *starting at the source*. Watershed and land-use controls are usually the best preventive measures in both the short term and long term, coupled with point and nonpoint source control.

Table 2.7 summarizes treatment methods for the removal of inorganic chemicals from drinking water. Several are discussed in some detail as follows.

Arsenic Removal   Inorganic arsenic in water occurs naturally in two oxidation states, arsenite [As(III)] and arsenate [As(V)]. Arsenate is a negatively charged molecule and is relatively easy to remove since it strongly adsorbs onto the surface of metal hydroxide particles. Arsenite, however, is more difficult to remove due to its neutral charge. If the water being treated contains only arsenite (or enough that achieving the MCL is questionable) the treatment should begin with an oxidation step to convert all inorganic arsenic to arsenate. Oxidation is effectively achieved using simple chlorination, ozonation, or use of potassium permanganate.

For surface water systems with a conventional process using either ferric or alum coagulation, arsenic is easily removed during the coagulation–semination process. Groundwater systems (with or without disinfection) will likely choose one of the following treatment processes, which will all achieve greater than 95 percent removal:

- Sorption process (ion exchange or adsorption onto an iron, aluminum, or copper media)
- Precipitation process (coagulation followed by filtration)
- Membrane process (nanofiltration or reverse osmosis)

The final choice of treatment technology will depend on several factors, including the availability of sewers to handle waste brine; the presence of competing ions such as nitrate and sulfate; the number of bed-volumes of an adsorbent media (i.e., useful life); and the presence of other constituents such as hardness or TDS. Following an assessment of the most appropriate technology, pilot studies should be conducted to confirm the treatment efficacy prior to full-scale implementation.[119]

Cadmium Removal   Cadmium removal of greater than 90 percent can be achieved by iron coagulation at about pH 8 and above. Greater percentage removal is obtained in higher turbidity water. Lime and excess lime softening remove nearly 100 percent cadmium at pH 8.7 to 11.3. Ion exchange treatment

TABLE 2.7 Most Effective Treatment Methods for Inorganic Contaminant Removal

Contaminant	Most Effective Methods
Arsenic As^{3+}	Ferric sulfate coagulation, pH 6–8
	Alum coagulation, pH 6–7
	Excess lime softening
	Oxidation before treatment required
As^{5+}	Ferric sulfate coagulation, pH 6–8
	Alum coagulation, pH 6–7
	Excess lime softening
Barium	Lime softening, pH 10–11
	Ion exchange
Cd^{3+}	Ferric sulfate coagulation, above pH 8
	Lime softening
	Excess lime softening
Chromium Cr^{3+}	Ferric sulfate coagulation, pH 6–9
	Alum coagulation, pH 7–9
	Excess lime softening
	Ferrous sulfate coagulation, pH 7–9.5
Cr^{6+}	
Fluoride	Ion exchange with activated alumina or bone char media
Lead	Ferric sulfate coagulation, pH 6–9
	Alum coagulation, pH 6–9
	Lime softening
	Excess lime softening
Mercury	
Inorganic	Ferric sulfate coagulation, pH 7–8
Organic	Granular activated carbon
Nitrate	Ion exchange
Selenium	Ferric sulfate coagulation, pH 6–7
Se^{4+}	Ion exchange
	Reverse osmosis
Se^{6+}	Ion exchange
	Reverse osmosis
Silver	Ferric sulfate coagulation, pH 7–9
	Alum coagulation, pH 6–8
	Lime softening
	Excess lime softening

Source: T. J. Sorg, "Treatment Techniques for the Removal of Inorganic Contaminants from Drinking Water," *Manual of Treatment Techniques for Meeting the Interim Primary Drinking Water Regulations*, U.S. Environmental Protection Agency, Cincinnati, OH, May 1977, p. 3.

with cation exchange resin should remove cadmium from drinking water. Powdered activated carbon is not efficient and granular activated carbon will remove 30 to 50 percent. Reverse osmosis may not be practical for cadmium removal.[120]

Lead Removal   Normal water coagulation and lime softening remove lead—99 percent for coagulation at pH 6.5 to 8.5 and for lime softening at pH 9.5 to 11.3. Turbidity in surface water makes particulate lead removal easier by coagulation, flocculation, settling, and filtration. Powdered activated carbon removes some lead; GAC effectiveness is unknown; and reverse osmosis, electrodialysis, and ion exchange should be effective.[121] Lead in the soluble form may be removed by reverse osmosis or distillation.

Nitrate Removal   Treatment methods for the removal of nitrates from drinking water include chemical reduction, biological denitrification, anion exchange, reverse osmosis, distillation, and electrodialysis. Ion exchange is the most practical method. At one community water system,* the water has approximately 200 mg/l total dissolved solids; the nitrate–nitrogen levels are reduced from 20 to 30 mg/l to less than 2 mg/l.[122] Little plant-scale data are otherwise available. Reverse osmosis and electrodialysis are effective (40 to 95 percent), but these methods are more costly than ion exchange.

Fluoride Removal   Treatment methods for the removal of fluorides from drinking water have been summarized by Sorg.[123] They include high (250–300 mg/l) alum doses, activated carbon at pH 3.0 or less; lime softening if sufficient amounts of magnesium (79 mg/l to reduce fluoride from 4 to 1.5 mg/l) are present or added for coprecipitation with magnesium hydroxide; ion exchange using activated alumina, bone char, or granular tricalcium phosphate; and reverse osmosis. Of these methods, alum coagulation and lime softening are not considered practical. Reverse osmosis has not been demonstrated on a full-scale basis for this purpose, but ion exchange has. Activated alumina and bone char have been successfully used, but the former is the method of choice for the removal of fluoride from drinking water.[124]

Selenium Removal   Selenium is predominantly found in water as selenite and selenate. Selenite can be removed (40–80 percent) by coagulation with ferric sulfate, depending on the pH, coagulant dosage, and selenium concentration. Alum coagulation and lime softening are only partially effective, 15 to 20 percent and 35 to 45 percent, respectively. Selenite and selenate are best removed by ion exchange, reverse osmosis, and electrodialysis, but the effectiveness of these methods in removing selenium has not been demonstrated in practice.[125]

---

*Garden City Park Water District, Garden City, NY. Nitrates have also been reduced in Bridge-water, MA, since 1979 and in McFarland, CA, since 1983. ("Letters," *AWWA MainStream*, January 1986, p. 2.)

Radionuclide Removal Coagulation and sedimentation are very effective in removing radioactivity associated with turbidity and are fairly effective in removing dissolved radioactive materials—with certain exceptions. The type of radioactivity, the pH of the treatment process, and the age of the fission products in the water being treated must be considered. For these reasons, jar-test studies are advised before plant-scale operation is initiated. A comprehensive summary of the effectiveness of different chemical treatment methods with various radionuclides is given by Straub.[126] The effectiveness of rapid and slow sand filtration, lime–soda ash softening, ion exchange, and other treatment processes is also discussed.

Studies for military purposes show that radioactive materials present in water as undissolved turbidity can be removed by coagulation, hypochlorination, and diatomite filtration. Soluble radioisotopes are then removed by ion exchange using a cation exchange column followed by an anion exchange column operated in series. Hydrochloric acid is used for regenerating the cation resin and sodium carbonate the anion resin. The standard Army vapor compression distillation unit is also effective in removing radioactive material from water. Groundwater sources of water can generally be assumed to be free of fallout radioactive substances and should, if possible, be used in preference to a surface-water source[127] in emergency situations. However, radionuclides can travel great distances in groundwater.

Kosarek[128] reviewed the water treatment processes used to reduce dissolved radium contamination to an acceptable level (5 pCi/l or less) in water for industrial and municipal purposes. Processes for industrial water uses are selective membrane mineral extraction, reverse osmosis, barium sulfate co-precipitation, ion exchange, activated alumina, lime–soda ash softening, and sand filtration. Processes for municipal water uses are reverse osmosis, ion exchange, lime–soda ash softening, aeration, greensand filtration, and sand filtration. Aeration, greensand filtration, and sand filtration have low radium removal efficiency. Lime–soda ash has a 50 to 85 percent efficiency; the other remaining processes have an efficiency of 90 to 95 percent or better. A manganese dioxide–coated fiber filter can effectively remove radium from drinking water by adsorption.[129]

Packed tower aerators can remove more than 95 percent of the radon and conventional cascading tray aerators better than 75 percent.[130] Radon is effectively removed from well water by GAC adsorption. However, as in other processes, the spent carbon and other solid and liquid wastes collected present a disposal problem because of the radioactive materials retained in the waste. Possible waste disposal options for treatment plant solid and liquid wastes containing radium, if approved by the regulatory authority, include sanitary sewers, storm sewers, landfills, and land spreading. Conditions for disposal must be carefully controlled.[131]

Uranium can be removed from well water to a level as low as 1 μg/l using conventional anion exchange resins in the chloride form. Gamma radiation build-up in the system does not appear to be significant.[132] Treatment methods to remove uranium from surface waters and groundwaters include iron coagulation (80–85 percent), alum coagulation (90–95 percent), lime softening (99 percent), cation exchange (70–95 percent), anion exchange (99 percent), activated alumina (99 percent), granular activated carbon (90+ percent), and reverse osmosis (99 percent).[133]

The EPA is considering the setting of MCLs for certain radionuclides in water and a proposal of best available treatment (BAT) technologies to achieve the MCLs and MCLGs. Radon MCLs may fall between 300 and 4,000 pCi/l in water, equivalent to about 0.03 to 0.4 pCi/l in air. The BATs given are aeration and GAC. Radium-226 and Ra-228 MCLs may fall between 2 and 20 pCi/l each. The BATs given are cation exchange, lime softening, and reverse osmosis. Uranium MCLs may fall between 5 and 40 pCi/l. The BATs given are coagulation/filtration, reverse osmosis, anion exchange, and lime softening. Beta particle and photon emitter MCL concentrations may be equal to the risk posed by a 4-mrem effective dose equivalent. The BATs given for betas are reverse osmosis and ion exchange (mixed bed).[134]

Prevention and Removal of Organic Chemicals

As noted for inorganic chemicals, the control of organic chemicals in drinking water should start with a sanitary survey to identify the sources, types, and amounts of pollutants, followed by their phased elimination as indicated by the associated hazard. Included would be watershed use regulation and protection, watershed management to minimize turbidity and organic and inorganic runoff, vigorous compliance with the national and state water and air pollution elimination objectives, enforcement of established water and air classification standards, and complete effective drinking water treatment under competent supervision. It is obvious that selection of the cleanest available protected source of water supply, for the present and the future, would greatly minimize the problems associated not only with organic chemicals but also with inorganic, physical, and microbiological pollution. In any case, water treatment plants must be upgraded where needed to consistently produce a water meeting the national drinking water standards.

Trihalomethanes, Removal and Control   The halogenated, chloro-organic compounds* include the trihalomethanes: trichloromethane (chloroform), bromodichloromethane, dibromochloromethane, and tribromomethane (bromoform). These chlorination byproducts are formed by the reaction of *free* chlorine with certain organic compounds in water. The major cause of trihalomethane (THM) formation in chlorinated drinking water is believed to be humic and fulvic substances (natural organic matter in soil, peat, other decay products of plants and animals, and runoff) and simple low-molecular-weight compounds not removed by conventional filtration treatment—all referred to as precursors. Treatment to remove turbidity should remove high-molecular-weight

---

*Halogenated organics are organic compounds that contain one or more halogens—fluorine, chlorine, bromide, iodine, and astatine.

compounds. Low-molecular-weight compounds are best reduced by GAC treatment. Chlorination of municipal wastewater also results in the formation of halo-organics, but their concentration is very low when combined chlorine is formed,[135] which is usually the case. However, chloramination produces other yet undefined chloro-organic compounds. The reaction is dependent on chlorine dose, pH, temperature, and contact time. The point of chlorination, to avoid precursors, is critical in drinking water treatment to minimize or prevent the formation of THMs. Total trihalomethane concentration in treated water has been found to be higher in the summer, after reservoir turnover, and lowest in the winter. It is also related to the presence of phytoplankton and correlates well with chlorine demand of untreated water, but not with organic carbon and chloroform extract.[136] The potential for THM formation in groundwater was found to be strongly correlated with total organic carbon (TOC) concentration, ammonia, iron, and manganese, but very few sources were found to exceed 100 μg/l.[137]

Prechlorination with long contact periods and sunlight increases the formation of THMs, as does increased chlorine dosage and the addition of chlorine prior to coagulation and settling. Preozonation is effective in oxidizing in part naturally present organic compounds, thereby reducing the potential for THM production after subsequent postchlorination. Alternative disinfectants are chloramines and chlorine dioxide as well as potassium permanganate and ultraviolet radiation if approved. Ozone and chloramine treatment is reported to produce only about 2 percent of the THMs produced by free chlorine.

Granular activated carbon has been found to be of limited effectiveness in removing precursor materials; GAC is effective for only a few weeks.[138] In contrast, GAC for taste and odor control needs regeneration every three to six years.[139] It is not efficient for the removal of THMs once formed. Treatment to remove suspended, colloidal, and dissolved materials by coagulation, flocculation, settling, and filtration should precede GAC treatment if used for taste and odor control. The same holds true for the removal of synthetic organic chemicals so as not to coat and reduce the adsorptive capacity of the carbon. Such treatment will also remove most THM precursors, as previously noted.

Recommended Standards for Water Works summarizes recommended practice in the "Policy Statement on Trihalomethane Removal and Control for Public Water Supplies."[140]

Trihalomethanes (THMs) are formed when free chlorine reacts with organic substances, most of which occur naturally. These organic substances (called *precursors*), are a complex and variable mixture of compounds. Formation of THMs is dependent on such factors as amount and type of chlorine used, temperature, concentration of precursors, pH, and contact time. Approaches for controlling THMs include:

1. Control of precursors at the source.
    a. Selective withdrawal from reservoirs—varying depths may contain lower concentrations of precursors at different times of the year.

    b. Plankton control—Algae and their oils, humic acid, and decay products have been shown to act as THM precursors.

    c. Alternative sources of water may be considered, where available.

2. Removal of THM precursors and control of THM formation.

    a. Moving the point of chlorination to minimize THM formation.

    b. Removal of precursors prior to chlorination by optimizing:

        (1) Coagulation/flocculation including sedimentation and filtration

        (2) Precipitative softening/filtration

        (3) Direct filtration

    c. Adding oxidizing agents such as potassium permanganate, ozone or chlorine dioxide to reduce or control THM formation potential.

    d. Adsorption by powdered activated carbon (PAC).

    e. Lowering the pH to inhibit the reaction rate of chlorine with precursor materials. Corrosion control may be necessary.

3. Removal of THM.

    a. Aeration—by air stripping towers.

    b. Adsorption by:

        (1) Granular Activated Carbon (GAC)

        (2) Synthetic Resins

4. Use of Alternative Disinfectants—Disinfectants that react less with THM precursors may be used as bacteriological quality of the finished water is maintained. Alternative disinfectants may be less effective than free chlorine, particularly with viruses and parasites. Alternative disinfectants, when used, must be capable of providing an adequate distribution system residual. Use of alternative disinfectants may also produce possible health effects and must be taken into consideration. The following alternative disinfectants may be used:

    a. Chlorine dioxide

    b. Chloramines

    c. Ozone

Using various combinations of THM controls and removal techniques may be more effective than a single control or a treatment method.

Any modifications to existing treatment process must be approved by the reviewing authority. Pilot plant studies are desirable.

The maximum contaminant level for total THMs in drinking water in the United States is 100 μg/l. The goal is 10 to 25 μg/l. The Canadian maximum acceptable level is 350 μg/l.[141] The WHO has set a guideline for chloroform only at 30 μg/l; several countries have set limits of 25 to 250 μg/l for the sum of four specific THMs.[142]

Synthetic Organic Chemicals and Their Removal The major sources of synthetic organic chemical pollution (also inorganic pollution in many places) are industrial wastewater discharges; air pollutants; municipal wastewater discharges; runoff from cultivated fields, spills, and waste storage sites; and leachate from sanitary landfills, industrial and commercial dump sites, ponds, pits, and lagoons. Illegal dumping and coal-tar-based pipe coating and linings may also contribute organics. Both surface waters and groundwaters may be affected. It cannot be emphasized enough that *control of all pollutants must start at the source*, including raw-material selection, chemical formulation, and manufacturing process control. Separation of floating oils and collection of low-solubility, high-density compounds in traps on building drains and improved plant housekeeping could reduce pollutant discharges and recover valuable products. Such actions would reduce the extent of needed plant upgrading, sophisticated wastewater treatment and control, burden on downstream aquatic life and water treatment plants, and hence risks to the consumer associated with the ingestion of often unknown hazardous or toxic chemicals.

Waters containing a mixture of organic chemicals and soluble metals are difficult to treat and require special study.

The more common water treatment methods considered to reduce the concentration of volatile organic chemicals (VOCs) and other synthetic organic chemicals (SOCs) in drinking water sources are aeration and adsorption through GAC. Other possible methods include ozonation, oxidation, osmosis, ion exchange, and ultrafiltration.[143] However, before a treatment method is selected and because of the many variables involved, characterization of the organic contaminants involved and bench-scale and pilot plant studies of aeration and GAC are generally required to be carried out with the actual water to be treated to determine the effectiveness of a process and the basis for design. This is also necessary to determine the GAC adsorption capacity before exhaustion and its reactivation cost. Organics have different adsorptive characteristics on GAC. It should also be noted that bench-scale tests using strongly basic anion exchange resins showed that most organics present in surface water can be removed.[144] Conventional coagulation, flocculation, sedimentation, and sand filtration treatment does not remove VOCs to any significant extent.

Aeration (air stripping) will remove many VOCs. Methods include diffused air in which air is forced up through the falling water spray, packed tower with forced or induced draft, waterfall, mechanical surface aerators, cascade aeration, tray aeration, and air-lift pump. The extent to which aeration is successful will depend on the concentration, temperature, solubility, and volatility of the compounds in the water. The rate of removal depends on the amount of air used, contact time, and temperature of the air and water. Removals of 95 to 99 percent have been reported. Very low efficiencies are obtained at freezing temperatures. Aeration is usually more effective for removing the lighter, more volatile SOCs such as found in groundwater. The GAC is more effective for removal of heavier SOCs found in surface water. Compounds reported to be removed by aeration include trichloroethylene, carbon tetrachloride, tetrachloroethylene, benzene, toluene, napthalene, biphenyl methyl bromide, bromoform, chloroform, dibromochloromethane, bromodichloromethane, methylene chloride, vinyl chloride, sodium fluoroacetate, dichloroethylene, dichloroethane, perchloroethylene, and others. The potential for air pollution and its control must be considered. Synthetic organic chemicals, referred to as refractory compounds, resist decomposition and removal. Corrosion control is usually required after aeration. Airborne contamination, including worm growth in the aerator, must be guarded against.

Granular activated carbon is considered the best available broad-spectrum adsorber of SOCs and appears to be indicated where nonvolatile organics are present. The carbon is similar in size to filter sand. Adsorption is a complex process. It is influenced by the surface area of the carbon grains, the material being adsorbed or concentrated (adsorbate), the pH and temperature of the water being treated, the mixture of compounds present, and the nature of the adsorbent—that is, the carbon grain structure, surface area, and pores. The smaller the grain size within the range of operational efficiency, the greater the rate of adsorption obtained.[145] Disposal of spent carbon may be a problem.

The EPA has designated packed-tower aeration and GAC filtration as the BAT for the removal of regulated VOCs. The exception is vinyl chloride, for which packed-tower aeration is the preferred technology. The GAC treatment is considered more costly than air stripping.[146]

Treatment consisting of coagulation, filtration, and powdered activated carbon is reported[147] to remove 85 to 98 percent endrin, 90 to 98 percent 2,4-D, and 30 to 99 percent lindane at dosages of 5 to 79 mg/l. Reverse osmosis is also effective in removing organics, including pesticides, with proper design and membrane selection. Highly colored waters and iron can coat GAC and interfere with its adsorption of VOCs.

WATER SYSTEM DESIGN PRINCIPLES*

Water Quantity

The quantity of water upon which to base the design of a water system should be determined in the preliminary planning stages. Future water demand is based on social, economic, and land-use factors, all of which can be expected to change with time. Population projections are a basic consideration. They are made using arithmetic, geometric, and demographic methods and with graphical comparisons with the growth of other comparable cities or towns of greater population.[148] Adjustments should be made for hospital and other institution populations, industries, fire protection, military reservations, transients, and tourists, as well as for leakage and unaccounted-for water, which may amount to 10 to 15 percent or more. Universal metering is necessary for an accounting.

*Refer to *Recommended Standards for Water Works*, Great Lakes–Upper Mississippi River Board of State Public Health and Environmental Managers, Health Research Inc., Health Education Services Division, Albany, NY, 1987. See also state and design publications.

Numerous studies have been made to determine the average per-capita water use for water system design. Health departments and other agencies have design guides, and standard texts give additional information. In any case, the characteristics of the community must be carefully studied and appropriate provisions made. See "Water Quantity and Quality" in Chapter 1, for average water uses.

Design Period

The design period (the period of use for which a structure is designed) is usually determined by the future difficulties to acquire land or replace a structure or pipeline, the cost of money, and the rate of growth of the community or facility served. In general, large dams and transmission mains are designed to function for 50 or more years; wells, filter plants, pumping stations, and distribution systems for 25 years; and water lines less than 12 inches in diameter for the full future life. When interest rates are high or temporary or short-term use is anticipated, a lesser design period would be in order. Fair et al.[149] suggest that the dividing line is in the vicinity of 3 percent per annum. Treatment of water, design, and operation control has been discussed earlier.[150]

Watershed Runoff and Reservoir Design

Certain basic information, in addition to future water demand, is needed upon which to base the design of water works structures. Long-term precipitation, stream flow data, and groundwater information are available from the U.S. Geological Survey and state sources, but these seldom apply to small watersheds. Precipitation data for specific areas are also available from the National Oceanic and Atmospheric Administration, local weather stations, airports, and water works. Unit hydrographs, maximum flows, minimum flows, mass diagrams,* characteristics of the watershed, precipitation, evaporation losses, percolation, and transpiration losses should be considered for design purposes and storage determinations when these are applicable.

Watershed runoff can be estimated in different ways. The rational method for determining the maximum rate of runoff is given by this formula:

$$Q = AIR$$

where

Q = runoff, ft^3/sec
A = area of the watershed, acres
I = imperviousness ratio, that is, the ratio of water that runs off the watershed to the amount precipitated on it
R = rate of rainfall on the watershed, in./hr

---

*A plot of the summation of accumulated stream inflow in million gallons vs. the summation of the mean daily demand in years (25 or more if stream flow data are available) to determine the required (available) storage to meet the daily demand.

The ratio I will vary from 0.01 to 0.20 for wooded areas; from 0.05 to 0.25 for farms, parks, lawns, and meadows depending on the surface slope and character of the subsoil; from 0.25 to 0.50 for residential semirural areas; from 0.05 to 0.70 for suburban areas; and from 0.70 to 0.95 for urban areas having paved streets, drives, and walks.[151] For maximum storms, use these equations:

$$R = 360/t + 30$$

for ordinary storms in eastern United States

$$R = 105/t + 15$$

for San Francisco

$$R = 7/\sqrt{t}$$

for New Orleans

$$R = 56/(t + 5)^{0.85}$$

and for St. Louis

$$R = 19/\sqrt{t}$$

where t is time (duration) of rainfall in minutes.[152]

Another formula for estimating the average annual runoff by Vermuelé may be written as

$$F = R - (11 + 0.29R)(0.035)T - 0.65)$$

where

F = annual runoff, in.
R = annual rainfall, in.
T = mean annual temperature, °F

This formula is reported to be particularly applicable to streams in northern New England and in rough mountainous districts along the Atlantic Coast.[153] For small water systems, the design should be based on the year of minimum rainfall or on about 60 percent of the average.

In any reservoir storage study, it is important to take into consideration the probable losses due to seepage, outflows, evaporation from water surfaces during the year, and loss in storage capacity due to sediment accumulation if the sediment cannot be released during high inflow. This becomes very significant in small systems when the water surfaces exceed 6 to 10 percent of the drainage area.[154] In the North Atlantic states, the annual evaporation from land surfaces averages about 40 percent, while that from water surfaces is about 60 percent of the annual rainfall.[155] The watershed water loss due to land evaporation and transpiration is significant and hence must be taken into consideration when determining precipitation minus losses.

The minimum stream flow in New England has been estimated to yield 0.2 to 0.4 cfs/mi^2 of tributary drainage and an annual yield of 750,000 gpd/mi^2 with storage of 200 to 250 × 10^6 gal/mi^2. New York City reservoirs located in upstate

New York have a dependable yield of about 1 mgd/mi^2 of drainage area. For design purposes, long-term rainfall and stream flows should be used and a mass diagram constructed. See Figure 2.9 later.

Groundwater runoff at the 70 percent point (where flow is equaled or exceeded 70 percent of the time) for the United States land area averaged a yield of 0.23 mgd/mi^2. In the Great Lakes Basin, 25 to 75 percent of the annual flow of streams is derived from groundwater seepage.[156]

The feasibility of implementing watershed rules and regulations should have a high priority in the selection of a water supply source. The management of land use and the control of wastewater discharges, including stormwater drainage on a watershed from urban, suburban, and rural areas, are necessary. Erosion and the input of sediment and organic and inorganic materials such as oils, pesticides, heavy metals, road salt, and other synthetic chemicals must be adequately minimized. Of course, these factors will affect the water quality and reservoir eutrophication, treatment required, and overall quality of the water source. Development of a reservoir should, if possible, include removal of rich organic topsoil from the site to conserve the resource and delay the development of anaerobic conditions.

Intakes and Screens

Conditions to be taken into consideration in design of intakes include high- and low-water stages; navigation or allied hazards; floods and storms; floating ice and debris; water velocities, surface and subsurface currents, channel flows, and stratification; location of sanitary, industrial, and storm sewer outlets; and prevailing wind direction.

Small communities cannot afford elaborate intake structures. A submerged intake crib, or one with several branches and upright tee fittings anchored in rock cribs 4 to 10 feet above the bottom, is relatively inexpensive. The inlet fittings should have a coarse strainer or screen with about 1-inch mesh. The total area of the inlets should be at least twice the area of the intake pipe and provide an inlet velocity less than 0.5 fps. Low-entrance velocities reduce ice troubles and are less likely to draw in fish or debris. Sheet ice over the intake structure also helps avoid anchor ice or frazil ice. If ice clogging of intakes is anticipated, provision should be made for an emergency intake or injecting steam, hot water, or compressed air at the intake. Backflushing is another alternative that may be incorporated in the design. Fine screens at intakes will become clogged; hence, they should not be used unless installed at accessible locations that will make regular cleaning simple. Duplicate stationary screens in the flow channel with 1/8- to 3/8-inch corrosion-resistant mesh can be purchased.

Some engineers have used slotted well screens in place of a submerged crib intake for small supplies. The screen is attached to the end of the intake conduit and mounted on a foundation to keep it off the bottom, and, if desired, crushed rock or gravel can be dumped over the screen. For example, a 10-foot section of a 24-inch-diameter screen with 0.25-inch openings is said to be able to handle

12 mgd at an influent velocity of less than 0.5 fps. Attachment to the foundation should be made in such a way that removal for inspection is possible.

In large installations, intakes with multiple-level inlet ports are provided in deep reservoirs, lakes, or streams to make possible depth selection of the best water when the water quality varies with the season of the year and weather conditions. Special bottom outlets should be provided in reservoirs to make possible the flushing out of sediment and accumulated organic matter during periods of high inflow.

For a river intake, the inlet is perpendicular to the flow. The intake structure is constructed with vertical slotted channels before and after the bar racks and traveling screens for the placement of stop planks if the structure needs to be dewatered. Bar racks, 1 × 6 inch vertical steel, spaced 2 to 6 inches apart, provided with a rake operated manually or mechanically, keep brush and large debris from entering. This may be followed by a continuous slow-moving screen traveling around two drums, one on the bottom of the intake and the other above the operating floor level. The screen is usually a heavy wire mesh with square openings 3/8 to 1 inches; it is cleaned by means of water jets inside that spray water through the screen, washing off debris into a wastewater trough. In cold-weather areas, heating devices such as steam jets are needed to prevent icing and clogging of the racks and screens. Intake velocities should be maintained at less than 5 fps.

Pumping

When water must be pumped from the source or for transmission, electrically operated pumps (at least two) should have gasoline or diesel standby units having at least 50 percent of the required capacity. If standby units provide power for pumps supplying chlorinators and similar units, the full 100 percent capacity must be provided where gravity flow of water will continue during the power failure.

The distribution of water usually involves the construction of a pumping station, unless one is fortunate enough to have a satisfactory source of water at an elevation to provide a sufficient flow and water pressure at the point of use by gravity. The size pump selected is based on whether hydropneumatic storage (steel pressure tank for a small system), ground level, or elevated storage is to be used; the available storage provided; the yield of the water source; the water usage; and the demand. Actual meter readings should be used, if available, with consideration being given to future plans, periods of low or no usage, and maximum and peak water demands. Metering can reduce water use by 25 percent or more. Average water consumption figures must be carefully interpreted and considered with required fire flows. If the water system is to also provide fire protection, then elevated storage is practically essential, unless ground-level storage with adequate pumps is available.

The capacity of the pump required for a domestic water system with elevated storage is determined by the daily water consumption and volume of the storage tank. Of course, where the topography is suitable, the storage tank can be located on high ground, although the hydraulic gradient necessary to meet the highest water demand may actually govern. The pump should be of such capacity as to deliver the average daily water demand to the storage tank in 6 to 12 hours. In very small installations, the pump chosen may have a capacity to pump in 2 hours all the water used in one day. This may be desirable when the size of the centrifugal pump is increased to 60 gpm or more and the size of the electric motor to 5 to 10 hp or more, since the efficiencies of these units then approach a maximum. On the other hand, larger transmission lines, if not provided, would be required in most cases to accommodate the larger flow, which would involve increased cost. Due consideration must also be given to the increased electrical demand and the effects this has. A careful engineering analysis should be made.

Pumping stations should be at least 3 feet above the 100-year flood level or the highest known level, whichever is higher. They should be secured and weather protected.

Distribution Storage Requirements

Water storage requirements should take into consideration the peak daily water use, the maximum-day demand plus the required fire flow, the capacity of the normal and standby pumping equipment, the availability and capacity of auxiliary power, the probable duration of power failure, and the promptness with which repairs can be made. Additional considerations include land use, topography, pressure needs, distribution system capacity, special demands, and the increased cost of electric power and pumps to meet peak demands.

Water storage is necessary to help meet peak demands, fire requirements, and industrial needs; to maintain relatively uniform water pressures; to eliminate the necessity for continuous pumping; to make pumping possible when the electric rate is low; and to use the most economical pipe sizes. Surges in water pressure due to water hammer are also dissipated. Other things being equal, a large-diameter shallow tank is preferable to a deep tank of the same capacity. It is less expensive to construct, and water pressure fluctuations on the distribution system are less. The cost of storage compared to the decreased cost of pumping, the increased fire protection and possibly lowered fire insurance rate, the greater reliability of water supply, and the decreased probability of negative pressures in the distribution system will be additional factors in making a decision.

In general, it is recommended that water storage equal not less than one-half the total daily consumption, with at least one-half the storage in elevated tanks. A preferred minimum storage capacity would be a two-day average use plus fire flow, or the maximum-day usage plus fire requirements less the daily capacity of the water plant and system for the fire-flow period.

Hudson[157] suggests the provision of two tank outlets, one to withdraw the top third of tank water for general purposes and a second outlet at the bottom of the tank to withdraw the remaining two-thirds of tank water if needed to supply building sprinkling systems in developed areas with high-rise apartments, industries, shopping centers, office complexes, and the like. In small communities, real estate subdivisions, institutions, camps, and resorts, elevated storage should be equal to at least a full day's requirements during hot and dry months when lawn sprinkling is heavy. Two or three days storage is preferred. The amount of water required during peak hours of the day may equal 15 to 25 percent of the total maximum daily consumption. This amount in elevated storage will meet peak demands, but not fire requirements. Some engineers provide storage equal to 20 to 40 gal/capita, or 25 to 50 percent of the total average daily water consumption. A more precise method for computing requirements for elevated storage is to construct a mass diagram. Two examples are shown in Figures 2.8 and 2.9. Fire requirements should be taken into consideration.[158]

It is good practice to locate elevated tanks near the area of greatest demand for water and on the side of town opposite from where the main enters. Thus, peak demands are satisfied with the least pressure loss and smallest main sizes. All distribution reservoirs should be covered; provided with an overflow that will not undermine the footing, foundation, or adjacent structures; and provided with a drain, water-level gauge, access manhole with overlapping cover, ladder, and screened air vent.

Water storage tanks are constructed of concrete, steel, or wood. Tanks may be constructed above or partly below ground, except that under all circumstances the manhole covers, vents, and overflows must be well above the normal ground level and the bottom of the tank must be above groundwater or floodwater. Good drainage should be provided around the tank. Tanks located partly below ground must be at a higher level than any sewers or sewage disposal systems and not closer than 50 feet. Vents and overflows should be screened and the tanks covered to keep out dust, rain, insects, small animals, and birds. A cover will also prevent the entrance of sunlight, which tends to warm the water and encourage the growth of algae. Manhole covers should be locked and overlap at least 2 inch over a 2- to 6-inch lip around the manhole. Partly below-ground storage is usually less costly and aesthetically more acceptable than elevated storage.*

Properly constructed reinforced concrete tanks ordinarily do not require waterproofing. If tanks are built of brick or stone masonry, they should be carefully constructed by experienced craftsmen and only hard, dense material laid with full Portland cement mortar joints should be used. Two 0.5-inch coats of 1:3 Portland cement mortar on the inside, with the second coat carefully troweled, should make such tanks watertight. A newly constructed concrete or masonry tank should be allowed to cure for about one month, during which time it should be wetted down frequently. The free lime in the cement can be neutralized by washing the interior with a weak acid, such as a 10 percent muriatic acid solution, or with a solution made up of 4 pounds of zinc sulfate per gallon of water and then flushed clean.

Wooden elevated storage tanks are constructed of cypress, fir, long-leaf yellow pine, or redwood. They are relatively inexpensive and easily assembled, and need

---

*For small concrete reservoir construction details, see *Manual of Individual Water Supply Systems*, U.S. EPA, Washington, DC, 1973, pp. 127–128.

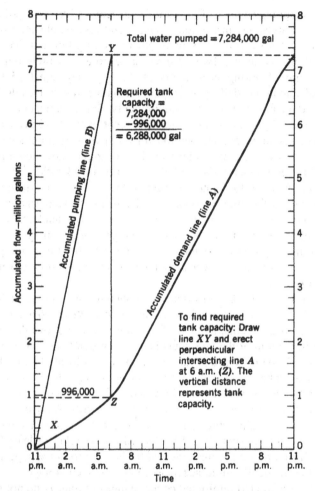

FIGURE 2.8 Mass diagram for determining capacity of tank when pumping 7 hours, from 11 p.m. to 6 a.m. (*Source*: J. E. Kiker, Jr., "Design Criteria for Water Distribution Storage," *Public Works* (March 1964): 102–104. This illustration originally appeared in the March 1964 issue of *Public Works* ®, published by Public Works Journal Corporation, 200 South Broad Street, Ridgewood, NJ 07450. © 2002 Public Works Journal Corporation. All rights reserved.)

not be painted or given special treatment; their normal life is 15 to 20 years. Wooden tanks are available with capacities up to 500,000 gallons. The larger steel tanks start at 5,000 to 25,000 gallons; they require maintenance in order to prolong their life. Reinforced prestressed concrete tanks are also constructed. Underground fiberglass reinforced plastic tanks are also available up to a capacity of 25,000 to 50,000 gallons. Tanks having exterior lead-based paint needing repair present special problems regarding removal and prevention of air pollution.

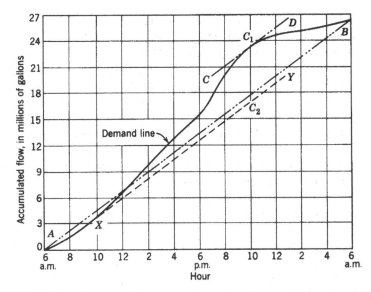

FIGURE 2.9 Mass diagram of storage requirements. The cumulative demand curve is plotted from records or estimates and the average demand line, *AB*, drawn between its extremities. Lines *CD* and *XY* are drawn parallel to line *AB* and tangent to the curve at points of greatest divergence from the average. At C_1 (the point of maximum divergence), a line is extended down the coordinate to line *XY*. This line, C_1C_2, represents the required peak-hour storage: in this case, it scales to 6.44×10^6 gal. (*Source*: G. G. Schmid, "Peak Demand Storage," *J. Am. Water Works Assoc.* (April 1956). Copyright 1956 by the American Water Works Association. Reprinted with permission.)

Steel standpipes, reservoirs, and elevated tanks are made in a variety of sizes and shapes. As normally used, a standpipe is located at some high point to make available most of its contents by gravity flow and at adequate pressure; a reservoir provides mainly storage. A standpipe has a height greater than its diameter; a reservoir has a diameter greater than its height. Both are covered, except when a reservoir is a natural body of water. The altitude of elevated tanks, standpipes, and reservoirs is usually determined, dependent on topography, to meet special needs and requirements. Elevated tanks rising more than 150 feet above the ground or located within 15,000 feet of a landing area and in a 50-mile-wide path of civil airways, must meet the requirements of the Civil Aeronautics Administration.

Peak Demand Estimates

The maximum hourly or peak-demand flow upon which to base the design of a water distribution system should be determined for each situation. A small residential community, for example, would have characteristics different from a new realty subdivision, central school, or children's camp. Therefore, the design flow to determine distribution system capacity should reflect the pattern of living or operation, probable water usage, and demand of that particular type of establishment or community. At the same time, consideration should be given to the location of existing and future institutions, industrial areas, suburban or fringe areas, highways, shopping centers, schools, subdivisions, and direction of growth. In this connection, reference to the city, town, or regional comprehensive or master plan, where available, can be very helpful. Larger cities generally have a higher per-capita water consumption than smaller cities, but smaller communities have higher percentage peak-demand flow than larger communities.

The maximum hourly domestic water consumption for cities with a population above 50,000 will vary from about 200 to 700 percent of the average-day annual hourly water consumption; the maximum hourly water demand in smaller cities will probably vary from 300 to 1,000 percent of the average-day annual hourly water consumption. The daily variation is reported to be 150 to 250 percent, and the monthly variation 120 to 150 percent of the average annual daily demand in small cities.[159] A survey of 647 utilities serving populations of 10,000 or more in 1970 found the mean maximum daily demand to be 1.78 times the average day, with a range of 1.00 to 5.22. Studies in England showed that the peak flow is about 10 times the average flow in cities with a population of 5,000.[160] It can be said that the smaller and newer the community, the greater the probable variation in water consumption from the average will be.

Various bases have been used to estimate the probable peak demand at real estate subdivisions, camps, apartment buildings, and other places. One assumption for small water plants serving residential communities is to say that, for all practical purposes, almost all water for domestic purposes is used in 12 hours.[161] The maximum hourly rate is taken as twice the maximum daily hourly rate, and the maximum daily hourly rate is 1.5 times the average maximum hourly rate. If the average maximum monthly flow is 1.5 times the average monthly annual flow, then the maximum hour's consumption rate is 9 times the average daily hourly flow rate.

Another basis used on Long Island is maximum daily flow rate = 4 times average daily flow rate; maximum 6-hour rate = 8 times average daily flow rate; and maximum 1-hour rate = 9.5 times average daily flow rate.[162]

A study of small water supply systems in Illinois seems to indicate that the maximum hourly demand rate is 6 times the average daily hourly consumption.[163]

An analysis by Wolff and Loos[164] showed that peak water demands varied from 500 to 600 percent over the average day for older suburban neighborhoods with small lots; to 900 percent for neighborhoods with 0.25- to 0.5-acre lots; and to 1,500 percent for new and old neighborhoods with 0.33- to 3-acre lots. Kuranz, Taylor, and many others have also studied the variations in residential water use.[165]

The results of a composite study of the probable maximum momentary demand are shown in Figure 2.10. It is cautioned, however, that for other than average conditions, the required supply should be supplemented as might be appropriate for fire flows, industries, and other special demands.

Peak flows have also been studied at camps, schools, apartment buildings, highway rest areas, and other places.

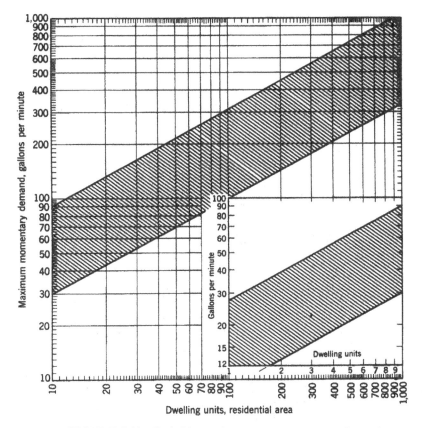

FIGURE 2.10 Probable maximum momentary water demand.

The design of water requirements at toll road and superhighway service areas introduces special considerations that are typical for the installation. It is generally assumed that the sewage flow equals the water flow. In one study of national turnpike and highway restaurant experience, the extreme peak flow was estimated at 1,890 gpd per counter seat and 810 gpd per table seat; the peak day was taken as 630 gpd per counter seat and 270 gpd per table seat.[166] In another study of the same problem, the flow was estimated at 350 gpd per counter seat plus 150 gpd per table seat.[167] The flow was 200 percent of the daily average at noon and 160 percent of the daily average at 6 p.m. It was concluded that 10 percent of the cars passing a service area will enter and will require 15 to 20 gallons per person. A performance study after 1 year of operation of the Kansas Turnpike service areas showed that 20 percent of cars passing service areas will enter; there will be 1.5 restaurant customers per car; average water usage will be 10 gallons per restaurant customer, of which 10 percent is in connection with gasoline service; and plant flows may increase four to five times in a matter of seconds.[168]

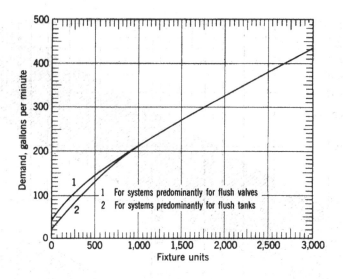

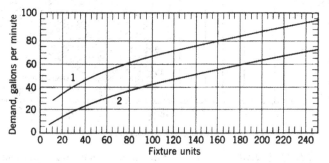

FIGURE 2.11 Estimate curves for demand load. (Source: R. B. Hunter, "Water-Distributing Systems for Buildings," Report BMS 79, National Bureau of Standards for Building Materials and Structures, November 1941.)

Peak flows for apartment-type buildings can be estimated using the curves developed by Hunter.[169] Figure 2.11 and Tables 2.8 and 2.9 can be used in applying this method. Additions should be made for continuous flows. This method may be used for the design of small water systems, but the peak flows determined will be somewhat high.

At schools, peak flows would occur at recess and lunch periods and after gym classes. At motels, peak flows would occur between 7 and 9 a.m. and between 5 and 7 p.m.

It must be emphasized that actual meter readings from a similar type establishment or community should be used whenever possible in preference to an estimate. Time spent to obtain this information is a good investment, as each installation has different characteristics. Hence, the estimates and procedures mentioned here should be used as a guide to supplement specific studies and aid in the application of informed engineering judgment. Peak demands and

TABLE 2.8 Demand Weight of Fixtures in Fixture Units[a]

Fixture or Group[b]	Occupancy	Type of Supply Control	Weight in Fixture Units[c]
Water closet	Public	Flush valve	10
		Flush tank	5
Pedestal urinal	Public	Flush valve	10
Stall or wall urinal	Public	Flush valve	5
		Flush tank	3
Lavatory	Public	Faucet	2
Bathtub	Public	Faucet	4
Shower head	Public	Mixing valve	4
Service sink	Office, etc.	Faucet	3
Kitchen sink	Hotel or restaurant	Faucet	4
Water closet	Private	Flush valve	6
		Flush tank	3
Lavatory	Private	Faucet	1
Bathtub	Private	Faucet	2
Shower head	Private	Mixing valve	2
Bathroom group	Private	Flush valve for closet	8
		Flush tank for closet	6
Separate shower	Private	Mixing valve	2
Kitchen sink	Private	Faucet	2
Laundry trays (1–3)	Private	Faucet	3
Combination fixture	Private	Faucet	3

[a]For supply outlets likely to impose continuous demands, estimate continuous supply separately and add to total for fixtures.

[b]For fixtures not listed, weights may be assumed by comparing the fixture to a listed one using water in similar quantities and at similar rates.

[c]The given weights are for total demand. For fixtures with both hot and cold water supplies, the weights for maximum separate demands may be taken as three-fourths the listed demand for supply.

Source: R. B. Hunter, *Water-Distributing System for Buildings*, Report No. BMS 79, National Bureau of Standards Building Materials and Structures, November 1941.

per-capita daily water use can be expected to decline as water-saving plumbing fixtures and devices come into general use.

Distribution System Design Standards

As far as possible, distribution system design should follow usual good water-works practice and provide for fire protection.[170] Mains should be designed on the basis of velocities of 4 to 6 fps with maximums of 10 to 20 fps, the rates of water consumption (maximum daily demand), and fire demand, plus a residual pressure of not less than 35 psi or more than 100 psi using the Hazen and Williams coefficient $C = 100$, with a normal working pressure of about 60 psi.

Air release valves or hydrants are provided as necessary, where air can accumulate in the transmission lines, and blowoffs are provided at low drain points.

TABLE 2.9 **Flow Rate and Required Pressure**

Fixture	Flow Pressure[a] (psi)	Flow Rate (gpm)
Ordinary basin faucet	8	3.0
Self-closing basin faucet	12	2.5
Sink faucet		
3/8 in.	10	4.5
1/2 in.	5	4.5
Bathtub faucet	5	6.0
Laundry-tub cock, 1/2 in.	5	5.0
Shower	12	5.0
Ball cock for closet	15	3.0
Flush valve for closet	10–20	15–40[b]
Flush valve for urinal	15	15.0
Garden hose, 50 ft and sill cock	30	5.0
Dishwashing machine, commercial	15–30	6–9

[a]Flow pressure is the pressure in the pipe at the entrance to the particular fixture considered. Some codes permit 8 psi for faucet fixtures and lesser flow rates.
[b]Wide range due to variation in design and type of flush-valve closets.
Source: Report of the Coordinating Committee for a National Plumbing Code, U.S. Department of Commerce, Washington, DC, 1951.

These valves must not discharge to below-ground pits unless provided with a gravity drain to the surface above flood level. As far as possible, dead ends should be eliminated or a blowoff provided, and mains should be tied together at least every 600 feet. Lines less than 6 inches in diameter should generally not be considered, except for the smallest system, unless they parallel secondary mains on other streets. In new construction, 8-inch pipe should be used. In urban areas 12-inch or larger mains should be used on principal streets and for all long lines that are not connected to other mains at intervals close enough for proper mutual support. Although the design should aim to provide a pressure of not less than 35 psi in the distribution system during peak-flow periods, 20 psi minimum may be acceptable. A minimum pressure of 60 to 80 psi is desired in business districts, although 50 psi may be adequate in small villages with one- and two-story buildings. Thrust blocks and joint restraints must be provided on mains where indicated, such as at tees, bends, plugs, and hydrants.

Valves are spaced not more than 500 feet apart in commercial districts and 800 feet apart in other districts and at street intersections. A valve book, at least in triplicate, should show permanent ties for all valves, number of turns to open completely, left- or right-hand turn to open, manufacturer, and dates valves operated. A valve should be provided between each hydrant and street main.

Hydrants should be provided at each street intersection and spacing may range generally from 350 to 600 feet, depending on the area served for fire protection and as recommended by the state Insurance Services Office. The connection to the street main should be not less than 6 inches in diameter. Operating nuts and direction of operation should be standard on all hydrants and should conform with AWWA standards. Hydrants should be set so that they are easily accessible to fire department pumpers; they should not be set in depressions, in cutouts, or on embankments high above the street. Pumper outlets should face directly toward the street. With respect to nearby trees, poles, and fences, there should be adequate clearance for connection of hose lines. Hydrants should be painted a distinguishing color so that they can be quickly spotted at night. Hydrant drains shall not be connected to or located within 10 feet of sanitary sewers or storm drains.

Main breaks occur longitudinally and transversely. Age is not a factor. Breaks are associated with sewer and other construction, usually starting with a leaking joint. The leak undermines the pipe, making a pipe break likely due to beam action. Sometimes poor quality control in pipe manufacture contributes to the problem. Good pipe installation practice, including bedding and joint testing, followed by periodic leak surveys, will minimize main leaks and breaks. Unavoidable leakage should not exceed 70 gallons per 24 hours per mile of pipe per inch of pipe diameter. A loss of 1,000 to 3,000 gallons per mile of main is considered reasonable.

Water lines are laid below frost, separated from sewers a minimum horizontal distance of 10 feet and a vertical distance of 18 inches. Water lines may be laid closer horizontally in a separate trench or on an undisturbed shelf with the bottom at least 18 inches above the top of the sewer line under conditions acceptable to the regulatory agency. It must be recognized that this type of construction is more expensive and requires careful supervision during construction. Mains buried 5 feet are normally protected against freezing and external loads.

The selection of pipe sizes is determined by the required flow of water that will not produce excessive friction loss. Transmission mains for small water systems more than 3 to 4 miles long should not be less than 10 to 12 inches in diameter. Design velocity is kept under 5 fps and head loss under 3 ft/1,000 ft. On the one hand, if the water system for a small community is designed for fire flows, the required flow for domestic use will not cause significant head loss. On the other hand, where a water system is designed for domestic supply only, the distribution system pipe sizes selected should not cause excessive loss of head. Velocities may be 1.5 to 5.5 fps. In any case, a special allowance is usually necessary to meet water demands for fire, industrial, and other special purposes.

Design velocities as high as 10 to 15 fps are not unusual, particularly in short runs of pipe. The design of water distribution systems can become very involved and is best handled by a competent sanitary engineer. When a water system is carefully laid out, without dead ends, so as to divide the flow through several pipes, the head loss is greatly reduced. The friction loss in a pipe connected at both ends is about one-quarter the friction loss in the same pipe with a dead end. The friction loss in a pipe from which water is being drawn off uniformly along its length is about one-third the total head loss. Also, for example, an 8-inch line will carry 2.1 times as much water as a 6-inch line for the same loss of head.*

Where possible, a water system that provides adequate fire protection is highly recommended. This is discussed further below. The advantages of fire protection should at the very least be compared with the additional cost of increased pipe size and plant capacity, water storage, and the possible reduced fire insurance rate. If, for example, the cost of 8-inch pipe is only 20 percent more per foot than 6-inch pipe, the argument for the larger diameter pipe, where needed, is very persuasive, since the cost of the trench would be the same. In any case, only pipes and fittings that have a permanent-type lining or inner protective surface should be used.

Small Distribution Systems

In some communities where no fire protection is provided, small-diameter pipe may be used. In such cases, a 2-inch line should be no more than 300 feet long, a 3-inch line no more than 600 feet, a 4-inch line no more than 1,200 feet, and a 6-inch line no more than 2,400 feet. If lines are connected at both ends, 2- or 3-inch lines should be no longer than 600 feet; 4-inch lines are not more than 2,000 feet.

Transmission lines for rural areas have been designed for peak momentary demands of 2 to 3 gpm per dwelling unit and for as low as 0.5 gpm per dwelling unit with storage provided on the distribution system to meet peak demands. Adjustments are needed for constant or special demands and for population size. For example, Figure 2.10 shows a probable maximum demand of 3 to 9 gpm per dwelling unit for 10 dwelling units, 1 to 3.2 gpm per dwelling unit for 100 dwelling units, and 0.33 to 1.1 gpm per dwelling unit for 1,000 dwelling units.

A general rule of thumb is that a 6-inch main can be extended only 500 feet if the average amount of water of 1000 gpm is to be supplied for fire protection, or about 2,000 feet if the minimum amount of 500 gpm is to be supplied.

The minimum pipe sizes and rule-of-thumb guides are not meant to substitute for distribution system hydraulic analysis but are intended for checking or rough approximation. Use of the equivalent pipe method, the Hardy Cross method, or one of its modifications should be adequate for the small distribution system. Computer analysis methods are used for large-distribution-system analysis.[171]

Fire Protection

Many factors enter into the classification of municipalities (cities, towns, villages, and other municipal entities) for fire insurance rate-setting purposes.

---

*A 6-in. line carries 2.9 times as much as a 4-in. line; an 8-in. line carries 6.2 times as much as a 4-in. line; a 12-in. line carries 18 times as much as a 4-in. line, 6.2 times as much as a 6-in. line, and 2.9 times as much as an 8-in. line. The discharges vary as the 2.63 power of the pipe diameters being compared, based on the Hazen-Williams formula. See flow charts, nomograms, or Table 2.14.

The Insurance Services Office, their state representatives, and other authorized offices use the *Fire Suppression Rating Schedule*[172] to classify municipalities with reference to their fire defenses. This is one of several elements in the development of property fire insurance rates.

The municipal survey and grading work formerly performed by the National Board of Fire Underwriters, then by the American Insurance Association, as well as that formerly performed by authorized insurance-rating organizations are continued under the Insurance Services Office. Credit is given for the facilities provided to satisfy the needed fire flows of the buildings in the municipality.[173] (Since this discussion is intended only for familiarization purposes, the reader interested in the details of the grading system is referred to the references cited in this section for further information.)

An adequate water system provides sufficient water to meet peak demands for domestic, commercial, and industrial purposes as well as for firefighting. For fire suppression rating, the water supply has a weight of 40 percent; the fire department, 50 percent; and receiving and handling fire alarms, 10 percent. The water system rating considers the adequacy of the supply works, mains and hydrant spacing, size and type of hydrants, and inspection and condition of hydrants.

To be recognized for fire protection, a water system must be capable of delivering at least 250 gpm at 20 psi at a fire location for at least 2 hours with consumption at the maximum daily rate. The method of determining the needed fire flow for a building is given in the *Fire Suppression Rating Schedule*.[174] The needed fire flow will vary with the class of construction, its combustibility class, openings and distance between buildings, and other factors. Table 2.10 shows the needed duration for fire flow. The needed fire flow for a community of one- and two-family dwellings varies from 500 gpm for buildings over 100 feet apart, to 1,500 gpm where buildings are less than 11 feet apart.[175] There should be sufficient hydrants within 1,000 feet of a building to supply its needed fire flow. Each hydrant with a pumper outlet and within 300 feet of a building is credited at 1,000 gpm; 301 to 600 feet, 670 gpm; and 601 to 1,000 feet, 250 gpm.

Where possible, water systems should be designed to also provide adequate fire protection, and old systems should be upgraded to meet the requirements. This will also help ensure the most favorable grading, classification, and fire insurance rates. Improvements in a water system resulting in a better fire protection grade and classification would generally be reflected in a reduced fire

TABLE 2.10   Needed Duration for Fire Flow

Needed Fire Flow (gpm)	Needed Duration (hr)
$\leq$2,500	2
3,000	3
3,500	3
$\geq$4,000	4

Source: Fire Suppression Rating Schedule, Insurance Services Office, New York, 1980.

insurance rate on specifically rated commercial properties, although other factors based on individual site evaluation may govern. However, this is not always the case in "class-rated properties" such as dwellings, apartment houses, and motels. It generally is not possible to justify the cost to improve the fire protection class *solely* by the resulting savings in insurance premiums.[176] Nevertheless, the greater safety to life and property makes the value of improved fire protection more persuasive.

It is prudent for the design engineer to follow the state Insurance Services Office requirements.[177]

One must be alert to ensure that fire protection programs do not include pumping from polluted or unapproved sources into a public or private water system main through hydrants or blowoff valves. Nor should bypasses be constructed around filter plants or provision made for "emergency" raw-water connections to supply water in case of fire. In *extreme emergencies*, the health department might permit a temporary connection under certain conditions, but in any case, the water purveyor must immediately notify every consumer not to drink the water or use it in food or drink preparation unless first boiled or disinfected as noted at the end of this chapter.

Cross-Connection Control

There have been numerous instances of illness caused by cross-connections.[178] A discussion of water system design would not be complete without reference to cross-connection control and backflow prevention. The goal is to have no connection between a water of drinking water quality (potable) and an unsafe or questionable (nonpotable) water system or between a potable system and any plumbing, fixture, or device whereby nonpotable water might flow into the potable water system.

A *cross-connection* is any physical connection between a potable water system and a nonpotable water supply; any waste pipe, soil pipe, sewer, drain; or any direct or indirect connection between a plumbing fixture or device whereby polluted water or contaminated fluids including gases or substances might enter and flow back into the potable water system. Backflow of nonpotable water and other fluids into the potable water system may occur by backpressure or backsiphonage. In *backpressure* situations, the pressure in the nonpotable water system exceeds that in the potable water system. In *backsiphonage*, the pressure in the potable water system becomes less than that in the nonpotable water system due to a vacuum or reduced pressure developing in the potable water system.

Negative or reduced pressure in a water distribution or plumbing system may occur when a system is shut off or drained for repairs, when heavy demands are made on certain portions of the system causing water to be drawn from the higher parts of the system, or when the pumping rate of pumps installed on the system (or of fire pumps or fire pumpers at hydrants) exceeds the capacity of the supply line to the pump. Backpressure may occur when the pressure in a nonpotable water system exceeds that in the potable water system, such as when

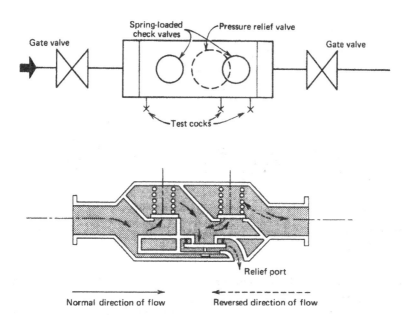

FIGURE 2.12   Reduced pressure zone backflow preventer—principle of operation. Malfunctioning of check or pressure-relief valve is indicated by discharge of water from relief port. Preferred for hazardous facility containment. (*Source: Cross-Connection Control*, EPA-430/9-73-002, U.S. EPA, Water Supply Division (WSD), Washington, DC, 1976, p. 25.)

a fire pumper at a dock or marina pumps nonpotable water into a hydrant or when a boiler chemical feed pump is directly connected to the potable water system.

The more common acceptable methods or devices to prevent backflow are air gap separation, backpressure units as shown in Figures 2.12 and 2.13, and vacuum breakers.[179] The non-pressure-type vacuum breaker is always installed on the atmospheric side of a valve and is only intermittently under pressure, such as when a flushometer valve is activated. The pressure-type vacuum breaker is installed on a pressurized system and will function only when a vacuum occurs. It is spring loaded to overcome sticking and is used only where authorized. The vacuum breaker is not designed to provide protection against backflow resulting from backpressure and should not be installed where backpressure may occur.

The barometric or atmospheric loop that extends 34 to 35 feet above the highest outlet is not acceptable as a backflow preventer because backpressure due to water, air, steam, hot water, or other fluid can negate its purpose. The swing joint, four-way plug valve, three-way two-port valve, removable pipe section, and similar devices are not reliable because nonpotable water can enter the potable water system at the time they are in use.[180]

An elevated or ground-level tank providing an air gap, the reduced pressure zone backflow preventer, and the double-check-valve assembly are generally used on public water system service connections to prevent backflow into the

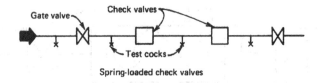

Spring-loaded check valves

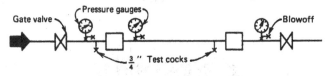

All bronze special-type (factory mutual) check valves

FIGURE 2.13   Double check valve–double gate valve assembly. For aesthetically objectionable facility containment.

distribution system. The vacuum breaker is usually used on plumbing fixtures and equipment.

An approved backflow preventer or air break should be required on the water service line to every building or structure using or handling any hazardous substance that might conceivably enter the potable water system. In addition, building and plumbing codes should prohibit cross-connections within buildings and premises and require approved-type backflow preventers on all plumbing, fixtures, and devices that might cause or permit backflow. It is the responsibility of the designing engineer and architect, the building and plumbing inspector, the waterworks official, and the health department to prevent and prohibit possibilities of pollution of public and private water systems.

There are two major aspects to a cross-connection control program. One is protection of the water distribution system to prevent its pollution. The other is protection of the internal plumbing system used for drinking and culinary purposes to prevent its pollution.

The water purveyor has the responsibility to provide its customers with water meeting drinking water standards. This requires control over unauthorized use of hydrants, blowoffs, and main connections or extensions. It also means requirement of a backflow prevention device at the service connection (containment) of all premises where the operations or functions on the premises involve toxic or objectionable chemical or biological liquid substances or use of a nonpotable water supply, which may endanger the safety of the distribution system water supply through backflow. However, although these precautions may protect the water system, it is also necessary to protect the consumers on the premises using the water for drinking and culinary purposes. This responsibility is usually shared by the water purveyor, the building and plumbing department, the health department, and the owner of the structure, depending on state laws and local ordinances. The AWWA Policy Statement on Cross-Connection states, in part, that the "water purveyor must take reasonable precaution to protect the community distribution system from the hazards originating on the premises of its customers that may degrade the water in the community distribution system."[181] The water purveyor has been held legally responsible for the delivery of safe water to the consumer and the Safe Drinking Water Act bases compliance with federal standards on the quality of water coming out of the consumer's tap. Under these circumstances, a cross-connection control program is needed in every community having a public water system to define and establish responsibility and ensure proper installation and adequate inspection, maintenance, testing, and enforcement.

A comprehensive cross-connection control program should include the following six components:[182]

1. An implementation ordinance that provides the legal basis for the development and complete operation of the program
2. The adoption of a list of devices acceptable for specific types of cross-connection control
3. The training and certification of qualified personnel to test and ensure devices are maintained
4. The establishment of a suitable set of records covering all devices
5. Public education seminars wherein supervisory, administrative, political, and operating personnel, as well as architects, consulting engineers, and building officials, are briefed and brought up-to-date on the reason for the program as well as on new equipment in the field
6. An inspection program with priority given to potentially hazardous connections

In some states, the legal basis for the adoption of a local cross-connection ordinance is a state law or sanitary code; hence, consultation with the state health department or other agency having jurisdiction is advised in the development of a local ordinance and program. Model ordinances and instruction manuals are available.[183]* Enforcement is best accomplished at the local level.[184]

Implementation of a control program requires, in addition to the above, that a priority system be established. Grouping structures and facilities served as "Hazardous," "Aesthetically Objectionable," and "Nonhazardous" can make inspection manageable and permit concentration of effort on the more serious conditions. Estimating the cost of installing backflow prevention devices is helpful in understanding what is involved and obtaining corrections. Some devices are quite costly. An inspection program, with first priority to hazardous situations, is followed by review of findings with the local health department public health engineer or sanitarian, official notification of the customer, request for submission and approval of plans, establishment of a correction timetable, inspection and testing of the backflow device when installed, enforcement action if indicated, follow-up inspections, and testing of installed devices. The program progress should be reviewed and adjusted as needed every six months.[185]

---

*See also local building and plumbing codes.

Hydropneumatic Systems

Hydropneumatic or pressure-tank water systems are suitable for small communities, housing developments, private homes and estates, camps, restaurants, hotels, resorts, country clubs, factories, and institutions and as booster installations. In general, only about 10 to 20 percent of the total volume of a pressure tank is actually available. Hydropneumatic tanks are usually made of 3/16-inch or thicker steel and are available in capacities up to 10,000 or 20,000 gal. Tanks should meet American Society of Mechanical Engineers (ASME) code requirements. Small commercial-size tanks are 42, 82, 120, 144, 180, 220, 315, 525, and 1,000 gallons. Smaller tanks are available precharged with air.

The required size of a pressure tank is determined by peak demand, the capacity of the pump and source, the operating pressure range, and air volume control (available water). The capacity of well and pump should be at least 10 times the average daily water consumption rate, and the gross tank volume in gallons should be at least 10 times the capacity of the pump in gallons per minute.[186] The EPA suggests that the pump capacity for private dwellings be based on the number of fixtures in a dwelling, as shown in Table 2.11. The Water System Council recommends a 7-minute peak-demand usage for one- to four-bedroom homes and suggests a storage of 15 gpd per dwelling unit.[187]

A simple and direct method for determining the recommended volume of the pressure storage tank and size pump to provide is given by Figure 2.14. This figure is derived from Boyle's law and is based on the following formula:

$$Q = \frac{Q_m}{1 + P_1/P_2}$$

where

Q = pressure-tank volume, gal
Q_m = 15-minutes storage at the maximum hourly demand rate
P_1 = minimum absolute operating pressure (gauge pressure plus 14.7 lb/in.^{2})
P_2 = maximum absolute pressure[188]

TABLE 2.11 Recommended Pump Capacity for Private Dwellings

Number of Fixtures	Recommended Pump Capacity (gpm)
2–7	7–8
8	8–9
10	9–11
12	10–12
14–16	11–13
18–20	12–14

Source: Manual of Individual Water Systems, EPA-570/9-82-004, U.S. Environmental Protection Agency, Office of Drinking Water, Washington, DC, October 1982, p. 99.

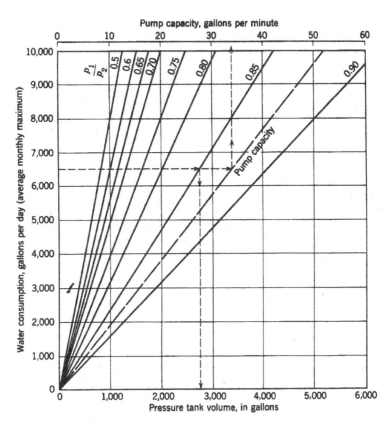

FIGURE 2.14   Chart for determining pressure storage tank volume and pump size. Pressure tank volume provides 15-minutes of storage. (*Source*: J. A. Salvato Jr., "The Design of Pressure Tanks for Small Water Systems," *J. Am. Water Works Assoc.*, June 1949, pp. 532–536. Reprinted by permission. Copyright © 1949 by the American Water Works Association.)

The pump capacity given on the curve is equal to 125 percent of the maximum hourly demand rate. The maximum hourly demand is based on the following but should be determined for each situation:

$$\text{Average daily rate} = \frac{\text{Average water use per day}}{1440 \text{ min/day in gpm}} \text{based on annual water use}$$

$$\text{Average maximum monthly rate} = 1.5 \times \text{average daily rate}$$

$$\text{Maximum hourly demand rate} = 6 \times \text{average maximum monthly rate}$$

or

$$9 \times \text{average daily rate}$$

$$\text{Instantaneous rate (pump capacity)} = 1.25 \times \text{maximum hourly demand rate}$$

or

$$11.25 \times \text{average daily rate}$$

The pressure tank is assumed to be just empty when the pressure gauge reads zero. Figure 2.14 can also be used for larger or smaller flows by dividing or multiplying the vertical and horizontal axes by a convenient factor. The required pressure tank volume can be reduced proportionately if less than 15-minutes of storage is acceptable. For example, it can be reduced to one-third if 5-minutes storage is adequate, or to 1/15 if 1-minutes storage is adequate. Also, if the water consumption in Figure 2.14 is 1/10 of 6,500 gpd, that is 650 gpd, the corresponding pressure tank volume would be 1/10 of 2,800 gallons, or 280 gallons. The pump capacity would be 1/10 of 34 gpm, or 3.4 gpm. But if all water is used in 12 hours, as in a typical residential dwelling, double the required pump capacity, which in this case would be 6.8 gpm. The larger pump is usually provided in small installations for faster pressure tank recovery and to meet momentary demands that are more likely to vary widely than in large installations. See previous text and Table 2.11. Also see Figures 2.10 and 2.11. An example for a larger system is given under "Design of Small Water Systems," this chapter.

The water available for distribution is equal to the difference between the dynamic head (friction plus static head) and the tank pressure. Because of the relatively small quantity of water actually available between the usual operating pressures, a higher initial (when the tank is empty) air pressure and range are sometimes maintained in a pressure tank to increase the water available under pressure. When this is done, the escape of air into the distribution system is more likely. Most home pressure tanks come equipped with a pressure switch and an automatic air volume control (Figure 2.15), which is set to maintain a definite air-water volume in the pressure tank at previously established water pressures, usually 20 to 40 psi. Air usually needs to be added to replace that absorbed by the water to prevent the tank from becoming waterlogged. Small pressure tanks are available with a diaphragm inside that separates air from the water, thereby minimizing this problem. Some manufacturers, or their representatives, increase the pressure tank storage slightly by precharging the tank with air. With deep-well displacement and submersible pumps, an excess of air is usually pumped with the water, causing the pressure tank to become airbound unless an air-release or needle valve is installed to permit excess air to escape.

In large installations an air compressor is needed, and an air-relief valve is installed at the top of the tank. A pressure-relief valve should also be included on the tank. See Figure 2.16.

Where a well yield (source) is inadequate to meet water demand with a pressure tank, then gravity or in-well storage, an additional source of water, or double pumping with intermediate storage, may be considered. Intermediate ground-level storage can be provided between the well pump and the pressure-tank pump. The well pump will require a low-water cutoff, and *its capacity must be related to the dependable well yield*. The intermediate storage tank (tightly covered) should have a pump stop-and-start device to control the well pump and a low-water sensor to signal depletion of water in the intermediate storage tank. A centrifugal

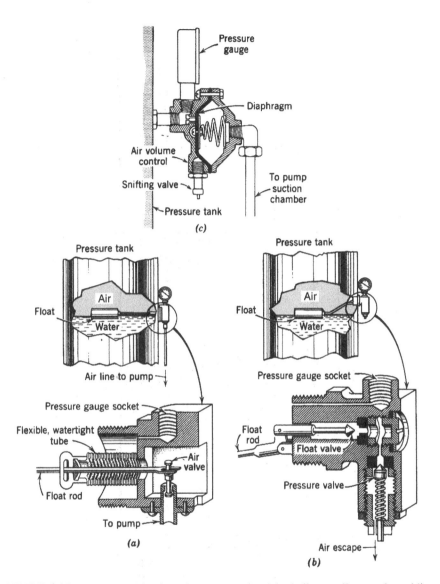

FIGURE 2.15 Pressure-tank air volume controls: (*a*) shallow-well type for adding air; (*b*) deep-well type for air release—used with submersible and piston pumps; (*c*) diaphragm-type in position when pump is not operating (used mostly with centrifugal pump). Small air precharged pressure tanks with a diaphragm to separate air and water are replacing air-volume controls. (*Source: Pumps and Plumbing for the Farmstead*, Tennessee Valley Authority, Agriculture and Engineering Development Division, November 1940.)

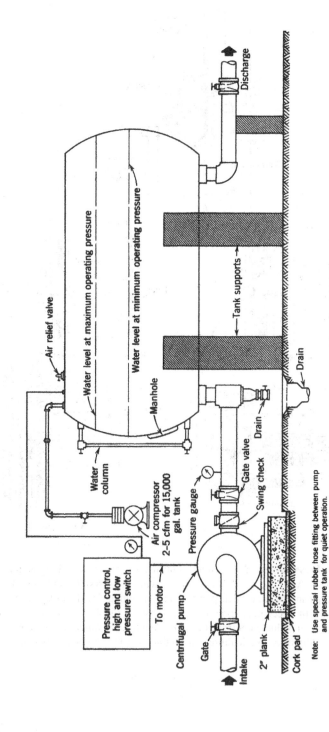

FIGURE 2.16 Typical large pressure tank installation.

pump would pump water from the intermediate tank to a pressure tank, with a pressure switch control, and thence to the distribution system.

Low-rate pumping to elevated storage, a deeper well to provide internal storage, or an oversize pressure tank may be possible alternatives to intermediate ground-level storage, depending on the extent of the problem and relative cost.

Pumps

The pump types commonly used to raise and distribute water are referred to as positive displacement, including reciprocating, diaphragm, and rotary; centrifugal, including turbine, submersible, and ejector jet; air lift; and hydraulic ram. Pumps are classified as low lift, high lift, deep well, booster, and standby. Other types for rural and developing areas include the chain and bucket pump and hand pump.

Displacement Pump

In reciprocating displacement pumps, water is drawn into the pump chamber or cylinder on the suction stroke of the piston or plunger inside the pump chamber and then the water is pushed out on the discharge stroke. This is a simplex or single-acting reciprocating pump. An air chamber (Figure 2.17) should be provided on the discharge side of the pump to prevent excessive water hammer caused by the quick-closing flap or ball valve; by the quick closing or opening of a gate valve, float valve, or pressure-reducing valve; and the sudden shutdown of a pump. The air chamber or other surge suppressor will protect piping and equipment on the line and will tend to even out the intermittent flow of water. See "Water Hammer," this chapter. Reciprocating pumps are also of the duplex type wherein water is pumped on both the forward and backward stroke, and of the triplex type, in which three pistons pump water. The motive power may be manual; a steam, gas, gasoline, or oil engine; an electric motor; or a windmill. The typical hand pump and deep-well plunger or piston pumps over wells are displacement pumps.

A rotary pump is also a displacement pump, since the water is drawn in and forced out by the revolution of a cam, screw, gear, or vane. It is not used to any great extent to pump water.

Displacement pumps have certain advantages over centrifugal pumps. The quantity of water delivered does not vary with the head against which the pump is operating but depends on the power of the driving engine or motor. A pressure-relief valve is necessary on the discharge side of the pump to prevent excessive pressure in the line and possible bursting of a pressure tank or water line. They are easily primed and operate smoothly under suction lifts as high as 22 feet. Practical suction lifts at different elevations are given in Table 2.12.

Displacement pumps are flexible and economical. The quantity of water pumped can be increased by increasing the speed of the pump, and the head can vary within wide limits without decreasing the efficiency of the pump. A displacement pump can deliver relatively small quantities of water as high as

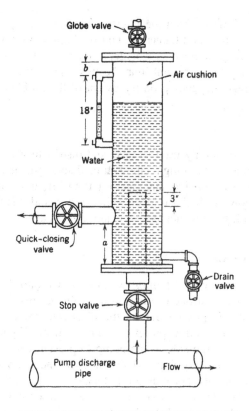

Globe valve

b

Air cushion

18″

Water

3″

Quick-closing valve

a

Drain valve

Stop valve

Pump discharge pipe

Flow

Air Chamber Dimensions				
Discharge Pipe	Inside Diameter of Air Chamber	Total Height	*a*	*b*
2″	8″	3′0″	4″	9″
2¼″	8″	3′6″	4″	12″
3″	10″	4′0″	5″	15″
4″	10″	5′0″	6″	21″
5″	12″	6′0″	6″	27″
6″	16″	7′0″	6″	33″

FIGURE 2.17 Air chamber dimensions for reciprocating pumps. (*Source: Water Supply and Water Purification*, T.M. 5-295, War Department, Washington, DC, 1942.)

800 to 1,000 feet. Its maximum capacity is 300 gpm, although horizontal piston pumps are available in sizes of 500 to 3,000 gpm. The overall efficiency of a plunger pump varies from 30 percent for the smaller sizes to 60 to 90 percent for the larger sizes with electric motor drive. It is particularly suited to pumping small quantities of water against high heads and can, if necessary, pump air with water. This type of pump is no longer widely used.

TABLE 2.12 Atmospheric Pressure and Practical Suction Lift

Elevation Above Sea level		Atmospheric Pressure		Design Suction Life (ft)		
ft	miles	lb/in.^{2}	ft of water	Displacement Pump	Centrifugal Pump	Turbine Pump
0	—	14.70	33.95	22	15	28
1,320	$\frac{1}{4}$	14.02	32.39	21	14	26
2,640	$\frac{1}{2}$	13.33	30.79	20	13	25
3,960	$\frac{3}{4}$	12.66	29.24	18	11	24
5,280	1	12.02	27.76	17	10	22
6,600	$1\frac{1}{4}$	11.42	26.38	16	9	20
7,920	$1\frac{1}{2}$	10.88	25.13	15	8	19
10,560	2	9.88	22.82	14	7	18

Note: The possible suction lift will decrease about 2 ft for every 10°F increase in water temperature above 60°F; 1 lb/in.^{2} = 2.31 ft head of water.

Centrifugal Pump, Also Submersible and Turbine

There are several types of centrifugal pumps; the distinction lies in the design of the impeller. They include radial, mixed, and axial flow, turbine, close-coupled, submersible, and adjustable blade impeller pumps. Water is admitted into the suction pipe or pump casing and is rotated in the pump by an impeller inside the pump casing. The energy is converted from velocity head primarily into pressure head. In the submerged multistage, turbine-type pump used to pump water out of a well, the centrifugal pump is in the well casing below the drawdown water level in the well; the motor is above ground. In the submersible pump, the pump and electric motor are suspended in the well attached to the discharge pipe, requiring a minimum 3-inch- (preferably 4-inch-) diameter casing. It is a multistage, centrifugal pump unit.

If the head against which a centrifugal pump operates is increased beyond that for which it is designed and the speed remains the same, then the quantity of water delivered will decrease. By contrast, if the head against which a centrifugal pump operates is less than that for which it is designed, then the quantity of water delivered will be increased. This may cause the load on the motor to be increased, and hence overloading of the electric or other motor, unless the motor selected is large enough to compensate for this contingency.

Sometimes two centrifugal pumps are connected in series so that the discharge of the first pump is the suction for the second. Under such an arrangement, the capacity of the two pumps together is only equal to the capacity of the first pump, but the head will be the sum of the discharge heads of both pumps. At other times, two pumps may be arranged in parallel so that the suction of each is connected to the same pipe and the discharge of each pump is connected to the same discharge line. In this case, the static head will be the same as that of the individual pumps, but the dynamic head, when the two pumps are in operation, will increase because of the greater friction and may exceed the head for which the pumps are designed. It may be possible to force only slightly more water through the same line when using two pumps as when using one pump, depending on the pipe size. Doubling the speed of a centrifugal pump impeller doubles the quantity of water pumped, produces a head four times as great, and requires eight times as much power to drive the pump. In other words, the quantity of water pumped varies directly with the speed, the head varies as the square of the speed, and the horsepower varies as the cube of the speed. It is usual practice to plot the pump curves for the conditions studied on a graph to anticipate operating results.

The centrifugal pump has no valves or pistons; there is no internal lubrication; and it takes up less room and is relatively quiet. A single-stage centrifugal pump is generally used where the suction lift is less than 15 feet and the total head not over 125 to 200 feet. A single-stage centrifugal pump may be used for higher heads, but where this occurs, a pump having two or more stages, that is, two or more impellers or pumps in series, should be used. The efficiency of centrifugal pumps varies from about 20 to 85 percent; the higher efficiency can be realized in the pumps with a capacity of 500 gpm or more. The peculiarities of the water system and effect they might produce on pumping cost should be studied from the pump curve characteristics. A typical curve is shown in Figure 2.18. All head and friction losses must be accurately determined in arriving at the total pumping head.

Centrifugal pumps that are above the pumping water level should have a foot valve on the pump suction line to retain the pump prime. However, foot valves sometimes leak, thereby requiring a water connection or other priming

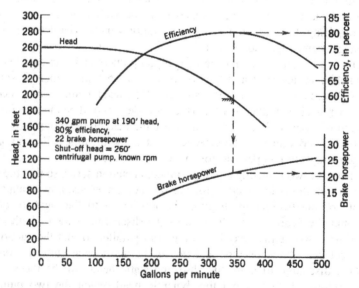

FIGURE 2.18 Typical centrifugal pump characteristic curves.

device or a new check valve on the suction side of the pump. The foot valve should have an area equal to at least twice the suction pipe. It may be omitted where an automatic priming device is provided. In the installation of a centrifugal pump, it is customary to install a gate valve on the suction line to the pump and a check valve followed by a gate valve on the pump discharge line near the pump. An air chamber, surge tank, or similar water-hammer suppression device should be installed just beyond the check valve, particularly on long pipelines or when pumping against a high head. Arrangements should be made for priming a centrifugal pump, unless the suction and pump are under a head of water, and the suction line should be kept as short as possible. The suction line should be sloped up toward the pump to prevent air pockets.

Pump maintenance items to check include cavitation, bearings, coupling alignments, packings, and mechanical seals. Pump manufacturers' catalog efficiencies do not include lift, friction losses in suction and discharge lines, elbow and increaser, or coupling, bearing frame, packing, or mechanical seal losses. Catalog efficiencies should be confirmed. Pump efficiency and capacity will vary with time—wear of bearings, disks or rings, stuffing box, impeller, and casing—as well as with pump and driver misalignment, change in pump speed, and increased pipe friction.

Jet Pump

The jet pump is actually a combination of a centrifugal pump and a water ejector down in a well below or near the water level. The pump and motor can be located some distance away from the well, but the pipelines should slope up to the pump about 1.5 inches in 20 feet. In this type of pump, part of the water raised is diverted back down into the well through a separate pipe. This pipe has attached to it at the bottom an upturned ejector connected to a discharge riser pipe that is open at the bottom. The water forced down the well passes up through the ejector at high velocity, causing a pressure reduction in the venturi throat, and with it draws up water from the well through the riser or return pipe. A jet pump may be used to raise small quantities of water 90 to 120 feet, but its efficiency is lowered when the lift exceeds 50 feet. Efficiency ranges from 20 to 25 percent. The maximum capacity is 50 gpm. There are no moving parts in the well. Jet pumps are shallow-well single-pipe-type (ejector at pump) and deep-well single- and multistage types (ejector in well). Multistage pumps may have impellers horizontal or vertical.

The air ejector pump is similar in operation to a water ejector pump except that air is used instead of water to create a reduced pressure in the venturi throat to raise the water.

Air-Lift Pump

In an air-lift pump, compressed air is forced through a small air pipe extending below the pumping water level in a well and discharged in a finely diffused state in a larger (education) pipe. The air–water mixture in the eduction pipe, being lighter than an equal volume of water, rises. The rise (weight of column of water) must at least equal the distance (weight of column of the same cross-sectional area) between the bottom of the eduction pipe and the water level in the well. A 60 percent submergence is best. For maximum efficiency, the distance from the bottom of the eduction pipe to the water level in the well should equal about twice the distance from the water surface to the point of discharge. The depth of submergence of the eduction pipe is therefore critical, as are the relative sizes of the air and eductor pipes. The area (in square inches) of the eduction pipe is

$$A = Q/20$$

where Q is the volume of water discharged (in gallons per minute) and depends on V, the rate at which air is supplied (in cubic feet per minute):

$$V = Qh/125$$

where h is the distance between the water surface and the point of discharge (in feet).

Efficiencies vary from about 20 to 45 percent. The eduction pipe is about 1 inch smaller in diameter than the casing.[189] The well casing itself can be used as the eductor pipe, provided it is not too much larger than the air pipe.

Hydraulic Ram

A hydraulic ram is a type of pump where the energy of water flowing in a pipe is used to elevate a smaller quantity of water to a higher elevation. An air chamber and weighted check valve are integral parts of a ram. Hydraulic rams are suitable where there is no electricity and the available water supply is adequate to furnish the energy necessary to raise the required quantity of water to the desired level. A battery of rams may be used to deliver larger quantities of water provided the supply of water is ample. Double-acting rams can make use of a nonpotable water to pump a potable water. The minimum flow of water required is 2 to 3 gpm with a fall of 3 feet or more. A ratio of lift to fall of 4 to 1 can give an efficiency of 72 percent, a ratio of 8 to 1 an efficiency of 52 percent, a ratio of 12 to 1 an efficiency of 37 percent, and a ratio of 24 to 1 an efficiency of 4 percent.[190] Rams are known to operate under supply heads up to 100 feet and a lift, or deliver heads, of 5 to 500 feet. In general, a ram will discharge from to of the water delivered to it. From a practical standpoint, it is found that the pipe conducting water from the source to the ram (known as the drive pipe) should be at least 30 to 40 feet long for the water in the pipe to have adequate momentum or energy to drive the ram. It should not, however, be on a slope greater than about 12 degrees with the horizontal. If these conditions cannot be met naturally, it may be possible to do so by providing an open stand pipe on the drive pipeline, so that the pipe beyond it meets the conditions given. The diameter of the delivery pipe is usually about one-half the drive pipe diameter. The following formula may be used to determine the capacity of a ram:

$$Q = \frac{\text{supply to ram} \times \text{power head} \times 960}{\text{pumping head}}$$

where

Q = Gallons delivered per day
Supply to ram = Water delivered to and used by the ram, gpm
Power head = Available supply head of water, ft or fall
Pumping head = Head pumping against, ft, or delivery head

Note: This information plus the length of the delivery pipe and the horizontal distance in which the fall occurs are needed by manufacturers to meet specific requirements.

Pump and Well Protection

A power pump located directly over a deep well should have a watertight well seal at the casing as illustrated in Chapter 1, Figures 1.11 and 1.12. An air vent is used on a well that has an appreciable drawdown to compensate for the reduction in air pressure inside the casing, which is caused by a lowering of the water level when the well is pumped. The vent should be carried 18 inches above the floor and flood level and the end should be looped downward and protected with screening. A downward-opening sampling tap located at least 12 inches above the floor should be provided on the discharge side of the pump. In all instances, the top of the casing, vent, and motor are located above possible flood level.

The top of the well casing or pump should not be in a pit that cannot be drained to the ground surface by gravity. In most parts of the country, it is best to locate pumps in some type of housing above ground level and above any high water. Protection from freezing can be provided by installing a thermostatically controlled electric heater in the pump house. Small, well-constructed, and insulated pump housings are sometimes not heated but depend on heat from the electric motor and a light bulb to maintain a proper temperature. Some type of ventilation should be provided, however, to prevent the condensation of moisture and the destruction of the electric motor and switches. See Figure 1.11.

Use of a submersible pump in a well would eliminate the need for a pump-house but would still require that the discharge line be installed below frost. See Figure 1.7.

Pump Power and Drive

The power available will usually determine the type of motor or engine used. Electric power, in general, receives first preference, with other sources used for standby or emergency equipment.

Steam power should be considered if pumps are located near existing boilers. The direct-acting steam pump and single, duplex, or triplex displacement pump can be used to advantage under such circumstances. When exhaust steam is available, a steam turbine to drive a centrifugal pump can also be used.

Diesel-oil engines are good, economical pump-driving units when electricity is not dependable or available. They are high in first cost. Diesel engines are constant low-speed units.

Gasoline engines are satisfactory portable or standby pump power units. The first cost is low, but the operating cost is high. Variable-speed control and direct connection to a centrifugal pump are common practice. Natural gas, methane, and butane can also be used where these fuels are available.

When possible, use of electric motor pump drive is the usual practice. Residences having low lighting loads are supplied with single-phase current, although this is becoming less common. When the power load may be 3 hp or more, three-phase current is needed. Alternating-current (AC) two- and three-phase motors are of three types: the squirrel-cage induction motor, the wound-rotor or slip-ring induction motor, and the synchronous motor. Single-phase motors are the repulsion–induction type having a commutator and brushes; the capacitator or condenser type, which does not have brushes and commutator; and the split-phase type. The repulsion–induction motor is, in general, best for centrifugal pumps requiring 0.75 hp or larger. It has good starting torque. The all-purpose capacitor motor is suggested for sizes below 0.75 hp. It is necessary to ensure the electric motor is grounded to the pump and to check the electrical code.

The *squirrel-cage motor* is a constant-speed motor with low starting torque but heavy current demand, low power factor, and high efficiency. Therefore, this type of motor is particularly suited where the starting load is large. Larger power lines and transformers are needed, however, with resultant greater power use and operating cost.

The *wound-rotor motor* is similar to the squirrel-cage motor. The starting torque can be varied from about one-third to three times that of normal, and the speed can be controlled. The cost of a wound-rotor motor is greater than a squirrel-cage motor, but where the pumping head varies, power saving over a long-range period will probably compensate for the greater first cost. Larger transformers and power lines are needed.

The *synchronous motor* runs at the same frequency as the generator furnishing the power. A synchronous motor is a constant-speed motor even under varying loads, but it needs an exciting generator to start the electric motor. Synchronous motors usually are greater in size than 75 to 100 hp.

An electric motor starting switch is either manually or magnetically operated. Manually operated starters for small motors (less than 1 hp) throw in the full voltage at one time. Overload protection is provided, but undervoltage protection is not. Full-voltage magnetic starters are used on most jobs. Overload and under-voltage control to stop the motor is generally included. Clean starter controls and proper switch heater strips are necessary. Sometimes a reduced voltage starter must be used when the power company cannot permit a full voltage starter or when the power line is too long. A voltage increase or decrease of more than 10 percent may cause heating of the equipment and winding and fire.

Lightning protection should be provided for all motors. Electric motors can be expected to have efficiencies of about 84 percent for motors under 7.5 hp to about 92 percent for motors of 60 hp or larger. Overall pump and motor efficiency of 65 percent can be achieved.[191] It is important to check with manufacturer and the National Electrical Manufacturers Association (NEMA).

Automatic Pump Control

One of the most common automatic methods of starting and stopping the operation of a pump on a hydropneumatic system is the use of a pressure switch. This switch is particularly adaptable for pumps driven by electric motors, although it can also be used to break the ignition circuit on a gasoline-engine-driven pump. The switch consists of a diaphragm connected on one side with the pump discharge line and on the other side with a spring-loaded switch. This spring switch makes and breaks the electric contact, thereby operating the motor when the water pressure varies between previously established limits.

Water-level control in a storage tank can be accomplished by means of a simple float switch. Other devices are the float with adjustable contacts and the electronic or resistance probes control and altitude valve. Each has advantages for specific installations.[192]

When the amount of water to be pumped is constant, a time cycle control can be used. The pumping is controlled by a time setting.

In some installations, the pumps are located at some distance from the treatment plant or central control building. Remote supervision can be obtained through controls to start or stop a pump and report pressure and flow data and faulty operation.

Another type of automatic pump switch is the pressure flow control. This equipment can be used on ground-level or elevated water storage tanks.

When pumps are located at a considerable distance from a storage tank and pressure controls are used to operate the pumps, heavy drawoffs may cause large fluctuations in pressure along the line. This will cause sporadic pump starting and stopping. In such cases and when there are two or more elevated tanks on a water system, altitude valves should be used at the storage tanks. An altitude valve on the supply line to an elevated tank or standpipe is set to close when the tank is full; it is set to open when the pressure on the entrance side is less than the pressure on the tank side of the valve. In this way, overflowing of the water tank is prevented, even if the float or pressure switch fails to function properly.

Water Hammer

Water hammer is the change in water pressure in a closed conduit (pipe) flowing full due to a very rapid acceleration or cessation of flow, resulting in very large momentary positive and negative pressure changes (surges) from normal.

Causes are pump startup, pump power failure, valve operation, and failure of the surge protection device. Control devices used include vacuum breaker–air relief valves, controlled shutoff valves, flywheel on a pump motor, a surge tank, and a reservoir or standpipe floating on the distribution system. See Figure 2.17. Vacuum breaker–air relief valves are usually located at high points of distribution system pipelines. Pressure-relief valves are usually found in pump stations to control pressure surges and protect the pump station. Air chambers may have a diaphragm to separate the air–water interface to prevent absorption and loss of air in the chamber or an inert gas in place of air. They are used on short pipelines. Each pipeline system should be studied for possible water hammer problems and protected as indicated. Selection of the proper devices requires careful analysis and proper sizing.[193]

Rural Water Conditions in the United States

A national assessment of rural water conditions made between May 1978 and January 1979 of a 2,654 sample of 21,974,000 rural households (places with a population of less than 2,500 and in open country) in the United States is shown in Table 2.13.[194]

Ninety percent of the individual systems were well-water supplies, mostly drilled wells. The remainder relied on driven, bored, jetted, or dug wells as well as springs (275,000), cisterns (133,000), surface water (93,000), or hauled water (269,000). The median rural household system consisted of a 6-gpm pump and a 30-gallon pressure tank with an effective volume of 0.3 gallons. Systems in the west had a larger capacity. Ninety-one percent had piped water and an electric pump. Of the intermediate systems, 90 percent had two or three connections; 88 percent were drilled well supplies. Eighty-eight percent of the community systems had a median of 59 connections and 1.5 miles of distribution system. Ninety percent had groundwater sources and used an average of 36,000 gpd.

TABLE 2.13   Rural Water Service and Water Quality

Water Service	Number of Systems	Water Number of Households[a]	Quality E. coli[b]	% Fecal E. coli
Individual systems	8,765,000	8,765,000	42.1	12.2
Intermediate systems[c]	845,000	2,228,000	43.3	12.2
Community systems[d]	34,000	10,981,000	15.5	4.5

[a]Median of 2.65 persons per household.
[b]More than one per 100 ml.
[c]2 to 14 connections.
[d]15 or more connections.

Source: J. D. Francis et al., *National Assessment of Rural Water Conditions*, EPA 570/9-84-004, U.S. Environmental Protection Agency, Office of Drinking Water, Washington, DC, June 1984.

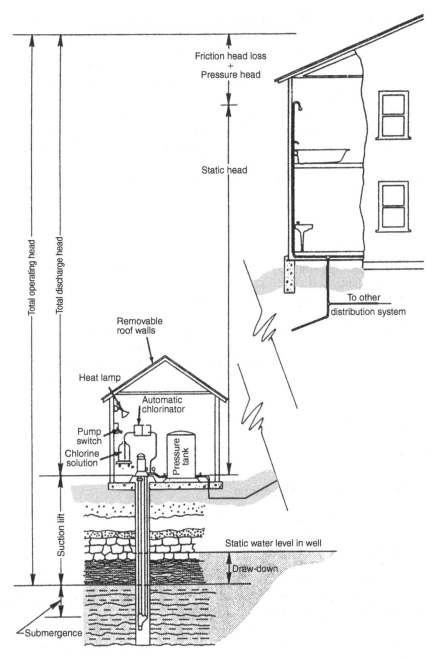

FIGURE 2.19 Components of total operating head in well pump installations. (*Source: Manual of Individual Water Systems*, EPA-570/9-82-004, U.S. EPA, Office of Drinking Water, Washington, DC, October 1982, p. 102.)

Consolidated systems had a median of 153 connections. The median for average daily use was 43,000 gpd.

Nationally, 28.9 percent of all rural household water supplies had coliform concentrations exceeding the standard of 1/100 ml. Individual and intermediate systems were more often contaminated than community systems. Dug, driven, and jetted wells and springs were more likely to be contaminated. However, even the rural community systems showed significantly higher levels of coliform contamination than the larger public water systems. This points out sharply an unresolved problem and the need for greater attention to rural water supplies, including the 34,000 rural community systems. The need for improved well and spring location, construction, and protection and competent intermediate and community system operation is apparent.

Design of a Household Water System

Major considerations in the design of a well-water system for a private dwelling are a dependable well yield and a well pump of adequate capacity and operating head. Chapter 1, Figure 1.4 shows a well log and well yield testing. Figure 2.10 shows the probable range of the maximum momentary water demand for one or more dwelling units. See also Figure 2.11 for fixture unit basis for demand load in gallons per minute and Figure 2.14 for pressure tank and pump size. Figure 2.19 shows the components that make up the total well pump operating head. Table 2.11 gives recommended pump capacities and supplements text suggestions.

EXAMPLES

Design of Small Water Systems

Small water systems, serving less than 10,000 households, supplied about 42 million people in the United States (1980 statistics). Many of these systems are marginally designed and poorly operated and maintained due to insufficient budgets, very low water rates, poorly paid and trained operators, and uninformed management. Such systems are frequently inadequately monitored and fail to meet drinking water standards. Some are too small to provide sufficient revenue to support proper operation, maintenance, and management. Very often, small water systems are the only alternative for small isolated communities and developments. A partial answer, where feasible, is the consolidation of small water systems or connection to a large municipal system. Other alternatives include regional management of several small systems, including professional supervision, administration, and technical and financial assistance. Rural water associations, local water works associations, and regulatory agencies can, and in many areas do, provide training programs, seminars, and speakers to meet some of the needs. Compliance with drinking water standards, operational problems, and maintenance can be discussed. The opportunity to share experiences is provided and made accessible to the small water system operator.

Experiences in new subdivisions show that peak water demands of 6 to 10 times the average daily consumption rate are not unusual. Lawn-sprinkling demand has made necessary sprinkling controls, metering, or the installation of larger distribution and storage facilities and, in some instances, ground storage and booster stations. As previously stated, every effort should be made to serve a subdivision from an existing public water supply. Such supplies can afford to employ competent personnel and are in the business of supplying water, whereas a subdivider is basically in the business of developing land and does not wish to become involved in operating a public utility.

In general, when it is necessary to develop a central water system to serve the average subdivision, consideration should first be given to a drilled well-water supply. Infiltration galleries or special shallow wells may also be practical sources of water if their supply is adequate and protected. Such water systems usually require a minimum of supervision and can be developed to produce a known quantity of water of a satisfactory sanitary quality. Simple chlorination treatment will normally provide the desired factor of safety. Test wells and sampling will indicate the most probable dependable yield and the chemical and bacterial quality of the water. Well logs should be kept in duplicate.

Where a clean, clear lake supply or stream is available, chlorination and slow sand filtration can provide reliable treatment with daily supervision for the small development. The turbidity of the water to be treated should not exceed 30 NTU. Preliminary settling may be indicated in some cases.

Other more elaborate types of treatment plants, such as rapid sand filters, are not recommended for small water systems unless specially trained operating personnel can be assured. Pressure filters have limitations, as explained earlier in this chapter.

The design of small slow sand filter and well-water systems is explained and illustrated earlier in Figures 2.2 and 2.3 and shown in Figure 2.22.

An example (Figure 2.20) will serve to illustrate the design bases previously discussed. The design population at a development consisting of 100 two-bedroom dwellings, at two persons per bedroom, is 400. The average water use at 75 gallons per person or 150 gallons per bedroom is 30,000 gpd for the development. From Figure 2.10, the peak demand can vary from 100 to 320 gpm. An average conservative maximum or peak demand would be 210 gpm. Adjustment should be made for local conditions. This design provides no fire protection.

Examples showing calculations to determine pipe diameters, pumping head, pump capacity, and motor size follow.

In one instance, assume that water is pumped from a lake at an elevation of 658 feet to a slow sand filter and reservoir at an elevation of 922 ft. See Figure 2.20. The pump house is at an elevation of 665 feet and the intake is 125 feet long. The reservoir is 2,000 feet from the pump. All water is automatically

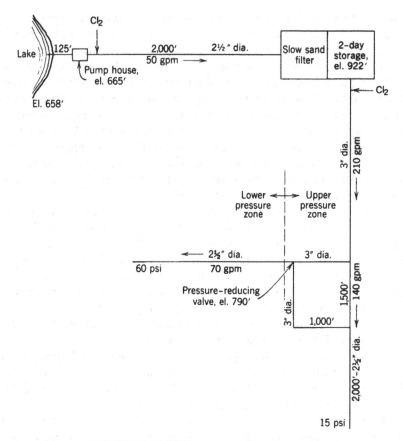

FIGURE 2.20   Water system flow diagram.

chlorinated as it is pumped. The average water consumption is 30,000 gpd. With the reservoir at an elevation of 922 feet, a pressure of at least $15 \, lb/in.^2$ is to be provided at the highest fixture. Find the size of the intake and discharge pipes, the total pumping head, the size pump, and motor. The longest known power failure is 14 hours and repairs can be made locally. Assume that the pump capacity is sufficient to pump 30,000 gallons in 10 hours, or 50 gpm. Provide one 50-gpm pump and one 30-gpm standby, both multistage centrifugal pumps, one to operate at any one time and one generator.

From the above, with a flow of 50 gpm, a 2-inch pipeline to the storage tank is indicated.

The head losses, using Tables 2.14 and Table 2.15, are as follows:

Intake, 125 ft of 2-in. pipe$(3.3 \times 1.25) = 4.1$ ft

TABLE 2.14 Friction Due to Water Flowing in Pipe

Capacity (gpm)	Friction Head Loss by Pipe Diameter[a]														
	1/2	3/4	1	1-1/4	1-1/2	2	2-1/2	3	4	5	6	8	10	12	14
1	2.1														
2	7.4	1.9													
3	15.8	4.1	1.3												
4	27.0	7.0	2.1	0.57											
5	41.0	10.5	3.2	0.84	0.40										
8	98.0	25.0	7.8	2.0	0.95										
10		38.0	11.7	3.0	1.4	0.50									
15		80.0	25.0	6.5	3.1	1.1									
20		136.0	42.0	11.1	5.2	1.8	0.61								
25			64.0	16.6	7.8	2.7	0.95	0.40							
30			89.0	23.5	11.0	3.8	1.3	0.54							
35			119.0	31.2	14.7	5.1	1.7	0.75							
40			152.0	40.0	18.8	6.6	2.2	0.91							
50				60.0	28.4	9.9	3.3	1.4							
60				85.0	39.6	13.9	4.6	1.9	0.47						
70				113.0	53.0	18.4	6.2	2.6	0.63						
80					68.0	23.7	7.9	3.3	0.81						
90					84.0	29.4	9.8	4.1	1.0						
100					102.0	35.8	12.0	5.0	1.2	0.41					
125						54.0	18.2	7.6	1.9	0.64					
150						76.0	26.0	10.5	2.6	0.87					
175						102.0	33.8	14.0	3.4	1.2					
200						129.0	43.1	17.8	4.4	1.5	0.62				
225							54.0	22.0	5.3	1.8	0.72				

(continues)

TABLE 2.14 (continued)

| | Friction Head Loss by Pipe Diameter[a] | | | | | | | | | | | | | | |
Capacity (gpm)	1	3/4	1	1-1/4	1-1/2	2	2-1/2	3	4	5	6	8	10	12	14
250							65.0	27.1	6.7	2.2	0.92				
300							92.0	38.0	9.3	3.1	1.3				
400								65.0	16.0	5.4	2.2				
500								98.0	24.0	8.1	3.3	0.83			
600									33.8	11.7	4.7	1.2			
700									45.0	15.2	6.2	1.5	0.52		
800									57.6	19.4	8.0	2.0	0.67		
900									71.6	24.2	10.0	2.5	0.83		
1000									87.0	29.4	12.1	3.0	1.0	0.42	
1500										62.2	25.6	6.3	2.1	0.88	0.42
2000											43.6	10.8	3.6	1.5	0.71
3000												22.8	7.7	3.2	1.5
4000													13.1	5.4	2.6
5000													19.8	8.2	3.8

[a] In inches.

TABLE 2.15 Friction of Water in Fittings

Pipe Fitting	Friction Head Loss as Equivalent Number of Feet of Straight Pipe by Pipe Size (in.) (nominal diameter)											
	1/2	3/4	1	1-1/4	1-1/2	2	2-1/2	3	3-1/2	4	5	6
Open gate valve	0.4	0.5	0.6	0.8	0.9	1.2	1.4	1.7	2.0	2.3	2.8	3.5
Three-quarters closed gate valve	40.0	60.0	70.0	100.0	120.0	150.0	170.0	210.0	250.0	280.0	350.0	420.0
Open globe valve	19.0	23.0	29.0	38.0	45.0	58.0	70.0	85.0	112.0	120.0	140.0	170.0
Open angle valve	8.4	12.0	14.0	18.0	22.0	28.0	35.0	42.0	50.0	58.0	70.0	85.0
Standard elbow or through reducing tee	1.7	2.2	2.7	3.5	4.3	5.3	6.3	8.0	9.3	11.0	13.0	16.0
Standard tee	3.4	4.5	5.8	7.8	9.2	12.0	14.0	17.0	19.0	22.0	27.0	33.0
Open swing check	4.3	5.3	6.8	8.9	10.4	13.4	15.9	19.8	24.0	26.0	33.0	39.0
Long elbow or through tee	1.1	1.4	1.7	2.3	2.7	3.5	4.2	5.1	6.0	7.0	8.5	11.0
Elbow 45°	0.75	1.0	1.3	1.6	2.0	2.5	3.0	3.8	4.4	5.0	6.1	7.5
Ordinary entrance	0.9	1.2	1.5	2.0	2.4	3.0	3.7	4.5	5.3	6.0	7.5	9.0

Note: The frictional resistance to flow offered by a meter will vary between that offered by an open angle valve and globe valve of the same size. See manufacturers for meter and check valve friction losses; also Table 3-32. See manufacturer for head loss in butterfly, rotary, and special valves.

For this calculation, assume the entrance loss and loss through pump are negligible. Say that you have the following:

Four long elbows = 16.8 ft
Two globe valves = 140 ft
One check valve = 16 ft
Three standard elbows = 18.9 ft
Two 45° elbows = 6 ft
Total equivalent pipe = 198 ft of 2-1/2-in. pipe; head loss = 3.3 × 1.98:

Discharge pipe, 2,000 ft of 2-1/2-in. = 3.3 × 20 = 66.0 ft

Total friction head loss = 76.6 ft

Suction lift = 7 ft to center of pump = 7.0 ft

Static head, difference in elevation = 922−655 = 257.0 ft

No pressure at point of discharge; and total head = 340.6 ft

Add for head loss through meters if used (Table 2.16).

If a 3-inch intake and discharge line is used instead of a 2-1/2-inch intake, the total head can be reduced to about 300 ft. The saving thus effected in power consumption would have to be compared with the increased cost of 3-in. pipe over 2-1/2-in. pipe. The additional cost of power would be approximately $76.65

TABLE 2.16 Head Loss Through Meters

Flow (gpm)	Head Loss through Meter by Meter Size[a] (psi)								
	5/8	3/4	1	1-1/4	1-1/2	2	3	4	6
4	1								
6	2								
8	4	1							
10	6	2	1						
15	14	5	2	2					
20	25	9	4	3	1				
30		20	8	7	2	1			
40			15	12	4	2			
50			23	18	6	3			
75					14	5	1		
100					25	10	3	1	
200							10	4	1
300							24	9	2
400								16	4
500								25	6

[a] In inches.

Note: Flows of less than 1/4 gpm are not usually registered by domestic meters.

Source: Adapted from G. Roden, "Sizing and Installation of Service Pipes," *J. Am. Water Works Assoc.* **38**, 5 (May 1946). Copyright 1946 by the American Water Works Association. Reprinted with permission.

per year, with the unit cost of power at $0.02/kW-hr. This calculation is shown below using approximate efficiencies.*

For a 340-foot head,

$$\text{Horsepower to motor} = \frac{\text{gpm X total head in ft}}{3960 \times \text{pump efficiency} \times \text{motor efficiency}}$$

$$= \frac{50 \times 340}{3960 \times 0.45 \times 0.83} = 11.5$$

$$11.5\,\text{hp} = 11.5 \times 0.746\,\text{kW} = 8.6\,\text{kW}^\dagger$$

In 1 hr, at 50 gpm, 50 × 60, or 3,000 gallons, will be pumped and the power used will be 8.6 kW-hr. To pump 30,000 gallons of water will require 8.6 × (30,000/3,000) = 86 kW-hr.

If the cost of power is $0.02/kW-hr, the cost of pumping 30,000 gallons of water per day will be 86 × 0.02 = 1.72, or $1.72.

For a 300-ft head,

$$\text{Horsepower} = \frac{50 \times 300}{3960 \times 0.45 \times 0.83} = 110.1\,\text{hp}$$

$$= 10.1 \times 0.746\,\text{kW} = 7.55\,\text{kW}$$

In 1 hour, 3,000 gallons will be pumped as before, but the power used will be 7.55 kW-hr. To pump 30,000 gallons will require 7.55 × 10 = 75.5 kW-hr.

If the cost of power is $0.02 per kW-hr, the cost of pumping 30,000 gal will be 75.5 × 0.02 = 1.51, or $1.51.

The additional power cost due to using 2-inch pipe is 1.72 –1.51 = $0.21 per day, or $76.65 per year.

At 4 percent interest (i), compounded annually for 25 years (n), $76.65 ($D$) set aside each year would equal about $3,200 ($S$):

$$D = \frac{i}{(1+i)^n - 1} \times S \quad 76.65 = \frac{0.04}{(1+0.04)^{25} - 1}$$

$$S = \frac{76.65}{0.024} = \$3194, \text{ say } \$3200$$

This assumes that the life of the pipe used is 25 years and the value of money 4 percent. If the extra cost of 3-inch pipe over 2-1/2-in. pipe plus interest on the difference minus the saving due to purchasing a smaller motor and lower head pump is less than $1,200 (present worth of $3,200), then 3-inch pipe should be used.

---

*Adjust costs and interest rates to current conditions.
†Check pump and motor efficiencies with manufacturers.

The size of electric motor to provide for the 50-gpm pump against a total head of 340 feet was shown to be 11.5 hp. Since this is a nonstandard size, the next larger size, a 15-hp motor, will be provided. If a smaller motor is used, it might be overloaded when pumping head is decreased.

The size of electric motor to provide for the 30-gpm pump is

$$\text{Horsepower to motor drive} = \frac{\text{gpm} \times \text{total head in fit}}{3960 \times \text{pump efficiency} \times \text{motor efficiency}}$$

But the total head loss through 2-inch pipe when pumping 30 gpm would be as follows:

Intake, 125 ft of 2-1/2-in. pipe   125
Fittings, total equivalent pipe   198
Discharge pipe, 2,000 ft of 2-1/2-in.   2000

Total friction head loss $2323 \times \dfrac{1.3}{100} = 32\,\text{ft}$

Suction lift = 9
Static head = 255
Total head = -296 ft
Horsepower to motor drive

$$\frac{30 \times 296}{3960 \times 0.35 \times 0.85} = 7.55 \,(\text{use } 7\tfrac{1}{2}\text{-hp electric motor})$$

Because of the great difference in elevation (658 to 922 feet), it is necessary to divide the distribution system into two zones so that the maximum pressure in pipes and at fixtures will not be excessive. In this problem, all water is supplied the distribution system at elevation 922 feet. A suitable dividing point would be at elevation 790 feet. All dwellings above this point would have water pressure directly from the reservoir, and all below would be served through a pressure-reducing valve to provide not less than $15\,\text{lb/in.}^2$ at the highest fixture or more than $60\,\text{lb/in.}^2$ at the lowest fixture. If two-thirds of the dwellings are in the upper zone and one-third is in the lower zone, it can be assumed that the peak or maximum hourly demand rate of flow will be similarly divided (Figure 2.20).

Assume the total maximum hourly or peak demand rate of flow for an average daily water consumption of 30,000 gpd to be 210 gpm. Therefore, 70 + 140 gpm can be taken to flow to the upper zone and 70 gpm to the lower zone. If a 3-inch pipe is used for the upper zone and water is uniformly drawn off, the head loss at a flow of 210 gpm would be about 0.33 × 20 feet per 100 feet of pipe. And if 2-1/2-inch pipe is used for the lower zone and water is uniformly drawn off in its length, the head loss at a flow of 70 gpm would be about 0.33 × 6.2 feet per 100 feet of pipe. If the pipe in either zone is connected to form a loop, thereby eliminating dead ends, the frictional head loss would be further reduced to one-fourth of that with a dead end for the portion forming a loop. Check all head losses.

In all of these considerations, actual pump and motor efficiencies obtained from and guaranteed by the manufacturer should be used whenever possible. Their recommendations and installation detail drawings to meet definite requirements should be requested and followed if it is desired to fix performance responsibility.

In another instance, assume that all water is pumped from a deep well through a pressure tank to a distribution system. See Figure 2.21. The lowest pumping water level in the well is at elevation 160, the pump and tank are at elevation 200, and the highest dwelling is at elevation 350. Find the size pump, motor and pressure storage tank, operating pressures, required well yield, and size mains to supply a development consisting of 100 two-bedroom dwellings using an average of 30,000 gpd.

Use a deep-well turbine pump. The total pumping head will consist of the sum of the total lift plus the friction loss in the well drop pipe and connection to the pressure tank plus the friction loss through the pump and pipe fittings plus the maximum pressure maintained in the pressure tank. The maximum pressure in the tank is equal to the friction loss in the distribution system plus the static head caused by the difference in elevation between the pump and the highest plumbing fixture plus the friction loss in the house water system, including meter if provided, plus the residual head required at the highest fixture.

With the average water consumption at 30,000 gpd, the maximum hourly or peak demand was found to be 210 gpm. The recommended pump capacity is taken as 125 percent of the maximum hourly rate, which would be 262 gpm. This assumes that the well can yield 262 gpm, which frequently is not the case. Under such circumstances, the volume of the storage tank can be increased two or three times, and the size of the pump correspondingly decreased to one-half or one-third the original size to come within the well yield. Another alternative would be to pump water out of the well, at a rate equal to the safe average yield of the well, into a large ground-level storage tank from which water can be pumped through a pressure tank at a higher rate to meet maximum water demands. This would involve double pumping and, hence, increased cost. Another arrangement, where possible, would be to pump out of the well directly into the distribution system, which is connected to an elevated storage tank. Although it may not be

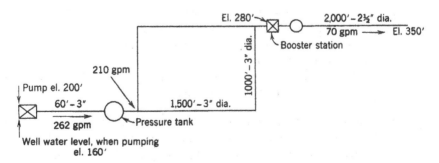

FIGURE 2.21 A water system flow diagram with booster station.

economical to use a pressure-tank water system, it would be of interest to see just what this would mean.

The total pumping head would be

$$\text{Lift from elevation } 160 - 200 \, \text{ft} = 40 \, \text{ft} \quad 40 \, \text{ft}$$

Figure 2.21 shows a distribution system that forms a rectangle $1,000 \times 1,500$ feet with a 2,000-foot dead-end line serving one-third of the dwellings taking off at a point diagonally opposite the feed main. The head loss in a line connected at both ends is approximately one-fourth that in a dead-end line. The head loss in one-half the rectangular loop, from which water is uniformly drawn off, is one-third the loss in a line without drawoffs. Therefore, the total head loss in a 3-inch pipeline with a flow of about 210 gpm is equal to

$$\frac{1}{4} \times \frac{1}{3} \times \frac{20}{100} \times 2500 = 42 \, \text{ft}$$

and the head loss through a 2,000-foot dead-end line, with water being uniformly drawn off, assuming a flow of 70 gpm through 2-1/2-inch pipe is equal to

$$\frac{1}{3} \times \frac{6.2}{100} \times 2000 = 41 \, \text{ft}$$

This would make a total of 42 + 41, or 83, ft. $= 83 \, \text{ft}$

(For a more accurate computation of the head loss in a water distribution grid system by the equivalent pipe, Hardy Cross, or similar method, the reader is referred to standard hydraulic texts. However, the assumptions made here are believed sufficiently accurate for our purpose.)

The static head between pump and the curb of the highest dwelling plus the highest fixture is $(350 -200) + 12 = 162 \, \text{ft}$.	162 ft
The friction head loss in the house plumbing system (without a meter) is equal to approximately 20 ft.	20ft
The residual head at highest fixture is approximately 20 ft.	20 ft
The friction loss in the well drop pipe and connections to the pressure tank and distribution system with a flow of 262 gpm in a total equivalent length of 100 ft of 3-in. pipe is 30 ft.	30 ft
The head loss through the pump and fittings is assumed negligible. Total pumping head	355 = 147 psi

Because of the high pumping head and so as not to have excessive pressures in dwellings at low elevations, it will be necessary to divide the distribution system into two parts, with a booster pump and pressure storage tank serving the upper half.

If the booster pump and storage tank are placed at the beginning of the 2,000 feet of 2-1/2-inch line, at elevation 280 feet, only one-third of the dwellings need be served from this point. The total pumping head here would be as follows.

Friction loss in 2,000 feet of 2-1/2-inch pipe with water withdrawn uniformly 41 feet along its length and a flow of 70 gpm is

$$\frac{1}{3} = 210 \times \frac{6.2}{100} = 41 \text{ ft}$$

The static head between the booster pump and the curb of the highest dwelling plus the highest fixture is

$(350 - 280) + 12 = 82$ ft	82 ft
The head loss in the house plumbing is 20 ft	20 ft
The residual pressure at highest fixture is 20 ft	20 ft
Booster pumping station total head	163 ft
	$= 71$ psi

The total pumping head at the main pumping station at the well would be as follows:

Lift in well is 40 ft	40 ft
Friction loss in distribution system forming loop* is 42 ft	42 ft
The static head between the pump and the booster station, which is also adequate to maintain a 20-ft head at the highest fixture, is $280 - 200 = 80$ ft	80 ft
Friction loss in well-drop pipe and connections to the distribution system is 30 ft	30 ft
Main pumping station, total head	192 ft
	$= 84$ psi

With an average daily water consumption of 30,000 gallons, the average daily maximum demand, on a monthly basis, would be $30,000 \times 1.5 = 45,000$ gallons. The ratio of the absolute maximum and minimum operating pressures at the main pumping station, using a 10-pound differential, would be

$$\frac{84 + 14.7}{94 + 14.7} = \frac{98.7}{108.7} = 0.908, \text{ say } 0.90$$

From Figure 2.14 the pressure tank volume should be (28,500 gal) about 30,000 gallons if 15 minutes of storage is to be provided at the maximum demand rate, 10,000 gallons if 5 minutes storage is acceptable, or 2,000 gallons if 1 minutes storage is acceptable, with a standby pump and well of adequate

---

*At 210 gpm, peak flow $(\frac{1}{4} \times \frac{1}{3} \times \frac{20}{100} \times 2500) = 42$ ft.

capacity. When the average monthly maximum water consumption exceeds that in Figure 2.14, multiply the vertical *and* horizontal axis by 5 or 10 (or other suitable factor) to bring the reading within the desired range. The pump capacity, as previously determined, should be 262 gpm. Use a 260-gpm pump. If a 20-lb pressure differential is used, $P_1/P_2 = 0.83$, and Figure 2.14 indicates an 18,000-gal pressure tank could be used to provide 15 minutes storage at the probable maximum hourly demand rate of flow, or 6,000 gallons for 5 minutes storage.

The booster pumping station serve one-third of the population; hence, the average daily maximum demand on a monthly basis would be 1/3 (45,000), or 15,000 gallons. The ratio of the absolute maximum and minimum operating pressures at the booster pumping station using a 10-lb differential would be

$$\frac{71 + 14.7}{81 + 14.7} = \frac{85.7}{95.7} = 0.90$$

From Figure 2.14, the pressure tank should have a volume of about 10,000 gallons to provide 15 minutes storage at times of peak demand. The pump capacity should be 78 gpm. Use a 75-gpm pump. By contrast, if the operating pressure differential is 20 pounds and only 5 minutes storage at peak demand is desired, the required pressure tank volume would be 1,600 gallons.

To determine the required size of motor for the main pumping station and booster pumping stations, use the average of the maximum and minimum operating gauge pressures as the pumping head. The size motor for the main pumping station using manufacturer's pump and motor efficiencies is:

$$\frac{262 \text{ gpm} \times (192 + 11\frac{1}{2}) \text{ ft avg. had}}{3960 \times 0.57 \times 0.85} = 27.8$$

Use a 30-hp motor. The size motor for the booster station is

$$\frac{70 \text{ gpm} \times (163 + 11\frac{1}{2}) \text{ ft avg. head}}{3960 \times 0.50 \times 0.80} = 7.7$$

Use a 7-1/2- or 10-hp motor.

In the construction of a pumping hydropneumatic station, provision should be made for standby pump and motive power equipment.

The calculations are based on the use of a multistage centrifugal-type pump. Before a final decision is made, the comparison should include the relative merits and cost using a displacement-type pump. Remember that price and efficiency, although important when selecting a pump, are not the only factors to consider. The requirements of the water system and peculiarities should be anticipated and a pump with the desirable characteristics selected.

Design of a Camp Water System

A typical hydraulic analysis and design of a camp water system is shown in Figure 2.22.

Water System Cost Estimates

Because of the wide variations in types of water systems and conditions under which they are constructed, it is impractical to give reliable cost estimates. Some approximations are listed to provide insight into the costs involved. Adjust costs use Engineering News Record (ENR) or other appropriate construction cost index:

1. The approximate costs (1990) of water pipes, valves, and hydrants, including labor and material but not including engineering, legal, land, and administrative costs are as follows*:

3/4-in. copper pipe, per ft	$ 10
1-in. copper pipe, per ft	12
1-1/4-in. copper pipe, per ft	15
1-1/2-in. copper pipe, per ft	19
3/4-in. service taps and curb boxes	150
6-in. ductile iron pipe, per ft	$15–20
8-in. ductile iron pipe, per ft	18–23
12-in. ductile iron pipe, per ft	25–30
16-in. ductile iron pipe, per ft	32–35
6-in. ABS or PVC pipe, per ft	$15–17
8-in. ABS or PVC pipe, per ft	18–20
10-in. ABS or PVC pipe, per ft	24–28
12-in. ABS or PVC pipe, per ft	32–40
6-in. double-gate valve and box	$300–400
8-in. double-gate valve and box	500–600
12-in. butterfly valve and box	600–800
6-in. hydrant assembly including valve, and tee on main	2,300

2. Elevated storage, small capacity—20,000-gal capacity, $50,000 to $56,000; 50,000 gal, $75,000 to $134,000; 100,000 gal, $124,000 to $200,000. Ground-level storage—41,000 gal, $45,000; 50,000 gal, $54,000; 72,000 gal, $57,000; 92,000 gal, $63,000 (1990 cost).[195] For larger installations, standpipe costs may run $90,000 for capacity of 0.15×10^6 gal; $210,000 for 0.5×10^6 gal; $350,000 for 1.0×10^6 gal; and $750,000 for 3.0×10^6 gal. For elevated tanks, cost may run

---

*The assistance of Kestner Engineers, P. C., Troy, NY, is gratefully acknowledged in arriving at the cost estimates.

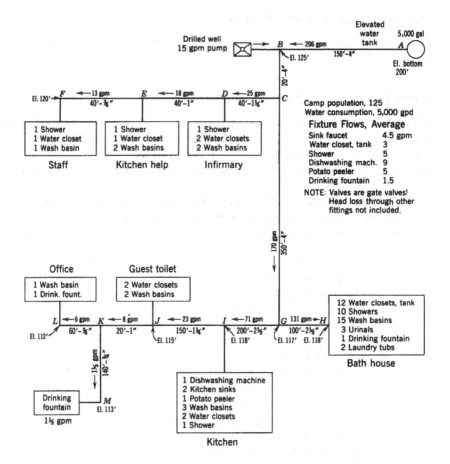

Distance			Gpm Flow			Head Available (Ft)				Pipe	Head Loss		Head	
From	To	Ft	Max.	%	Prob-able	Initial	+ Fall	– Rise	Total	Size (in.)	Ft per 100 ft	Total	(Ft Remain-ing)	Facility Served
A	C	170	295	70	206	0	75	0	75	4	4.5	7.6	67.4	
C	D	40	50	50	25	67.4	1.6	0	69	1¼	16.6	6.6	62.4	Inf., kitch., staff
D	E	40	30	60	18	62.4	1.0	0	63.4	1	36	14.4	49.0	Kitch., staff
E	F	40	13	90	12	49.0	1.6	0	50.6	¾	52	21	29.6	Staff cabin
C	G	350	245	70	170	67.4	8.0	0	75.4	4	3.4	11.9	63.5	
G	H	100	174	75	131	63.5	0	1	62.5	2½	20	20	42.5	Bath house
G	I	200	71	100	71	63.5	0	1	62.5	2½	6	12	50.5	Kitch. guest, off.
I	J	150	23	100	23	50.5	3.0	0	53.5	1¼	15	22	31.5	
J	K	20	8	100	8	31.5	0	0	31.5	1	7.8	1.6	29.9	
K	L	60	6	100	6	29.9	3.0	0	32.9	¾	12	7.2	25.7	Office
K	M	140	1½	100	1½	25.7	2.0	0	27.7	½	1.0	1.4	26.3	Drink. fount.

FIGURE 2.22 Typical hydraulic analysis of camp water system.

$180,000 for 0.15×10^6 gal; $460,000 for 0.5×10^6 gal; $815,000 for 1.00×10^6 gal; and $1,900,000 for 3.0×10^6 gal (1990 adjusted cost).

3. A complete conventional rapid sand filter plant including roads, landscaping, lagoons, laboratory, and low-lift pumps may cost $450,000 for a 0.3-mgd plant; $660,000 for a 0.5-mgd plant; $1,120,000 for a 1.0-mgd plant; $2,500,000 for a 3.0-mgd plant; $3,700,000 for a 5.0-mgd plant; $6,000,000 for a 10.0-mgd plant; and $10,300,000 for a 20.0-mgd plant (1990 adjusted cost).

4. The annual cost of water treatment plants (at 7 percent, 20 years) has been estimated at $63,000, $126,000, and $188,000 for 70-, 350-, and 700-gpm complete treatment package plants; $600,000 for 5-mgd plant; $240,000 and $728,000 for 1- and 10-mgd direct filtration plants; and $376,000 and $1,600,000 for a 2- and 20-mgd GAC plants (1990 cost).[196]

5. Iron and manganese removal plant, well supply, 3 mgd $1,700,000, including new well pumps and disinfection equipment, site work and treatment building (1990 adjusted cost).

6. Well construction costs including engineering, legal, and site development have been estimated[197] as follows:

Cost—Adjusted to 1990 ENR Construction Cost Index

Yield, gpm	70	350	500	600	700
Type	Drilled	Gravel Pack	Gravel Pack	Drilled	Gravel Pack
Diameter, in.	10	16–12	18–12	16	16–12
Depth, ft	40	50	80	68	50
Pump	Submersible	Turbine	Turbine	Turbine	Turbine 2 wells
Average Cost	$186,000	$285,600	$276,000	$300,000	$560,000

7. The National Water Well Association reported it costs $3,000 to drill a private domestic well, $12,000 to drill an irrigation well, and $45,000 to drill a municipal or industrial well.[198] The average cost of a 6-inch drilled well is estimated at $7 to $15 per foot plus $7 to $10 per foot for steel casing. A shallow well pump may cost $270 to $450 and a deep well pump $530 to $1,900, plus installation (1990 adjusted cost).

CLEANING AND DISINFECTION

Special precautions must be taken before entering an open or covered well, spring basin, reservoir, storage tank, manhole, pump pit, or excavation to avoid accidents due to lack of oxygen (and excess carbon dioxide) or exposure to hazardous gases such as hydrogen sulfide or methane, which are found in groundwater and underground formations. Hydrogen sulfide, for example, is explosive and very toxic. Methane is flammable and in a confined space displaces oxygen. Open flames and sparks from equipment or electrical connections can cause explosion and hence must be prevented. Wells, tanks, and other confined spaces should be well ventilated before entering. Mechanical ventilation should be on and the atmosphere tested for oxygen and toxic gases *before* entering. In any case, the person entering should use a safety rope and full-body harness, and two strong persons above the ground or the tank should be ready to pull the worker out should dizziness or other weakness be experienced. Self-contained positive-pressure breathing apparatus should be available and used. It is essential to comply with state and federal occupational safety and health requirements. These include, in addition to confined space entry, such matters as hazardous operations and chemical handling, respiratory protection, electrical safety, excavations, and construction safety.

Wells and Springs

Wells or springs that have been altered, repaired, newly constructed, flooded, or accidentally polluted should be thoroughly cleaned and disinfected after all the work is completed. The sidewalls of the pipe or basin, the interior and exterior surfaces of the new or replaced pump cylinder and drop pipe, and the walls and roof above the water line, where a basin is provided, should be scrubbed clean with a stiff-bristled broom or brush and detergent, insofar as possible, and then washed down or thoroughly sprayed with water followed by washing or thorough spraying with a strong chlorine solution. A satisfactory solution for this purpose may be prepared by dissolving 1 ounce of 70 percent high-test calcium hypochlorite made into a paste, 3 ounce of 25 percent chlorinated lime made into a paste, or 1 pint of 5-1/4 percent sodium hypochlorite in 25 gallon of water. The well or spring should be pumped until clear and then be disinfected.

To disinfect the average well or spring basin, mix 2 quarts of 5-1/4 percent "bleach" in 10 gallons of water. Pour the solution into the well; start the pump and open all faucets. When the chlorine odor is noticeable at the faucets, close each faucet and stop the pump. It will be necessary to open the valve or plug in the top of the pressure tank, where provided, just before pumping is stopped in order to permit the strong chlorine solution to come into contact with the entire inside of the tank. Air must be readmitted and the tank opening closed when pumping is again started. Mix one more quart of bleach in 10 gallons of water and pour this chlorine solution into the well or spring. Allow the well to stand idle at least 12 to 24 hours; then pump it out to waste, away from grass and shrubbery, through the storage tank and distribution system, if possible, until the odor of chlorine disappears. Bypass or disconnect the carbon filter if it is part of the system; do not drain into the septic tank. *It is advisable to return the heavily chlorinated water back into the well, between the casing and drop pipe where applicable, during the first 30 minutes of pumping to wash down and disinfect the inside of the casing and the borehole, insofar as possible.* A day or two after the disinfection, *after the well has been pumped out and all the chlorine has dissipated*, a water sample may be collected for bacterial examination to determine whether all contamination has been removed. If the well is not pumped out, chlorine may persist for a week or longer and give a very misleading bacteriological result if a sample is collected and examined. It is not unusual to repeat well disinfection several times, particularly where contaminated water has been used during drilling and where the well has not been adequately surged, cleaned, and pumped out.

A more precise procedure for the disinfection of a well or spring basin is to base the quantity of disinfectant needed on the volume of water in the well or spring. This computation is simplified by making reference to Table 2.17.

Although a flowing well or spring tends to cleanse itself after a period of time, it is advisable nevertheless to clean and disinfect all wells and springs that have had any work done on them before they are used.

Scrub and wash down the spring basin and equipment. Place twice the amount of calcium hypochlorite, swimming pool chlorine erosion tablets, or granular chlorine indicated in Table 2.17 in a weighted plastic container fitted with a cover. Punch holes in the container and fasten a strong line to the container and secure the cover. Suspend the can near the bottom of the well or spring, moving it up and down or around in order to distribute the strong chlorine solution formed throughout the water entering and rising up through the well or spring.

It should be remembered that disinfection is no assurance that the water entering a well or spring will be pollution free. The cause for the pollution, if present, should be ascertained and removed. Until this is done, all water used for drinking and culinary purposes should first be boiled.

TABLE 2.17   Quantity of Disinfectant Required to Give a Dose of 50 mg/l Chlorine

Diameter of Well, Spring, or Pipe (in.)	Gallons of Water per feet of Water Depth	Ounces of Disinfectant/10-ft Depth of Water		
		70% Calcium Hypochlorite[a]	25% Calcium Hypochlorite[b]	5-1/4% Sodium Hypochlorite[c]
2	0.163	0.02	0.04	0.20
4	0.65	0.06	0.17	0.80
6	1.47	0.14	0.39	1.87
8	2.61	0.25	0.70	3.33
10	4.08	0.39	1.09	5.20
12	5.88	0.56	1.57	7.46
24	23.50	2.24	6.27	30.00
36	52.88	5.02	14.10	66.80
48	94.00	9.00	25.20	120.00
60	149.00	14.00	39.20	187.00
72	211.00	20.20	56.50	269.00
96	376.00	35.70	100.00	476.00

[a]$Ca(OCl)_2$, also known as high-test calcium hypochlorite. A heaping teaspoonful of calcium hypochlorite holds approximately 1/2 oz. One liquid ounce = 615 drops = 30 ml.
[b]$CaCl(OCl)$.
[c]$Na(OCl)$, also known as bleach, Clorox, Dazzle, Purex, Javel Water, and Regina, can be purchased at most supermarkets and drugstores.

Pipelines

The disinfection of new or repaired pipelines can be expedited and greatly simplified if special care is exercised in the handling and laying of the pipe during installation. Trenches should be kept dry and a tight-fitting plug provided at the end of the line to keep out foreign matter. Lengths of pipe that have soiled interiors should be cleansed and disinfected before being connected. Each continuous length of main should be separately disinfected with a heavy chlorine dose or other effective disinfecting agent. This can be done by using a portable hypochlorinator, a hand-operated pump, or an inexpensive mechanical electric or gasoline-driven pump throttled down to inject the chlorine solution at the beginning of the section to be disinfected through a hydrant, corporation cock, or other temporary valved connection. Hypochlorite tablets can also be used to disinfect small systems, but water must be introduced very slowly to prevent the tablets being carried to the end of the line.

The first step in disinfecting a main is to shut off all service connections, then flush out the line thoroughly by opening a hydrant or drain valve below the section to be treated until the water runs clear. A velocity of at least 3 fps should be obtained. (Use a hydrant flow gauge.) After the flushing is completed, the valve is partly closed so as to waste water at some known rate. The rate of flow can be estimated with a flow gauge (the formula is in Appendix I) or by running the water into a can, barrel, or other container of known capacity and measuring the time to fill it. With the rate of flow known, determine from Table 2.18 the strength of chlorine solution to be injected into the main at the established rate of 1 pint in 3 minutes to give a dose of 50 mg/l. The rate of water flow can be adjusted and should be kept low for small-diameter pipe. It is a simple matter to approximate the time, in minutes, it would take for the chlorine to reach the open

TABLE 2.18 Hypochlorite Solution to Give a Dose of 50 mg/l Chlorine for Main Sterilization

Rate of Water Flow in Pipeline (gpm)	Quarts of 5-1/4% Sodium Hypochlorite Made Up to 10 gal with Water	Quarts of 14% Sodium Hypochlorite Made Up to 10 gal with Water	Pounds of 25% Chlorinated Lime to 10 gal Water	Pounds of 70% Calcium Hypochlorite to 10 gal Water
5	4.6	1.7	2.0	0.7
10	9.1	3.4	4.0	1.4
15	13.7	5.1	6.0	2.1
20	18.3	6.8	8.0	2.9
25	22.8	8.5	10.0	3.5
40	36.6	13.7	16.0	5.7

[a]*Notes*: Add hypochlorite solution at rate of 1 pt in 3 min. The 10-gal solution will last 4 hr if fed at rate of 1 pt in 3 min. Mix about 50% more solution than is theoretically indicated to allow for waste. A 100-mg/l available chlorine solution is recommended by some agencies.

hydrant or valve at the end of the line being treated by dividing the capacity of the main in gallons by the rate of flow in gallons per minute. In any case, injection of the strong chlorine solution should be continued at the rate indicated until samples of the water at the end of the main show at least 50 mg/l residual chlorine. The hydrant should then be closed, chlorination treatment stopped, and the water system let stand at least 24 hours. At the end of this time the treated water should show the presence of 25 mg/l residual chlorine. If no residual chlorine is found, the operation should be repeated. Following disinfection, the water main should be thoroughly flushed out, to where it will do no harm, with the water to be used and samples collected for bacterial examination for a period of several days. If the laboratory reports the presence of coliform bacteria, the disinfection should be repeated until two consecutive satisfactory results are received. Where poor installation practices have been followed, it may be necessary to repeat the main flushing and disinfection several times. The water should not be used until all evidence of contamination has been removed as demonstrated by the test for coliform bacteria.

If the pipeline being disinfected is known to have been used to carry polluted water, flush the line thoroughly and double the strength of the chlorine solution injected into the mains. Let the heavily chlorinated water stand in the mains at least 48 hours before flushing it out to waste and proceed as explained in the preceding paragraph. Cleansing of heavily contaminated pipe by the use of a nontoxic, biodegradable, nonfoaming detergent and a "pig," followed by flushing and then disinfection, may prove to be the quickest method. Tubercles found in cast-iron pipe in water distribution systems protect microorganisms against the action of residual chlorine.

Where pipe breaks are repaired, flush out the isolated section of pipe thoroughly and dose the section with 200 mg/l chlorine solution and try to keep the line out of service at least 2 to 4 hours before flushing out the section and returning it to service.

Potassium permanganate can also be used as a main disinfectant. The presence and then the absence of the purple color can determine when the disinfectant is applied and then when it has been flushed out.[199] See also AWWA Standard for Disinfecting Water Mains, C651-86.

Storage Reservoirs and Tanks

Make sure the tank is adequately and continuously ventilated before entering. Check with an oxygen deficiency meter. Wear protective clothing during the work, including self-contained breathing apparatus with full-face piece. *Insist on all safety precautions.*

Before disinfecting a reservoir or storage tank, it is essential to first remove from the walls (also bottom and top) all dirt, scale, and other loose material. The interior should then be flushed out (a fire hose is useful) and disinfected by one of the methods explained below.

If it is possible to enter the reservoir or tank, prepare a disinfecting solution by dissolving 1 ounce of 70 percent calcium hypochlorite (e.g., HTH, Perchloron,

Pitt-Chlor) made into a paste, 3 ounces of 25 percent calcium hypochlorite (chlorinated lime) made into a paste, or 1 pint of 5-1/4 percent sodium hypochlorite (e.g., bleach, Clorox, Dazzle,) in 25 gallons of water. Apply this strong 250-mg/l chlorine solution to the bottom, walls, and top of the storage reservoir or tank using pressure-spray equipment. Let stand for at least 2 hours. *Follow safety precautions given in this chapter*. See also AWWA Standard C652.

Another method is to compute the tank capacity. Add to the empty tank 1.25 pounds of 70 percent calcium hypochlorite, 4 pounds of 25 percent chlorinated lime completely dissolved, or 1 gallon of 5-1/4 percent sodium hypochlorite for each 1000-gallon capacity. Fill the tank with water and let it stand for 12 to 24 hours. This will give a 100+-mg/l solution. Then drain the water to waste, where it will do no harm. Dechlorinate if necessary.

A third method involves the use of a chlorinator or hand-operated force pump. Admit water to the storage tank at some known rate and add at the same time twice the chlorine solution indicated in Table 2.18 at a rate of 1 pint in 3 minutes. Let the tank stand full for 24 hours and then drain the chlorinated water to waste. Rinse the force pump immediately after use.

It should be remembered, when disinfecting pressure tanks, that it is necessary to open the air-relief or other valve at the highest point so that the air can be released and the tank completely filled with the heavily chlorinated water. Air should be readmitted before pumping is commenced. In all cases, a residual chlorine test should show a distinct residual in the water drained out of the tanks. If no residual can be demonstrated, the disinfection should be repeated.

Coliform bacteria, klebsiella, and enterobacter have been a problem in redwood water tanks. Klebsiella have been isolated from water samples extracted from redwood, which are apparently leached from the wood (especially new tanks) when the tank is filled with water. Tanks are treated with soda ash to leach out wood tannins (7 days duration) and disinfected with 200 mg/l chlorine water prior to use. A free chlorine residual of 0.2 to 0.4 mg/l in the tank water when in use will keep bacterial counts under control.[200]

EMERGENCY WATER SUPPLY AND TREATMENT

Local or state health departments should be consulted when a water emergency arises. Their sanitary engineers and sanitarians are in a position to render valuable, expert advice based on their experience and specialized training.

The treatment to be given a water used for drinking purposes depends primarily on the extent to which the water is polluted and the type of pollution present. This can be determined by making a sanitary survey of the water source to evaluate the significance of the pollution that is finding its way into the water supply. Bear in mind that all surface waters, such as from ponds, lakes, streams, and brooks, are almost invariably contaminated and hence must be treated. The degree of treatment required is based on the pollution present. However, under emergency conditions it is not practical to wait for the results of microbiological analyses. Be guided by the results of sanitary surveys, diseases endemic and epidemic in the watershed area, and such reliable local data as may be available. Using the best information on hand, select the cleanest and most attractive water available and give it the treatment necessary to render it safe. Prior approval of the regulatory agency is usually required. Water passing through inhabited areas is presumed to be polluted with sewage and industrial wastes. It must be boiled or given complete treatment, including filtration and disinfection, to be considered safe to drink. However, all chemical wastes may not be removed by conventional treatment.

Backpacker-type water filters with hand pump, manufactured by Katadyn[201] and First Need,[202] were found to be 100 percent effective in removing *Giardia* cysts when operated and maintained in accordance with the manufacturer's directions.[203] The Katadyn filter is also reported to remove bacteria and helminths from small quantities of water.

Boiling

In general, boiling clear water vigorously for 1 to 2 minutes will kill most disease-causing bacteria and viruses, including *E. histolytica* and *Giardia* cysts. Heating water to 158°F (70°C) will completely inactivate the *Giardia* cyst.[204] If sterile water is needed, water should be placed in a pressure cooker at 250°F (121°C) for 15 minutes.[205] A pinch of salt or aerating the water from one container to another will improve the taste of the water, but be careful not to recontaminate the water in the process.

Chlorination

Chlorination treatment is a satisfactory method for disinfecting water that is not grossly polluted. It is particularly suitable for the treatment of a relatively clean lake, creek, or well water that is of unknown or questionable quality. Chlorine for use in hand chlorination is available in supermarkets, drugstores, grocery stores, and swimming pool supply stores and can be purchased as a powder, liquid, or tablet. Store solutions in the dark. Chlorine is more effective in water at 68°F (20°C) than at 36°F (2°C) as well as at low pH and turbidity.

The powder is a calcium hypochlorite and the liquid a sodium hypochlorite. Both these materials deteriorate with age. The strength of the chlorine powder or liquid is on the container label and is given as a certain percent available chlorine. The quantity of each compound to prepare a stock solution, or the quantity of stock solution to disinfect 1 gallon or 1,000 gallons of water, is given in Table 2.19. When using the powder, make a paste with a little water, then dissolve the paste in a quart of water. Allow the solution to settle and then use the clear liquid, without shaking. The stock solution loses strength and hence should be made up fresh weekly. It is important to allow the treated water to stand for 30 minutes after the chlorine is added before it is used. Double the chlorine dosage if the water is turbid or colored.

TABLE 2.19 Emergency Disinfection of Small Volumes of Water

Product	Available Chlorine (%)	Stock Solution[a]	Quantity of Stock Solution to Treat 1 gal of Water[b]	Quantity of Stock Solution to Treat 1000 gal of Water[b]
Zonite	1	Use full strength	30 drops	2 qt
S.K., 101 solution	2-1/2	Use full strength	12 drops	1 qt
Clorox, White Sail,	5-1/4	Use full strength	6 drops	1 pt
Dazzle, Rainbow, Rose-X, bleach sodium	10	Use full strength	3 drops	1/2 pt
Hypochlorite sodium	15	Use full strength	2 drops	1/4 pt
Hypochlorite calcium hypochlorite, "bleaching powder," or chlorinated lime	25	6 heaping tablespoonfuls (3 oz) to 1 qt of water	1 teaspoonful or 75 drops	1 qt
Calcium hypochlorite	33	4 heaping tablespoonfuls to 1 qt of water	1 teaspoonful	1 qt
HTH, Perchloron, Pittchlor, calcium	70	2 heaping tablespoonfuls (1 oz)	1 teaspoonful	1 qt

[a] One quart contains 135 ordinary teaspoonfuls of water.
[b] Let stand 30 min before using. To dechlorinate, use sodium thiosulfate in same proportion as chlorine. One jigger = 1-1/2 liquid oz. Chlorine dosage is approximately 5–6 mg/l. (1 liquid oz = 615 drops.) Make sure chlorine solution or powder is fresh; check by making residual chlorine test. Double amount for turbid or colored water.

Chlorine-containing tablets suitable for use on camping, hunting, hiking, and fishing trips are available at most drugstores. The tablets contain 4.6 grains of chlorine; they deteriorate with age. Since chloramines are slow-acting disinfectants, the treated water should be allowed to stand at least 60 minutes before being used. Iodine tablets (Globaline) and halazone tablets are also available at most sporting goods stores and drugstores. Check the expiration date.

Homemade chlorinators may be constructed for continuous emergency treatment of a water supply where a relatively large volume of water is needed. Such units require constant observation and supervision as they are not dependable. Figures 2.23 and 2.24 show several arrangements for adding hypochlorite solution. In some parts of the country, it may be possible to have a commercial hypochlorinator delivered and installed within a very short time. Some health departments have a hypochlorinator available for emergency use. Communicate with the local or state health department for assistance and advice relative to the manufacturers of approved hypochlorinators. Simple erosion-type chlorinators can also be purchased or improvised for very small places. A daily report should

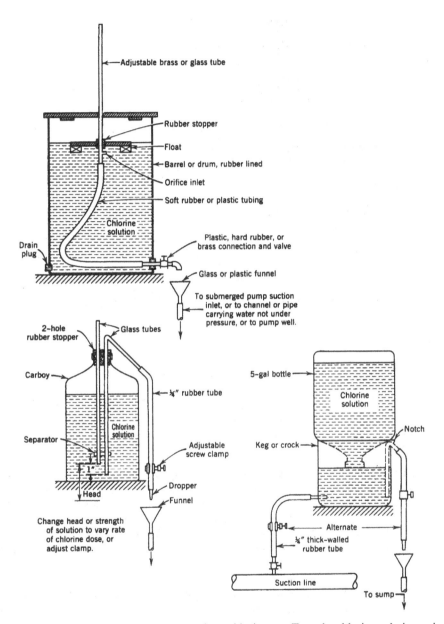

FIGURE 2.23 Homemade emergency hypochlorinators. To make chlorine solution, mix 4 pt of 5 percent hypochlorite to 5 gal of water.

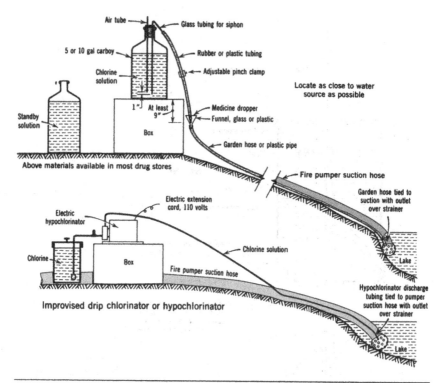

Chlorine	Quantity to 5 gal Water	Rate of Feed 500 gpm Pumper
Perchloron or HTH, 70% available chlorine	4 lb*	4 oz or ¼ pt per min
Chlorinated lime, 24% available chlorine	12 lb*	4 oz or ¼ pt per min
Sodium hypochlorite, 14% available chlorine	20 pt	4 oz or ¼ pt per min

FIGURE 2.24 Emergency chlorination for fire supply, under health department supervision, for pumping into a hydrant on the distribution system, if necessary. Data for preparation and feed of chlorine solution: The asterisk denotes that the paste should be made in a jar. Add water and mix; let settle for a few minutes; then pour into carboy or other container and make up to 5 gal. Discard white deposit; it has no value. Dosage is 5 mg/l chlorine. Double solution strength if necessary to provide residual of 4 to 5 mg/l.

be kept showing the gallons of water treated, the amount of chlorine solution used, and the results of hourly residual chlorine tests.

Iodine

Eight drops of 2 percent tincture of iodine may be used to disinfect 1 quart of clear water (8 mg/l dose). Allow the water to stand at least 30 minutes before it is used. (Bromine can also be used to disinfect water, although its use has been restricted to the disinfection of swimming pool water.) Studies of the usefulness of elemental iodine show it to be a good disinfectant over a pH range of 3 to 8, even in the presence of contamination.[206] Combined amines are not formed to use up the iodine. A dosage of 5 to 10 mg/l, with an average of 8 mg/l for most waters, is effective against enteric bacteria, *Giardia* and amoebic cysts, cercariae, leptospira, and viruses within 30 minutes. Tablets that can treat about 1 quart of water may be obtained from the National Supply Service, Boy Scouts of America, large camping supply centers, drugstores, and the Army, in emergency. These tablets dissolve in less than 1 minute and are stable for extended periods of time. They are known as iodine water purification tablets, of which Globaline, or tetraglycine hydroperiodide, is preferred. They contain 8.0 mg of active iodine per tablet. The treated water is acceptable. Iodine is toxic. It should be reserved for emergency use only.[207]

Filtration in an Emergency

Portable pressure filters are available for the treatment of polluted water. These units can produce acceptable water, provided they are carefully operated by trained personnel. Preparation of the untreated water by settling, prechlorination, coagulation, and sedimentation may be necessary, depending on the type and degree of pollution in the raw water. Pressure filters contain special sand, crushed anthracite coal, or diatomaceous earth. Diatomite filtration, or slow sand filtration, should be used where diseases such as amebic dysentery, giardiasis, ascariasis, schistosomiasis, or paragonimiasis are prevalent, in addition, of course, to chlorination.

Slow sand filters (consisting of barrels or drums) may be improvised in an emergency. Their principles are given in Figure 2.2 and Table 2.2. It is most important to control the rate of filtration so as not to exceed 50 gpd/ft^2 of filter area and to chlorinate each batch of water filtered, as shown in Table 2.19, in order to obtain reliable results.

Bottled, Packaged, and Bulk Water

The bottled water industry has shown a large growth in many parts of the world because of public demand for a more palatable and "pure" water. It is not uncommon to find a wide selection of waters from various sources in the United States and Europe and in supermarkets and small grocery stores in almost all parts of the world. A major bottler in France was reported to have a capacity of 800 million bottles per year. The 1989 production in the United States was estimated at 1,384.4 million gallons per year.[208] There were an estimated 700 water-bottling plants in the United States in 1972.[209] In addition, self-service water vending machines that dispense water into an individual's container are available.* Per-capita consumption increased from 5.2 gallons in 1985 to 6.2 gallons in 1989, with a 1989 sales value of $2.375 billion, compared to $1.5

---

*Water vending maching sanitary design details are given in ref. 211.

billion in 1985 and \$275 million in 1975.[210] The demand for bottled water is of course minimized where a safe, attractive, and palatable public water supply is provided.

In an emergency, it is sometimes possible to obtain bottled, packaged, or tank-truck water from an approved source that is properly handled and distributed. Such water should meet the federal and state drinking water standards as to source, protection, and microbiological, chemical, radiological, and physical quality. Water that is transported in tank trucks from an approved source should be batch chlorinated at the filling point as an added precaution. The tank truck, hoses, fittings, and connections should, of course, be thoroughly cleaned and disinfected (not less than 100 mg/l chlorine solution) before being placed in service. Detergents and steam are sometimes needed, particularly to remove gross pollution, followed by thorough rinsing with potable water and disinfection. Only tank trucks with a dedicated use for hauling potable water should be used. Each tank of water should be dosed with chlorine at the rate of 1 to 2 mg/l and so as to yield a free chlorine residual of not less than 0.5 mg/l. See Table 2.19.

Milk pasteurization plants and beverage bottling plants have much of the basic equipment needed to package water in paper, plastic, or glass containers. Contamination that can be introduced in processing (filtration through sand and carbon filters) and in packaging (pipelines, storage tanks, fillers) should be counteracted by germicidal treatment of the water just prior to bottling. In any case, the source of water, equipment used, and operational practices must meet recognized standards.

Bottled water should meet EPA and state water quality standards for drinking water and comply with FDA regulations and industry standards for the processing and bottling of drinking water.[212] Many states also have detailed regulations or codes including water quality standards. The National Sanitation Foundation also has guidelines and makes plant inspections. The FDA microbiological quality standards are 9 of 10 samples less than 2.2 coliforms per 100 ml, with no sample showing 9.2 or more by the multiple-tube fermentation method, and the arithmetic mean of all samples should be not greater than 1 per 100 ml with not more than one having 4.0 or more coliform organisms by the membrane filter method.[212] The standard plate count of the bottled water at the retail outlet should be less than 500 per 100 per milliliter. The FDA considers bottled water a "food" and regulates its purity.* Mineral water is exempt; its definition is vague. Mineral water has been defined as bottled water containing at least 500 ppm dissolved solids and originating entirely underground. Nevertheless, mineral waters should meet the physical, microbiological, and radiological standards for drinking water. Bottled water should also be free of *Pseudomonas aeruginosa*.

Bottled water may be labeled natural water (no change), natural sparkling water (carbon dioxide added), spring water (groundwater), purified water (demineralized), mineral water (assumed 500 ppm or greater total dissolved solids), seltzer water (carbonated tap water), and club soda (carbonated with salt

---

*Other standards refer to turbidity, color, odor, chemical quality, fluoride, and radiological quality.

and minerals added). Up to 0.02 percent caffeine and 0.5 percent alcohol by weight may be added to natural sparkling water, club soda, seltzer, and soda water, according to the FDA.[214]

REFERENCES

1. M. N. Baker, The Quest for Pure Water, American Water Works Association, New York, 1948.

2. Policy Statement Adopted by the Board of Directors on June 19, 1988, and revised on June 11, 2000, AWWA, Denver, CO, 2001–02 Officers and Committee Directory, p. 238.

3. "Treatment, Coliform Rules Give States Flexibility", *AWWA MainStream* (July 1989): 3.

4. Primary Drinking Water Regulations, pursuant to the Safe Drinking Water Act of 1974 (PL93-523), as amended 1986.

5. *Recommended Standards for Water Works*, A Report of the Committee of the Great Lakes–Upper Mississippi River Board of State Public Health and Environmental Managers, Health Research, Albany, NY, 1987.

6. "Update", *J. Am. Water Works Assoc.* (July 1979): 9.

7. Recommended Standards for Water Works, p. 57.

8. *The Chlorine Manual*, 5th ed., Chlorine Institute, Washington, DC, 1986.

9. "Disinfection—Committee Report", *J. Am. Water Works Assoc.* (April 1978): 219–222.

10. A. D. Eaton, L. S. Clesceri, E. W. Rice and A. E. Greenberg, (Eds.), *Standard Methods for the Examination of Water and Wastewater*, 21st ed., APHA, AWWA, WEF, 2005.

11. W. J. Cooper, P. H. Gibbs, and E. M. Ott, "Equivalent Testing Procedures for Measuring Free Available Chlorine", *J. Am. Water Works Assoc.* (December 1983).

12. G. Gordon, W. J. Cooper, R. G. Rice, and G. E. Pacey, "Methods of Measuring Disinfectant Residuals", *J. Am. Water Works Assoc.* (September 1988): 94–108; J. N. Jensen and J. D. Johnson, "Specificity of the DPD and Amperometric Titration Method for Free Available Chlorine: A Review", *J. Am. Water Works Assoc.* (December 1989): 59–64.

13. C. T. Butterfield, "Bactericidal Properties of Chloramines and Free Chlorine in Water", *Public Health Rep.*, **63**, 934–940 (1948).

14. Committee Report, "Disinfection", *J. Am. Water Works Assoc.* (July 1982): 376–380.

15. A. E. Greenberg and E. Kupka, "Tuberculosis Transmission by Waste Water—A Review", *Sewage Ind. Wastes*, **29**(5), (May 1957): 524–537.

16. S. Kelly and W. W. Sanderson, "Viruses in Sewage", *Health News*, **36**, (June 1959): 14–17.

17. S. G. Lensen et al., "Inactivation of Partially Purified Poliomyelitis Virus in Water by Chlorination, II", *Am. J. Public Health*, **37**(7), (July 1947); 869–874.

18. S. Weidenkopf, *Virology*, 1958.

19. N. A. Clarke and P. W. Kabler, "The Inactivation of Purified Coxsackie Virus in Water", *Am. J. Hygiene*, **59**, (January 1954): 119–127.

20. Weidenkopf.

21. A. F. Bush and J. D. Isherwood, "Virus Removal in Sewage Treatment", *J. Sanit. Eng. Div., ASCE*, **92**(SA 1, Proc. Paper 4653), (February 1966): 99–107.

22. J. E. Malina Jr., *The Effect of Unit Processes of Water and Wastewater Treatment on Virus Removal*, in Borchardt, Cleland, Redman, and Oliver (Eds.), *Virus and Trace Contaminants in Water and Wastewater*, Ann Arbor Science, 1977, pp. 33–52.

23. V. C. Rao, J. M. Symons, A. Ling, P. Wang, T. G. Metcalf, J. C. Hoff, and J. L. Melnick, "Removal of Hepatitis A Virus and Rotavirus by Drinking Water Treatment", *J. Am. Water Works Assoc.* (February 1988): 59–67.

24. Rao et al., andG. G. Robeck et al., "Effectiveness of Water Treatment Processes in Virus Removal", *J. Am. Water Works Assoc.* (October 1962): 1275–1290.

25. C. P. Gerba et al., *"Virus Removal During Conventional Drinking Water Treatment,"* EPA, National Technical Information Service, Springfield, VA, September 1985.

26. *Guidelines for Drinking-Water Quality*, Vol. 1, WHO, Geneva, 1984, p. 28.

27. Malina, pp. 33–52.

28. E. C. Lippy, "Chlorination to Prevent and Control Waterborne Disease", *J. Am. Water Works Assoc.* (January 1986): 49–52.

29. R. M. Clark, E. J. Read, and J. C. Hoff, "Analysis of Inactivation of *Giardia lamblia* by Chlorine", *J. Environ. Eng.* (February 1989): 80–90.

30. A. J. Rubin, J. P. Engel, and O. J. Sproul, "Disinfection of Amoebic Cysts in Water with Free Chlorine", *J. Water Pollut. Control Fed.* (September 1983): 1174–1182.

31. S. D. Lin, *"Giardia lamblia* and Water Supply", *J. Am. Water Works Assoc.* (February 1985): 40–47.

32. Rubin et al.; and Lin.

33. Rubin et al.; W. S. Stone, "The Resistance of *Entamoeba histolytica* cysts to Chlorine in Aqueous Solutions", *Am. J. Trop. Med.*, **17**, 539 (1937); R. Stringer and C. W. Krusé, "Amoebic Cysticidal Properties of Halogens in Water", in *Proceedings of National Specialty Conference on Disinfection*, ASCE, New York, 1970.

34. D. F. Metzler, C. Ritter, and R. L. Culp, "Combined Effect of Water Purification Processes on the Removal of *Histoplasma capsulatum* from Water", *Am. J. Public Health*, **46**(12), (December 1956): 1571–1575.

35. E. L. Jarroll et al., "Effects of Chlorine on *Giardia lamblia* cyst viability", *Appl. Environ. Microbiol.*, **41**, 483 (1981).

36. J. A. Salvato, *Annual Report, Division of Environmental Hygiene*, Rensselaer County Department of Health, Troy, NY, 1956, p. 11; W. W. Sanderson and S. Kelly, in *Advances in Water Pollution Research Proceedings*, Vol. 2, Pergamon, London, 1962, pp. 536–541.

37. Committee on Environmental Quality Management of the Sanitary Engineering Division, "Engineering Evaluation of Virus Hazard in Water", *J. Sanit. Eng. Div.*, ASCE, **96**(SA 1, Proc. Paper 7112) (February 1970): 111–150.

38. G. C. White, "Chlorination and Dechlorination: A Scientific and Practical Approach", *J. Am. Water Works Assoc.* (May 1968): 540–561.

39. Recommended Standards for Water Works, p. 57.

40. *MMWR*, March 15, 1985, p. 143.

41. E. E. Geldreich and E. W. Rice, "Occurrence, Significance, and Detection of *Klebsiella* in Water Systems", *J. Am. Water Works Assoc.* (May 1987): 74–80.

42. J. T. O'Connor, S. K. Banerji, and B. J. Brazos, "Recommended Chemical and Microbiological Analyses for Distribution System Monitoring", *Public Works* (February 1989): 46–48.

43. *Water Quality and Treatment*, 2nd ed., AWWA, New York, 1950, p. 94;   P. K. Knoppert, G. Oskam, and E. G. H. Vreedenburgh, "An Overview of European Water Treatment Practice", *J. Am. Water Works Assoc.*, November 1980, pp. 592–599.

44. A. Amirtharajah, "Variance Analyses and Criteria for Treatment Regulations", *J. Am. Water Works Assoc.* (March 1986): 34–49.

45. G. J. Turre, "Use of Micro-Strainer Unit at Denver";   G. R. Evans, "Discussion", *J. Am. Water Works Assoc.* (March 1959): 354–362.

46. Amirtharajah.

47. C. R. Cox, *Water Supply Control*, Bull. 22, New York State Department of Health, Albany, NY, 1952, pp. 38–53;   C. R. Cox, *Operation and Control of Water Treatment Processes*, WHO, Geneva, 1969, pp. 60, 78.

48. D. R. Lawson, "Polymer Cuts Cost of Rochester Water", *Am. City County*, September 1977, pp. 97–98;   K. E. Carns et al., "Using Polymers with Direct Filtration", *J. Am. Water Works Assoc.* (March 1985): 44–49.

49. R. M. Willis, "Tubular Settlers—A Technical Review", *J. Am. Water Works Assoc.* (June 1978): 331–335.

50. R. D. Letterman and T. R. Cullen Jr., "*Slow Sand Filter Maintenance: Costs and Effects on Water Quality*," EPA, Water Engineering Laboratory, Cincinnati, OH, August 1985.

51. T. M. Riddick, "An Improved Design for Rapid Sand Filter", *J. Am. Water Works Assoc.* (August 1952): 733–744.

52. C. R. Schultz and D. A. Okun, "Treating Surface Waters for Communities in Developing Countries", *J. Am. Water Works Assoc.* (May 1983): 212–219.

53. G. M. Fair and J. C. Geyer, *Water Supply and Waste Water Disposal*, Wiley, New York, 1954, p. 677.

54. R. W. Abbett, *American Civil Engineering Practice*, Vol. II, Wiley, New York, 1956.

55. Amirtharajah.

56. J. T. O'Connor et al., "*Removal of Viruses from Public Water Supplies*," EPA, Cincinnati, OH, August 1982.

57. Amirtharajah.

58. R. L. Culp, "Direct Filtration", *J. Am. Water Works Assoc.* (July 1977): 375–378.

59. G. P. Westerhoff, A. F. Hess, and M. J. Barnes, "Plant Scale Comparison of Direct Filtration Versus Conventional Treatment of a Lake Erie Water", *J. Am. Water Works Assoc.* (March 1980): 148–155.

60. Recommended Standards for Water Works, p. 55.

61. Ibid., pp. 48–50.

62. "Water Treatment Plant Sludges—An Update of the State of the Art: Part 2", Report of the AWWA Sludge Disposal Committee, *J. Am. Water Works Assoc.* (October 1978): 548–554.

63. G. P. Westerhoff, *"Treating Waste Streams: New Challenge to the Water Industry,"* *Civil Eng.*, ASCE (August 1978): 77–83.

64. K. M. Mackenthum, *Toward a Cleaner Aquatic Environment*, EPA, U.S. Government Printing Office, Washington, DC, p. 231.

65. C. N. Sawyer, "Phosphates and Lake Fertilization", *Sewage Ind. Wastes* (June 1952): 768.

66. A. F. Bartsch, "Practical Methods for Control of Algae and Water Weeds", *Pub. Health Rep.* (August 1954): 749–757.

67. Recommended Standards for Water Works, pp. 56, 234.

68. P. Doudoroff and M. Katz, "Critical Review of Literature on the Toxicity of Industrial Wastes and Their Components to Fish", *Sewage Ind. Wastes*, **22** (November 1950): 1432–1458.

69. *J. Environ. Eng. Div.*, ASCE (December 1973): 761–772.

70. Mackenthum.

71. M. Goralski, "Zebra Mussels Invade U.S., Terrorize Treatment Plants", *Water Environ. Technol.*, May 1990, pp. 31–33; J. McCommons, *Clearwaters*, Summer 1990, pp. 11–13; C. R. O'Neill, Jr., *"Dreissena Polymorpha: An Unwelcome New Great Lakes Invader,"* *Sea Grant*, Cooperative Extension, Fact Sheet, Cornell University, New York State (February 1990): 1–10.

72. *"Spraying Your Reservoir for Weed Control,"* in *Opflow*, AWWA, Denver, CO, May 1978.

73. *Manual of Water Quality Control*, AWWA, New York, 1940, p. 49.

74. J. D. Johnson and J. N. Jensen, "THM and TOX Formation: Routes, Rates, and Precursors", *J. Am. Water Works Assoc.* (April 1986): 156–162.

75. N. V. Brodtmann, Jr., and P. J. Russo, "The Use of Chloramine for Reduction of Trihalomethanes and Disinfection of Drinking Water", *J. Am. Water Works Assoc.* (January 1979): 40–45.

76. G. R. Scott, *"Aeration,"* in *Water Quality and Treatment*, McGraw-Hill, New York, 1971, Chapter 2, pp. 52–65.

77. Recommended Standards for Water Works, p. 65.

78. G. Culp, *Physical-Chemical Wastewater Treatment Design*, EPA 625/4-73-002A, EAP, Washington, DC, August 1973, p. 19.

79. A. A. Stevens, C. J. Slocum, D. R. Seeger, and G. R. Robek, "Chlorination of Organics in Drinking Water", *J. Am. Water Works Assoc.* (November 1976): 615; G. Dallaire, "Are Cities Doing Enough to Remove Cancer-Causing Chemicals from Drinking Water" *Civil Eng.*, ASCE (September 1977).

80. Eaton, Clesceri, Rice, and Greenberg.

81. T. M. Riddick, Letters to the Editor, *Public Works* (May 1975): 144.

82. Cox, pp 121–122.

83. "Disinfection", Committee Report, *J. Am. Water Works Assoc.* (July 1982): 376–380.

84. *Water Treatment by Chemical Disinfection*, Workers Compensation Board, British Columbia, Canada, April 1986, p. 8.

85. G. E. Symons and K. W. Henderson, "Disinfection—Where Are We" *J. Am. Water Works Assoc.* (March 1977): 148–154.

86. J. T. Kirk, "Ozonation Reduces Manganese Concentration", *Public Works* (January 1984): 40.

87. H. M. Rosen, "Ozonation—Its Time Has Come", *Water & Wastes Eng.* (September 1978): 106–108.

88. J. Holluta, "Ozone in Water Chemistry", *J. Water Pollut. Control Fed.* (December 1973): 2507.

89. V. J. Elia et al., "Ozonation in a Wastewater Reuse System: Examination of Products Formed", *J. Water Pollut. Control Fed.* (July 1978): 1727–1732.

90. Holluta, p. 2507.

91. Literature Review Issue, *J. Water Pollut. Control Fed.* (June 1989): 835.

92. D. G. McBride et al., "Design of the L.A. Aqueduct Water Filtration Plant", *Public Works* (November 1982): 37.

93. "Update," p. 15.

94. "NIOSH Recommendations for Occupational Safety and Health Standards", *MMWR*, **35**(1S, Suppl.) (1986).

95. S. W. Wells, "Hydrogen Sulfide Problems of Small Water Systems", *J. Am. Water Works Assoc.*, **46**(2), (February 1954): 160–170.

96. J. Josephson, "Groundwater Strategies", *Environ. Sci. Technol.* (September 1980): 1030–1035.

97. "Air Force Cleans Up a California Aquifer", *Civil Eng.*, ASCE (June 1987): 6.

98. *Division of Water Annual Data Report, 1987–88 Fiscal Year*, New York State Department of Environmental Conservation, Albany, July 31, 1988, p. 12.

99. Anonymous, "Ozone Technology Removes Organics From Groundwater", *J. Am. Water Works Assoc.* (June 1989): 124.

100. "Massachusetts Town Is Seeking Funds to Save Its Water Supply", *New York Times*, November 19, 1978; *Protection of Public Water Supplies from Ground-Water Contamination*, Seminar Publication, EPA/ 625/4-85/016, EPA Center for Environmental Research Information, Cincinnati, OH, September 1985, pp. 175–181; *A Compendium of Technologies Used in the Treatment of Hazardous Wastes*, EPA, Center for Environmental Research Information, Cincinnati, OH, September 1987, p. 28.

101. A. K. Naj, "Super Microbes Offer Way to Treat Hazardous Waste", *Wall Street Journal*, January 25, 1989.

102. W. Ghiorse, *Public Works* (January 1983).

103. J. M. Thomas and C. H. Ward, "In-Situ Biorestoration of Organic Contaminants in the Subsurface", *Environ. Sci. Technol.* (July 1989): 760–765.

104. Recommended Standards for Water Works, pp. 68–70.

105. H. R. Fosnot, "7 Methods of Iron Removal", *Public Works* **86**(11), (November 1955): 81–83.

106. E. J. Laubusch, "Chlorination of Water", *Water Sewage Works*, **105**(10), (October 1958): 411–417.

107. O. Maessen, B. Freedman, and R. McCurdy, "Metal Mobilization in Home Well Water Systems in Nova Scotia", *J. Am. Water Works Assoc.* (June 1985): 73–80.

108. A. W. Kenny, "Modern Problems in Water Supply", *J. R. Soc. Health* (June 1978): 116–121.

109. *Guidelines for Drinking-Water Quality*, Vol. 1, p. 67;   C. D. Larson, O. T. Love, Jr., and G. Reynolds III, "Tetrachloroethylene Leached from Vinyl-Toluene Lined Asbestos-Cement Pipe into Drinking Water", *J. Am. Water Works Assoc.* (April 1983): 184–190.

110. *AWWA News*, AWWA, June 1987.

111. E. L. Filby, "The New Thomas H. Allen Pumping Station and Iron-Removal Plant of Memphis", *Water Sewage Works*, **99**(4), (April 1952); 133–138.

112. W. Wheeler, "Notes and Comments", *J. Am. Water Works Assoc.* (February 1979): 116.

113. D. T. Merrill and R. L. Sanks, "Corrosion Control by Deposition of $CaCO^3$ Films: Part 1, Part 2, and Part 3", *J. Am. Water Works Assoc.* (November 1977): 592–599; (December 1977): 634–640;   (January 1978): 12–18.

114. Eaton, Clesceri, Rice, and Greenberg.

115. J. W. Patterson and J. E. O'Brien, "Control of Lead Corrosion", *J. Am. Water Works Assoc.* (May 1979): 264–271.

116. O. J. Gottlieb and R. F. Blattert, "Concepts of Well Cleaning", *J. Am. Water Works Assoc.* (May 1988): 34–39.

117. D. C. Schafer, "Use of Chemicals to Restore or Increase Well Yield", *Public Works* (April 1975).

118. E. Bellacks, *Fluoridation Engineering Manual*, EPA, Water Supply Division (WSD), Washington, DC, 1974, Reprint.

119. T. D. Chinn, B. D. Black, and S. A. L. Perry, *"Implementation of Arsenic Treatment Systems, Volume 2: Design Considerations, Operation and Maintenance,"* AWWA Research Foundation, Denver, CO, 2002.

120. T. J. Sorg, M. Casanady, and G. S. Logsdon, "Treatment Technology to Meet the Interim Primary Drinking Water Regulations for Inorganics: Part 3", *J. Am. Water Works Assoc.* (December 1978): 680–691;   T. J. Sorg, "Compare Nitrate Removal methods", *Water & Wastes Eng.* (December 1980): 26–31.

121. Ibid.

122. M. Sheinker and J. P. Codoluto, "Making Water Supply Nitrate Removal Practicable", *Public Works* (June 1977): 71–73;   T. J. Sorg, "Treatment Technology to Meet the Interim Primary Drinking Water Regulations for Inorganics", *J. Am. Water Works Assoc.* (February 1978): 105–109.

123. Sorg, pp. 105–109.

124. F. Rubel, Jr., and R. D. Woolsey, "The Removal of Excess Fluoride From Drinking Water by Activated Alumina", *J. Am. Water Works Assoc.* (January 1979): 45–49.

125. Chin et al.

126. C. P. Straub, *"Radioactivity,"* *Water Quality and Treatment*, AWWA, McGraw-Hill, New York, 1971, pp. 443–462.

127. D. C. Lindsten, "From Fission, Fusion, and Confusion Can Come Potable Water", *Water & Wastes Eng.* (July 1977): 56–61.

128. L. J. Kosarek, "Radionuclide Removal from Water", *Environ. Sci. Technol.* (May 1979): 522–525.

129. M. Y. Menetrez et al, *"Manganese Dioxide Coated Filters for Removing Radium from Drinking Water,"* EPA/600-S2-88/057, EPA, Cincinnati, OH, January 1989.

130. K. L. Dixon and R. G. Lee, "Occurrence of Radon in Well Supplies", *J. Am. Water Works Assoc.* (July 1988): 65–70.

131. N. A. Hahn, Jr., "Disposal of Radium Removed from Drinking Water", *J. Am. Water Works Assoc.* (July 1988): 71–78.

132. R. T. Jelinek and T. J. Sorg, "Operating a Small Full-Scale Ion Exchange System for Uranium Removal", *J. Am. Water Works Assoc.* (July 1988): 79–83.

133. T. J. Sorg, "Methods for Removing Uranium from Drinking Water", *J. Am. Water Works Assoc.* (July 1988): 105–111.

134. *Fact Sheet, Drinking Water Regulations Under the Safe Drinking Water Act*, Office of Drinking Water, EPA, Washington, DC, May 1990, pp. 36–37.

135. P. E. Gaffney, "Chlorobiphenyls and PCBs: Formation during Chlorination", *J. Water Pollut. Control Fed.* (March 1977): 401–404.

136. R. C. Hoehm, C. W. Randall, F. A. Bell Jr., and P. T. B. Shaffer, "Trihalomethane and Viruses in a Water Supply", *J. Environ. Eng. Div.*, ASCE (October 1977): 803–814.

137. R. E. Miller et al., "Organic Carbon and THM Formation Potential in Kansas Groundwaters", *J. Am. Water Works Assoc.* (March 1990): 49–62.

138. A. A. Stevens, C. J. Slocum, D. R. Seeger, and G. R. Robek, "Chlorination of Organics in Drinking Water", *J. Am. Water Works Assoc.* (November 1976): 615–620.

139. G. Daillaire, "Are Cities Doing Enough to Remove Cancer-Causing Chemicals from Drinking Water" *Civil Eng.*, ASCE (September 1977).

140. Recommended Standards for Water Works, pp. xix–xx.

141. E. Somers, "Physical and Chemical Agents and Carcinogenic Risk", *Bull. Pan Am. Health Org.*, **14**(2), (1980): 172–184; *Canadian Water Quality Guidelines*, Canadian Council of Resource and Environment Ministers, Environment Canada, Ottawa, Canada, March 1987, pp. 1–2.

142. *Guidelines for Drinking-Water Quality,* Vol. 1, p. 77.

143. R. M. Clark et al., "Removing Organic Contaminants from Groundwater", *Environ. Sci. Technol.* (October 1988): 1126–1129.

144. C. T. Anderson and W. T. Maier, "Trace Organics Removal by Anion Exchange Resins", *J. Am. Water Works Assoc.* (May 1979): 278–283.

145. W. J. Weber, Jr., *Physicochemical Processes for Water Quality Control*, Wiley-Interscience, New York, 1972, pp. 199–255; W. J. Weber, Jr. and E. H. Smith, "Removing Dissolved Organic Contaminants from Water", *Environ. Sci. Technol.* (October 1986) 970–979.

146. *Protection of Public Water Supplies from Ground-Water Contamination*, pp. 157–171.

147. O. T. Love Jr., "Let's Drink to Cleaner Water", *Water & Wastes Eng.* (September 1977): 136; O. T. Love Jr., *"Treatment Techniques for the Removal of Organic Contaminants from Drinking Water,"* in *Manual of Treatment Techniques for Meeting the Interim Primary Drinking Water Regulations*, EPA, Cincinnati, OH, May 1977, pp. 53–61.

148. M. Zitter, "Population Projections for Local Areas", *Public Works* (June 1957).

149. G. M. Fair, J. C. Geyer, and D. A. Okun, *Water and Wastewater Engineering*, Vol. 1, Wiley, New York, 1966, pp. 9–13.

150. Ibid.

151. F. W. MacDonald and A. Mehn, Jr., "Determination of Run-off Coefficients", *Public Works* (November 1963): 74–76.

152. L. C. Urquhart, *Civil Engineering Handbook*, McGraw-Hill, New York, 1959.

153. W. A. Hardenbergh, *Water Supply and Purification*, International Textbook Co., Scranton, PA, 1953.

154. E. L. MacLeman, "Yield of Impounding Reservoirs", *Water Sewage Works* (April 1958): 144–149.

155. Merriman and Wiggin, *American Civil Engineering Handbook*, Wiley, New York, 1946.

156. R. M. Waller, "World's Greatest Source of Fresh Water", *J. Am. Water Works Assoc.* (April 1974): 245–247.

157. W. D. Hudson, "Elevated Storage vs Ground Storage", *OpFlow*, AWWA (July 1978).

158. *Distribution System Requirements for Fire Protection*, Manual M31, AWWA, Denver, CO, 1989.

159. C. P. Hoover, *Water Supply and Treatment*, Bull. 211, National Lime Association, Washington, DC, 1934.

160. Institution of Water Engineers, *Manual of British Water Supply Practice*, W. Heffer & Sons, Cambridge, England, 1950.

161. O. G. Goldman, "Hydraulics of Hydropneumatic Installation", *J. Am. Water Works Assoc.*, **40**, 144 (February 1948).

162. *Criteria for the Design of Adequate and Satisfactory Water Supply Facilities in Realty Subdivisions*, Suffolk County Department of Health, Riverhead, NY, 1956.

163. C. W. Klassen, *"Hydropneumatic Storage Facilities,"* Tech. Release 10–8, Division of Sanitary Engineering, Illinois Department of Public Health, October 28, 1952.

164. J. B. Wolff and J. F. Loos, "Analysis of Peak Demands", *Public Works*, September 1956.

165. A. P. Kuranz, "Studies on Maximum Momentary Demand", *J. Am. Water Works Assoc.* (October 1942); D. R. Taylor, "Design of Main Extensions of Small Size", *Water Sewage Works* (July 1951); E. P. Linaweaver, Jr., *Residential Water Use, Report II on Phase Two of the Residential Use Research Project*, Department of Sanitary Engineering Water Resources, Johns Hopkins University, Baltimore, MD, June 1965; *Maximum Design Standards for Community Water Supply Systems*, Federal Housing Administration, FHA No. 751, (July 1965), 52; E. P. Linaweaver, Jr., J. C. Geyer, and J. B. Wolff, *A Study of Residential Water Use*, HUD, Washington, DC, 1967.

166. G. G. Dixon and H. L. Kauffman, "Turnpike Sewage Treatment Plants", *Sewage Ind. Wastes* (March 1956).

167. C. Sharp, "Kansas Sewage Treatment Plants", *Pub. Works* (August 1957).

168. M. Gray, "Sewage from America on Wheels—Designing Turnpike Service Areas Plants to Serve Transient Flows", *Wastes Eng.* (July 1959).

169. R. B. Hunter, *Water-Distributing Systems for Buildings*, Rept. No. BMS 79, National Bureau of Standards for Building Materials and Structures, November 1941.

170. *Distribution System Requirements for Fire Protection; Grading Schedule for Municipal Fire Protection* (1974) and *Guide for Determination of Required Fire-Flow* (1972), Insurance Services Office, New York.

171. A. L. Davis and R. W. Jeppson, "Developing a Computer Program for Distribution System Analysis", *J. Am. Water Works Assoc.* (May 1979): 236–241.

172. *Fire Suppression Rating Schedule*, Insurance Services Office, New York, 1980.

173. Ibid.

174. Ibid.

175. *Distribution System Requirements for Fire Protection*, p. 8.

176. K. Carl, "Municipal Grading Classifications and Fire Insurance Premiums", *J. Am. Water Works Assoc.* (January 1978): 19–22.

177. Fire Suppression Rating Schedule.

178. *Cross-Connection Control Manual*, EPA, Water Supply Division (WSD), Washington, DC, 1973, pp. 3–8; "Final Report of the Committee on Cross-Connections", *J. New Engl. Water Works Assoc.* (March 1979); "Ethylene Glycol Intoxication Due to Contamination of Water Systems", *MMWR* (September 18, 1987): 611–614.

179. *Distribution System Requirements for Fire Protection*.

180. "Use of Backflow-Preventers for Cross-Connection Control", Joint Committee Report, *J. Am. Water Works Assoc.* (December 1958): 1589–1617; G. J. Angele Sr., *Cross-Connections and Backflow Prevention*, AWWA, Denver, CO, 1970.

181. AWWA Board of Directors, *AWWA MainStream* (April 1989): 8.

182. E. K. Springer, "Cross-Connection Control", *J. Am. Water Works Assoc.* (August 1976): 405–406; V. W. Langworthy, "Persistent, These Cross Connections", *Am. City* (October 1974): 19.

183. *Backflow Prevention and Cross-Connection Control*, Pub. No. M14, AWWA, Denver, CO, 2004.

184. F. Aldridge, "The Problems of Cross Connections", *J. Environ. Health* (July/August 1977): 11–15; J. A. Roller, "Cross-Connection Control Practices in Washington State", *J. Am. Water Works Assoc.* (August 1976): 407–409.

185. *Public Water Supply Guide, Cross-Connection Control*, Bureau of Public Water Supply, New York State Department of Health, Albany, 1979.

186. Recommended Standards for Water Works, p. 107.

187. B. W. Lewis, "Design Considerations and Operating Tips for Small Residential Systems", *J. Am. Water Works Assoc.* (November 1987): 42–45.

188. J. A. Salvato Jr., "The Design of Pressure Tanks for Small Water Systems", *J. Am. Water Works Assoc.* (June 1949): 532–536.

189. H. E. Babbitt and J. J. Doland, *Water Supply Engineering*, McGraw-Hill, New York, 1939, pp. 306–313.

190. R. T. Kent, *Mechanical Engineers' Handbook*, Vol. II, Wiley, New York, 1950; S. B. Watt, *A Manual on the Hydraulic Ram for Pumping Water*, Intermediate Technology Development Group, Water Development Unit, National College of Agricultural Engineering, Silsoe, Bedford, United Kingdom, July 1977.

191. H. J. Hansen and W. L. Trimmer, "Increase Pumping Plant Efficiency to Save Energy", *OpFlow*, AWWA (September 1989): 3–5.

192. L. Wolfe, "Automatic Control of Level, Pressure, and Flow", *J. Am. Water Works Assoc.* (October 1973): 654–662.

193. J. R. Kroon et al., "Water Hammer: Causes and Effects", *J. Am. Water Works Assoc.* (November 1984): 39–45; *OpFlow*, AWWA (September 1983).

194. J. D. Francis et al., *National Assessment of Rural Water Conditions*, EPA 570/9-84-004, EPA, Office of Drinking Water, Washington, DC, June 1984.

195. D. H. Stoltenberg, "Construction Costs of Rural Water Systems", *Public Works*, **94** (August 1969).

196. "Cost Estimates for Water Treatment Plant Operation and Maintenance", *Public Works* (August 1980): 81–83. Based on R. C. Gumerman, R. L. Culp, and S. P. Hansen, *"Estimating Water Treatment Costs,"* EPA Report 600/2-79-162a, EPA, Washington, DC, August 1979.

197. D. W. Stone, *Organic Chemical Contamination of Drinking Water*, New York State Health Department internal document, Albany, June 1979.

198. *Water Well Location by Fracture Trace Mapping*, Technology Transfer, Office of Water Research and Technology, Department of the Interior, U.S. Government Printing Office, Washington, DC, 1978.

199. J. J. Hamilton, "Potassium Permanganate as a Main Disinfectant", *J. Am. Water Works Assoc.* (December 1974): 734–735.

200. H. W. Talbot, Jr., J. E. Morrow, and T. J. Seidler, "Control of Coliform Bacteria in Finished Drinking Water Stored in Redwood Tanks", *J. Am. Water Works Assoc.* (June 1979): 349–353.

201. Katadyn Products, Wallisellen, Switzerland.

202. General Ecology, Lionville, PA.

203. J. E. Ongerth et al., "Backcountry Water Treatment to Prevent Giardiasis", *Am. J. Public Health* (December 1989: 1633–1637.

204. Ibid.

205. *"Emergency Disinfection of Drinking Water, Boiling,"* Water Supply Guidance No. 51, EPA Office of Drinking Water Policy Statement, Washington, DC, May 8, 1978.

206. S. L. Chang and J. C. Morris, "Elemental Iodine as a Disinfectant for Drinking Water", *Ind. Eng. Chem.*, **45**, 1009 (May 1953).

207. R. S. Tobin, "Testing and Evaluating Point-of-Use Treatment Devices in Canada", *J. Am. Water Works Assoc.* (October 1987): 44.

208. *Beverage World* (May 1990): 36.

209. *"News of Environmental Research in Cincinnati,"* Water Supply Research Laboratory, EPA, Washington, DC, May 23, 1975.

210. Beverage World, p. 36; *The New York Times*, May 25, 1986, p. F6.

211. *Vending Machine Evaluation Manual*, National Automatic Merchandising Association, Chicago, IL.

212. 21 CFR 103.35, Standards of Quality, Bottled Water, and 21 CFR 129.1–129.37, Processing and Bottling of Bottled Drinking Water, April 1989; "AFDO Model Bottled Water Regulation", *Food Drug Officials Assoc. Q. Bull. Proc.*, **48**, 81–89 (1984).

213. Ibid.

214. Anonymous, "Why Drink Bottled Water", *Dairy Food Sanit.* (1987): 347.

BIBLIOGRAPHY

Ameen, J. S., *Source Book of Community Water Systems*, Technical Proceedings, High Point, NC, 1960.

American Water Works Association, *Water Quality and Treatment*, 5th ed., McGraw-Hill, New York, 1999.

Berg, G. (Ed.), *Viruses in Waste, Renovated and Other Waters*, Federal Water Quality Administration, U.S. Department of the Interior, Cincinnati, OH, 1969.

Cairncross, S., and R. Feachem, *Small Water Supplies.*, Ross Institute of Tropical Hygiene, London, January 1978.

Chinn, T.D., et al., *Implementation of Arsenic Treatment Systems, Volume 2: Design Considerations, Operations and Maintenance,* AWWA Research Foundation, Denver, CO, 2002

Cleasby, J. L., et al., *Effective Filtration Methods for Small Water Supplies*, EPA-600/S2-84-088, Municipal Environmental Research Laboratory, Cincinnati, OH, May 1984.

Cox, C. R., *Operation and Control of Water Treatment Processes*, WHO, Geneva, 1969.

Cross-Connection Control Manual, EPA, Water Supply Division (WSD), U.S. Government Printing Office, Washington, DC, 1973.

Design Standards for Small Public Drinking Water Systems, State of New Hampshire Department of Environmental Services, Concord, NH, 1986.

Donaldson, D., "Overview of Rural Water and Sanitation Programs for Latin America", *J. Am. Water Works Assoc.* (May 1983): 224–231.

Drinking Water Treatment Units and Related Products, Components and Materials, Standards Nos. 42, 44, 53, 58, 62, National Sanitation Foundation, Ann Arbor, MI, June 1, 1990.

Driscoll, F. G., *Groundwater and Wells*, Johnson Division, St. Paul, MN, 1986.

Eaton, A. D., L. S. Clesceri, E. W. Rice, and A. E. Greenberg (Eds.) *Standard Methods for the Examination of Water and Wastewater,* 21st ed., APHA, AWWA, WEF, 2005.

Fair, G. M., J. C. Geyer, and D. A. Okun, *Water and Wastewater Engineering*, Vol. **1**, Wiley, New York, 1966.

Guidelines for Delineation of Wellhead Protection Areas, EPA, Office of Ground-Water Protection, Washington, DC, June 1987.

Guidelines for Drinking-Water Quality, Vol. **1**: Recommendations, 1984, 1998; Vol. **2**: Health Criteria and Other Supporting Information, 1984, 1998; Vol. **3**: Guidelines for Drinking-Water Quality, WHO, Geneva, 1985, 1997.

HDR Engineering, Inc., *Handbook of Public Water Systems,* 2nd ed., Wiley, New York, 2001.

Henderson, G. E., and E. E. Jones, *Planning for an Individual Water System*, Agricultural Research Service, U.S. Department of Agriculture, and EPA, Water Programs Branch, American Association for Vocational Instructional Materials, Engineering Center, Athens, GA, May 1973.

Koch, A. G., and K. J. Merry, *Design Standards for Public Water Supplies*, Department of Social and Health Services, Office of Environmental Programs, Olympia, WA, May 1973.

Komorita, J. J., and V. L. Snoeyink, "Technical Note: Monochloramine Removal From Water by Activated Carbon", *J. Am. Water Works Assoc.* (January 1985): 62–64.

Liguori, F. R., *Manual Small Water Systems Serving the Public*, Conference of State Sanitary Engineers, June 1978.

MWH, *Water Treatment: Principles and Design,* 2nd ed., Wiley, New York, 2005

Mackenthun, K. M., and W. M. Ingram, *Biological Associated Problems in Fresh Water Environments*, Federal Water Pollution Control Administration, U.S. Department of the Interior, Washington, DC, 1967.

Manual for Evaluating Public Drinking Water Supplies, PHS Pub. 1820, DHEW, Bureau of Water Hygiene, Cincinnati, OH, 1971.

Manual of Individual Water Supply Systems, EPA, Office of Drinking Water, Washington, DC, October 1982.

Manual of Instruction for Water Treatment Plant Operators, New York State Department of Health, Albany, 1965.

Military Sanitation, Department of the Army Field Manual 21–10, May 1957.

Okun, D. A., "Alternatives in Water Supply", *J. Am. Water Works Assoc.* (May 1969): 215–221.

"Point-of-Use and Point-of-Entry Home Treatment", *J. Am. Water Works Assoc.* (October 1987): 20–88.

Quality Criteria for Water, EPA, Washington, DC, July 1976.

Recommended Standards for Water Works, Great Lakes–Upper Mississippi River Board of State Public Health and Environmental Managers, Health Research, Albany, NY, 1987.

Safe Drinking Water Act, as amended, EPA, U.S. Government Printing Office, Washington, DC, 1986, 1996.

Schulz, C. R., and D. A. Okun, *Surface Water Treatment for Communities in Developing Countries*, Wiley, New York, 1984.

Schulz, C. R., and D. A. Okun, "Treating Surface Waters for Communities in Developing Countries", *J. Am. Water Works Assoc.* (May 1983): 212–219.

State of the Art of Small Water Treatment Systems, EPA, Office of Water Supply, Washington, DC, August 1977.

"Surveillance of Drinking Water Quality", *WHO Monogr. Ser.*, **63** (1976).

Van Dijk, J. L., and J. H. C. M. Oomen, *Slow Sand Filtration for Community Water Supply in Developing Countries*, A Design and Construction Manual, Technical Paper No. 11, WHO International Reference Centre for Community Water Supply, The Hague, The Netherlands, December 1978.

Vesilind, P. A., and J. J. Peirce, *Environmental Engineering*, Ann Arbor Science, Ann Arbor, MI, 1982.

Visscher, J. T., R. Paramasivam, A. Raman, and H. A. Heijnen, *Slow Sand Filtration for Community Water Supply, Planning, Design, Construction, Operation and Maintenance*, Technical Paper No. 24, International Reference Centre for Community Water Supply and Sanitation, The Hague, The Netherlands, June 1987.

Wagner, E. G., and J. N. Lanoix, "Water Supply for Rural Areas and Small Communities", *WHO Monogr. Ser.*, **42** (1959).

Wagner, E. G., "Simplifying Design of Water Treatment Plants for Developing Countries", *J. Am. Water Works Assoc.* (May 1983): 220–223.

Water Distribution Operator Training Manual, American Water Works Association, Denver, CO, 1976.

Water Quality Criteria, Federal Water Pollution Control Administration, U.S. Department of the Interior, Washington, DC, April 1, 1968.

Water Quality Criteria 1972, National Academy of Sciences, National Academy of Engineering, Washington, DC, 1972.

Water Systems Handbook, Water Systems Council, Chicago, IL, 1977.

Water Treatment Plant Design, 2nd ed., ASCE and AWWA, McGraw-Hill, New York, 1990.

Well Drilling Operations, Technical Manual 5–297, Department of the Army and Air Force, September, 1965.

Wisconsin Well Construction Code, Division of Well Drilling, Wisconsin State Board of Health, 1951.

Wright, F. B., *Rural Water Supply and Sanitation*, Robert E. Krieger Publishing, Huntington, NY, 1977.

CHAPTER 3

WASTEWATER TREATMENT AND DISPOSAL

JOHN R. KIEFER
Consulting Environmental Engineer, Greenbrae, California

DISEASE HAZARD

Improper disposal of human excreta and sewage is one of the major causes of disease in areas where satisfactory sewage treatment is not available. Because many pathogenic microorganisms are found in sewage, all sewage should be considered contaminated. The work of investigators who have studied the survival of enteric pathogens in soil, water, and wastewater has been summarized in a number of publications.[1] As an example, it is known that some pathogens can survive from less than a day in peat to more than 2 years in freezing moist soil.

The waterborne microbiological agents of greatest concern are pathogenic bacteria, viruses, helminths, protozoa, and spirochetes. Infectious bacterial agents are associated with shigellosis and salmonella infections, while viral agents are associated with infectious hepatitis A, viral gastroenteritis, and other enteric viral diseases. Helminths are associated with ascariasis, taeniasis, dracunculiasis, trichuriasis, toxocariasis, enterobiasis, and other illnesses. Protozoa are associated with amebiasis and giardiasis, while spirochetes are associated with leptospirosis.

The ability of treatment systems to remove these pathogens varies, depending on the system and the pathogen in question. For example, primary sedimentation can remove zero to 30 percent of all viruses; 50 to 90 percent of the bacteria, taenia ova and cholera vibrio; zero percent of the leptospires; 10 to 50 percent of the *Entamoeba histolytica*; 30 to 50 percent of the ascaris; and 80 percent of the schistosomes. By contrast, trickling filters can remove 90 to 95 percent of the viruses, bacteria, and cholera vibrio; zero percent of the leptospires; 50 percent of the *Entamoeba histolytica*; 70 to 100 percent of the ascaris; 50 to 99 percent of the schistosome ova; and 50 to 95 percent of the taenia ova. Activated sludge treatment can remove 90 to 99 percent of the viruses, bacteria, and cholera vibrio; zero percent of the leptospires; 50 percent of *Entamoeba histolytica*; 70 to 100 percent of the ascaris; and 50 to 99 percent of taenia ova. Stabilization ponds (not less than 25-day retention) can be expected to remove all viruses, bacteria, vibrio, leptospires, *Entamoeba histolytica*, ascaris, schistosome ova, and taenia ova. Septic tanks can be expected to remove 50 to 90 percent of the cholera vibrio, ascaris, schistosoma, and taenia present; 100 percent of the leptospires; and zero percent of the *Entamoeba histolytica*.[2] Waste stabilization ponds in series (three with 25-day retention) remove practically all enteric viruses, bacteria, protozoan cysts, and helminth eggs.[3] Chemical coagulation, flocculation, sedimentation, and filtration will remove nearly all viruses, bacteria, protozoa, and helminths, particularly if supplemented by chlorination or other effective disinfection treatment. Although sewage treatment does not in itself prevent the waterborne diseases caused by these pathogens (i.e., treatment of drinking water does that), proper sewage treatment is needed to minimize the pollution load on water treatment plants.

The improper treatment or disposal of untreated excreta, sewage, or other wastewater (including gray water) dramatically increases the possibility of disease transmission via direct contact with any of the following:

- Houseflies or other insects that land on excreta and then land on and/or bite people
- Inanimate objects such as children's toys that contact excreta and are subsequently handled by people
- Direct ingestion of contaminated water or food

Such occurrences are especially true in developing countries where enteric diseases are prevalent and clean water, sanitation, and personal hygiene are wanting. Studies have indicated that diarrheal illnesses, especially in young children, could be reduced 16 percent where water quality was improved, 25 percent where water, not necessarily potable, was made available, 37 percent where both water availability and quality were improved, and 22 percent where excreta disposal facilities were improved.[4]

In addition, sewage sludge accumulates heavy metals from wastewater. Because many of these metals are very toxic, the use of sewage sludge as a soil builder may result in higher levels of toxic metals in treated vegetation and animals that eat that vegetation. For that reason, sewage sludge should not be used as a soil builder or fertilizer supplement for forage crops unless it is known to be free of significant amounts of toxic metals or other pathogens. See "Biosolids Treatment and Disposal," this chapter.

Awareness of these dangers, coupled with adequate treatment of sewage, provision of potable water and sanitation, have been shown to be the primary reason why intestinal and waterborne diseases are at their present low level in many parts of the world.

Criteria for Proper Wastewater Disposal

Proper disposal of wastewater is necessary not only to protect the public's health and prevent contamination of groundwater and surface water resources, but also to preserve fish and wildlife populations and other beneficial uses (e.g., water-based recreation). To that end, the following six criteria should be used in designing and operating any wastewater disposal system:[5]

1. Prevention of microbiological, chemical, and physical pollution of water supplies and contamination of fish and shellfish intended for human consumption
2. Prevention of pollution of bathing and recreational areas
3. Prevention of nuisance, unsightliness, and unpleasant odors
4. Prevention of human wastes and toxic chemicals from coming into contact with man, grazing animals, wildlife, and food chain crops, or being exposed on the ground surface accessible to children and pets
5. Prevention of fly and mosquito breeding, and exclusion of rodents and other animals
6. Adherence to surface and groundwater protection standards as well as compliance with federal, state, and local regulations governing wastewater disposal and water pollution control

Failure to observe these criteria can result in the development of health hazards and the degradation of living conditions, recreational areas, and natural resources that are essential to the well being of the general public.

Definitions

Commonly used terms in wastewater treatment and disposal include the following:

Aerobic bacteria Bacteria that require free dissolved oxygen for their growth. Carbon, nitrogen, and phosphorus are required nutrients for growth.

Anaerobic bacteria Bacteria that grow only in the absence of free dissolved oxygen and obtain oxygen from breaking down complex organic substances.

Biochemical oxygen demand (BOD) The difference between the initial dissolved oxygen in a sample and the dissolved oxygen after a stated period of time, which is the amount of dissolved oxygen in milligrams per liter (mg/l) required during stabilization of the decomposable organic matter by aerobic bacterial action. Although incubation can be for either 5 days (carbonaceous demand satisfied) or 20 days (total carbonaceous plus nitrification demand satisfied), use of this term in this chapter refers to the 5-day demand test otherwise specified. The 20-day demand is used to measure the oxygen needed to oxidize inorganic materials such as sulfides and ferrous iron, as well as the oxygen needed to oxidize reduced nitrogen forms if the nitrifying organisms are present.

Biosolids The solid, semisolid, or liquid residue that is generated during the treatment of domestic waste. Examples of biosolids include sewage sludge and septage. Sewage sludge is the material that settles out of wastewater during its treatment in a treatment plant. Septage is the material that is pumped out of septic tanks.

Chemical oxygen demand (COD) A measure of the amount of oxygen, in mg/l, that is chemically—rather than biologically—consumed in the oxidation of organic and oxidizable inorganic materials in water. COD is usually higher than the BOD of the water. The test is relatively rapid and, while it does not oxidize some organic pollutants (pyridine, benzene, toluene), it does oxidize inorganic compounds that are not measured in the BOD analysis.

Coliform organisms Microorganisms found in the intestinal tract of humans and animals whose presence in water indicates fecal pollution and potentially adverse contamination by pathogens.

Dissolved oxygen (DO) The oxygen in water that is available to support aquatic life and that is used by wastewater discharged to a water body. Cold water holds more oxygen in solution than warm water. Game fish typically require at least 4 to 5 mg/l of DO. Absence of DO results in anaerobic conditions and foul odors.

Domestic sewage Wastewater from a home, which includes toilet, bath, laundry, lavatory, and kitchen-sink wastes. See Table 3.1. The strength of sewage is commonly expressed in terms of BOD, COD, and suspended solids. Normal domestic sewage will average less than 0.1 percent total solids in soft water regions.

Excreta The waste matter eliminated from the human body; about 27 grams per capita per day dry basis (100 to 200 grams wet). Mara[6] reports that human feces have an average weight per capita per day of 150 grams wet basis and contain 2,000 million fecal coliform, and 450 million fecal streptococci.

Facultative bacteria Bacteria that have the ability to live under both aerobic and anaerobic conditions.

Industrial waste Any liquid, gaseous, solid, or waste substance arising from industrial, manufacturing, trade, or business processes, which cause or might reasonably be expected to cause pollution of water.

National pollutant discharge elimination system (NPDES) The national system for the issuance of permits for the discharge of treated sanitary, industrial, and commercial wastes under the 1972 Federal Water Pollution Control Act. The NPDES permit specifies the treatment to be used by the discharger to protect water quality.

Nonpoint pollution Any source other than a point source that impacts the chemical, physical, biological, or radiological integrity of water.

TABLE 3.1 Characteristics of Wastewater[a]

Constituents	Domestic Wastewater (Community)[b]	Household Wastewater[c]	Septic Tank Household Effluent[d]	Gray Water[e,f]	Black Water[e,f]
Color					
nonseptic	Gray				
septic	Blackish		3.5		
Odor					
nonseptic	Musty				
septic	H_2S		4.5		
Temperature (°F)	55° to 90°[g]		63		
Total solids	800	968	820	528	621
Total volatile solids	425	514			
Suspended solids	200	376	101	162	77
Total nitrogen	40	84	36	11.3	153
Organic nitrogen	25				
Ammonia nitrogen	0.5	64	12	1.7	138
Nitrate nitrogen	0.5		0.12	0.12	0.22
Total phosphate	15	61	15	1.4	18.6
Total bacteria, per 100 ml	30×10^8		76×10^6		
Total coliform, MPN per 100 ml	30×10^6		110×10^6	24×10^{6}[g]	0.25×10^6
Fecal coliform, per 100 ml				1.4×10^6	0.04×10^6
BOD, 5-day	200	435	140	149	90
COD		709	675	366	258
Total organic carbon				125	97
Grease		65			
pH	7.5	8.1	7.4	6.8	7.8

[a] Average, in mg/l unless otherwise noted.

[b] Peter F. Atkins, "Water Pollution By Domestic Wastes," *Selection and Operation of Small Wastewater Treatment Facilities—Training Manual*. Charles E. Sponagle, U.S. EPA, Cincinnati, OH, April 1973, p. 3–3.

[c] K. S. Watson, R. P. Farrell, and J. S. Anderson, "The Contribution from the Individual Home to the Sewer System," *J. Water Pollution Control Fed.*, December 1967, pp. 2039–2054.

[d] J. A. Salvato, Jr., "Experience with Subsurface Sand Filters," *Sewage Ind. Wastes*, **27**, 8, 909–916 (August 1955).

[e] Septic tank effluent. The higher concentration of coliform bacteria in the gray water effluent are attributed to the large amounts of undigested organic matter in kitchen wastewater.

[f] M. Brandes, Characteristics of Effluents from Separate Septic Tanks Treating Grey Water and Black Water from the Same House, Ministry of the Environment, Toronto, Canada, October 1977, pp. 9 and 27.

[g] For the Central States zone in United States.

Pollution Water pollution is the addition of agricultural, domestic, industrial, and commercial wastes in concentrations or quantities that result in the measurable degradation of water quality. A *point source* of pollution is "any discernible, confined, or discrete conveyance from which pollutants are or may be discharged,"[7] such as a pipe, ditch, well, vessel, vehicle, and feedlot.

Refractory organics Manmade organic compounds, such as chlordane, endrin, DDT, and lindane, that degrade very slowly. Also, a material having the ability to retain its physical shape and chemical identity when subjected to high temperatures.[8]

Sewer, sewerage, sewage, or wastewater treatment plant, or sewage works When stormwater and domestic sewage enter a sewer, it is called a *combined sewer*. If domestic sewage and stormwater are collected separately (i.e., in a *sanitary sewer* and a *storm sewer*), it is called a *separate sewer system*. A *sewer system* is a combination of sewer pipes and appurtenances for the collection, pumping, and transportation of sewage, or *sewerage*. When facilities for treatment and disposal of sewage are included, it is called either a *sewage* or *wastewater treatment plant* or *sewage works*.

Suspended solids Solids that are visible and in suspension in water.

Total organic carbon (TOC) A measure of the carbon as carbon dioxide. Because inorganic carbon compounds present can interfere with the test, they must be removed before the analysis is made, or a correction is applied.

Wastewater Sewage from domestic, agricultural, commercial and/or industrial establishments, which can also include surface and groundwater infiltration.

SMALL WASTEWATER DISPOSAL SYSTEMS

The number of households in the United States served by public sewers, septic tank or cesspool systems, or other means is shown in Table 3.2. With an estimated average occupancy of 3.0 persons per household in 1970, 2.7 in 1980, and 2.4 in 1990, a total of 58.5 million people were dependent on individual on-site facilities in 1970, 59.9 million in 1980, and 62.7 million in 1990. This represents 28.8 percent of the population in 1970, 26.4 percent in 1980, and 25.2 percent in 1990. Although data from the U.S. Census Bureau are not available for 2000, it is likely that the observed trends (i.e., an increase in the number of people using on-site facilities and a decrease in the percent of the total population using such facilities) continue today. At an estimated water usage of 50 gallons per capita per day, more than 3 billion gallons of sewage is discharged each day to on-site sewage disposal systems and ultimately to the groundwater that underlies these systems. For that reason alone, it is obvious that there is a need for continued support and research into ways to improve on-site sewage treatment and disposal facilities.

The most common system for wastewater treatment and disposal at homes in rural areas is by using a septic tank for the settling and treatment of the

TABLE 3.2   Housing Units Served by Public Sewers and Individual Facilities

Sewage Disposal Facility	Number of Housing Units Served[a]		
	1970	1980	1990
Public sewer	48,187,675	64,569,886	76,455,211
Septic tank or cesspool	16,601,792	20,597,165	24,670,877
Other means	2,904,375	1,602,338	1,137,590
Total housing units	67,693,842	86,769,389	102,263,678
Total population	203,302,031	226,542,203	248,709,873

[a] All units were not occupied.

Source: U.S. Census of Housing and Census Bureau.

wastewater and a subsurface absorption field for the disposal of the septic tank overflow. A soil percolation test is commonly used to determine the capacity of a soil to absorb septic tank overflow. Where soils are unsuitable, sand filters, elevated systems in suitable fill, evapotranspiration-absorption systems, evapotranspiration beds, aeration systems, stabilization ponds or lagoons, recirculating toilets, and various types of toilets and privies can be used. These systems are discussed later in this chapter.

Wastewater Characteristics

As noted in Table 3.1, wastewater from residential communities is fairly uniform. Wastewater from toilets is referred to as *black water*, while all other domestic wastewater is referred to as *gray water*. From a public health standpoint, gray water should be considered sewage and treated as such because it can also be expected to contain pathogens from shower and washbasin use, as well as the washing of baby diapers and soiled clothing.

Soil Characteristics

Soils may be divided, for classification purposes, into gravel, sand, silt, and clay. As noted in Table 3.3, the particle size of soil components decreases as classification moves from gravel to clay.

Clay and clay loam do not drain well and are usually considered unsuitable for the disposal of sewage and other wastewater by subsurface means. The permeability of a soil (i.e., its ability to absorb and allow water to pass through) is related to the chemical composition and physical structure of the soil. Soil with good structure will break apart, with little pressure, along definite cleavage planes. Prismatic and columnar structure enhances vertical percolation, and blocky and granular structure enhances both horizontal and vertical water flow.[9] If the color of the soil is yellow, brown, or red, it would indicate that air and water pass through. However, if it is blue or grayish, the soil is likely to have been saturated for extended periods and is probably unsuitable for subsurface absorption of wastewater. Magnesium and calcium tend to

TABLE 3.3 U.S. Department of Agriculture Size Limits
for Soil Separates

Soil Separate	Size Range (mm)	Tyler Standard Sieve No.
Sand	2–0.05	10–270 mesh
Very coarse sand	2–1	10–16
Coarse sand	1–0.5	16–35
Medium sand	0.5–0.25	35–60
Fine sand	0.25–0.1	60–140
Very fine sand	0.1–0.05	140–270
Silt	0.25–0.002	—
Clay	<0.002	—

Source: Design Manual, Onsite Wastewater and Disposal System, U.S. EPA, Office of Water Program Operations, Washington, D.C., October 1980, pp. 367–374.

keep the soil loose, whereas sodium and potassium have the opposite effect. Sodium hydroxide, a common constituent of septic tank cleaners, would cause a breakdown of soil structure with resultant smaller pore space and reduced soil permeability.

Soil adsorptive capacity is an important consideration in the design of a septic tank system. Soils in which subsurface absorption fields are to be laid should have a low permeability (i.e., effective size 0.1 to 0.3 millimeter) and some adsorptive capacity to allow organic material to be retained. A minimum soil organic content of 0.5 to 1 percent is suggested and can be found in practically all agricultural soils, together with some clay and silt, which add to the adsorptive capacity. By contrast, a soil with low adsorption (e.g., coarse gravel) or a formation with solution channels, fractures, or fissures will permit pollution to travel long distances without purification.

Soil Suitability

It is necessary to have at least 2 feet of suitable soil between the bottom of absorption trenches and leaching pits and the highest groundwater level, clay, rock, or other relatively impermeable layer. This helps ensure removal of most of the pathogenic viruses, bacteria, protozoa, and helminths in septic tank effluent before they reach the groundwater. Some regulatory agencies require a minimum of 3 or 4 feet. Where the soil is relatively nonpermeable at shallow depths, an alternate treatment and disposal system is needed in place of a conventional leaching system. Preferably, construction should be postponed in such situations until sewers are available.

Pollutant Travel from Septic Systems

Groundwater contamination potential from a septic tank system is determined by the physical, chemical, and microbiological characteristics of the soil; the unsaturated soil depth to groundwater; the volume and strength of the septic tank effluent; and the biological slime or mat on the trench gravel, bottom, and sidewalls.

The percentage of clay and silt in a soil appears to be a major factor in bacteria retention and in bacteria die-off. For example, studies have shown that 4 feet of unsaturated coarse-grained sandy soil is not adequate to prevent bacteria from reaching saturated soils beneath septic tank absorption trenches. However, in loamy sandy soil, no bacteria traveled beyond 3.6 feet.[10] Soils containing loam (clay, silt, and sand) will remove most of the phosphorus, soluble orthophosphates, and microorganisms in sewage effluent. If the absorption trenches are kept shallow (top of gravel about 8 to 12 inches from the ground surface) as recommended, the vegetative cover root system over the absorption field penetrates and takes up much of the nitrogen during the growing season.

Laboratory studies have found that 40 to 50 cm of agricultural-type soil is very effective in removing viruses from water and that soil with reasonable amounts of silt and clay can remove viruses within the first 2 feet.[11] However, virus travel in sandy soil was reported at distances of 5 feet and was found in 6 and 20 foot-deep wells.[12] Also, Sorber et al. reported finding enteric viruses in soils at considerable distances from their point of application.[13]

Soil Percolation Test

The suitability of soil for the subsurface disposal of sewage can be determined by a study of soil characteristics and the soil percolation test. The test is a measure of the relatively constant rate at which clear water maintained at a depth of 6 inches will seep out of a standard size test hole that has been previously saturated and is at the same depth as the proposed absorption system. Henry Ryon first introduced this test in 1924,[14] and based on results for a wide range of soils developed "safe gallons per square foot per day subsurface irrigation rates." The work done by Ryon was confirmed by the U.S. Public Health Service (USPHS) in independent field tests and, despite of its limitations, serves as the basis for present-day design of septic systems.

Investigators have, however, differed regarding the interpolation of test results to determine the allowable rate of septic-settled sewage application per square foot of leaching area. This rate is a percentage of the actual amount volume of water a test hole accepts in 24 hours with rates ranging from 0.4 to 7.0 percent. Percolation test results are typically reported in terms of the minutes it takes for the water level in the test hole to drop 1 inch. Depending on percentage rate used, an allowable rate of settled sewage application in gallons per day per square foot (gpd/ft^2) of absorption trench bottom can be obtained from Table 3.4.

TABLE 3.4 Interpretation of Soil Percolation Test

Time for Water to Fall 1 in. (min)	Allowable Rate of Settled Sewage Application (gpd/ft^2)			
	U.S. PHS[a]	U.S. EPA[b]	GLUMR[c]	Ryon[g]
<1	5.0[d]	[b]	1.2	4.0 to 3.4
1	5.0[d]	1.2	1.2	3.3
2	3.5[d]	1.2	1.2	2.9
3	2.9[d]	1.2	1.2	2.7
4	2.5[d]	1.2	1.2	2.4
5	2.2[d]	1.2	1.2	2.2
6	2.0	0.8	0.9	
7	1.9	0.8	0.9	2.0
8	1.8	0.8	0.9	
9	1.7	0.8	0.9	
10	1.6	0.8	0.9	1.7
11	1.5	0.8	0.6	
12	1.4	0.8	0.6	
15	1.3	0.8	0.6	1.4
16	1.2	0.6	0.6	
20	1.1	0.6	0.6	1.1
25	1.0	0.6	0.6	
30	0.9	0.6	0.6	0.9
31	0.8	0.45	0.5	
35	0.8	0.45	0.5	
40	0.8	0.45	0.5	
45	0.7	0.45	0.5	0.7
46	0.7	0.45	0.45	
50	0.7	0.45	0.45	
60	0.6	0.45	0.45	0.4
61–120	[e]	0.2	[e]	0.2
>120	[f]	[e]	[e]	[f]

[a] USPHS, *Manual of Septic-Tank Practice*, USPHS Pub. 526, HEW, Washington, DC, 1967. Increase leaching area by 20 percent where a garbage grinder is installed and by additional 40 percent where a home laundry machine is installed. The required length of the absorption field may be reduced by 20 percent if 12 in. of gravel is placed under the distribution lateral, or by 40 percent if 24 in. of gravel is used, provided the bottom of the trench is at least 24 in. above the highest groundwater level.

[b] *Design Manual, Onsite Wastewater Treatment and Disposal Systems*, U.S. EPA, Cincinnati, OH, October 1980. Soils with percolation rates < 1 min/in. can be used if the soil is replaced with a suitably thick (>2 ft) layer of loamy sand or sand. Use 6 to 15 min/in. percolation rate. Reduce application rate where applied BOD and TSS is higher than domestic sewage. Additional area credit may be given for sidewall trench area if more than 6 in. of gravel is placed below the distributor. The EPA and GLUMR application rates are lower than the U.S. PHS rates. The former recognize the importance of settled sewage retention in the unsaturated zone to obtain maximum purification before it reaches the groundwater and results in a larger disposal system.

[c] GLUMR, *Recommended Standards for Individual Sewage Disposal Systems*, Great Lakes-Upper Mississippi River Board of State Sanitary Engineers, 1980 Edition. Absorption trench or bed shall not be constructed in soils having a percolation rate slower than 60 min/in., or where rapid percolation may result in contamination of water-bearing formation or surface water. The percolation rate is for *trench* bottom area. For absorption *bed*, use application rate of 0.6 gpd/ft^2 for percolation rate up to 6 min/in., then use 0.45 gpd/ft^2. Trench or bed bottom, or seepage pit bottom, should not be less than 3 ft above highest groundwater level. Maximum trench width credit shall be 24 in. for design purposes, even if trench is wider.

[d] Henry Ryon, *Notes on Sanitary Engineering*, New York State, Albany, 1924, p. 33. These and the U.S. PHS rates are given for historical perspective.

[e] Reduce rate to 2.0 gpd/ft^2 where a well or spring water supply is downgrade; increase protective distance, and place 6 to 8 in. sandy soil on trench bottom below gravel and between gravel and sidewalls.

[f] Soil not suitable.

[g] See Small Wastewater Disposal Systems for Unsuitable Soils or Sites, this chapter.

Many variations and refinements of the soil percolation test, including the use of a float gauge, inverted carboy as in a water cooler, and permeability test, have been proposed.[15] Whatever the case, enough tests should be conducted to give information representative of the soil, as indicated by a relatively constant rate of water drop in the test hole. This data should make it possible to determine an average percolation rate that can be used in design the septic tank system. A typical layout for such a system is shown in Figures 3.1 and 3.2.

Where a small subdivision is under consideration, at least three holes per acre should be tested. More holes should be tested if the percolation results vary widely, say by more than 20 percent. If rock, clay, hardpan, or groundwater is encountered within 4 feet of the ground surface, the property should be considered unsuitable for the disposal of sewage by means of conventional subsurface absorption fields.

Septic systems should not be constructed on filled-in ground until it has been thoroughly settled or otherwise stabilized. Percolation tests should be made in fill after at least a six-month settling period. Soil tests in fill often are not reliable as the soil structure, texture, moisture, and density will be quite variable and other disposal systems (see "Small Wastewater Disposal Systems for Unsuitable Soils or Sites" later in this chapter) should be considered.

When septic tank systems do fail, the cause is usually either improperly performed and interpreted soil percolation tests, high groundwater, poor construction, lack of maintenance, abuse of the system, or use of septic tanks where they were never intended. Inadequate design, lack of inspection by regulatory agencies, and failure to consider soil color, texture, and structure may also contribute to the problem.

Sewage Flow Estimates

The sewage flow to be expected from various establishments can vary, depending on the day of the week, season of the year, habits of the people, water pressure, type and number of plumbing fixtures, and type of place or business. For that reason, septic system design should be based on the average maximum flow rate to ensure adequate capacity. In the absence of actual figures, the per capita or unit estimated water flow given in Chapter 1 may be used as a guide. Alternatively, a fixture basis (see Table 3.5) can be used for estimating sewage flow rates. This approach assumes that all water used finds its way to the sewage disposal or treatment system. After adjusting for lawn watering and car washing, the total number of fixtures is multiplied by the flow from each fixture to obtain a rough estimate of the probable flow in gallons per day.

The design flow and sewage application rate for subsurface absorption systems also should take into consideration the strength of the septic tank effluent (BOD and TSS) in addition to the hydraulic loading rates already presented.

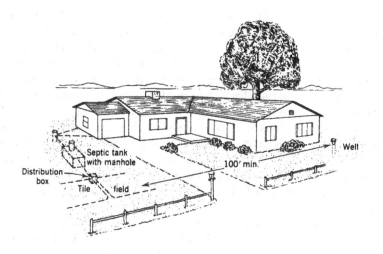

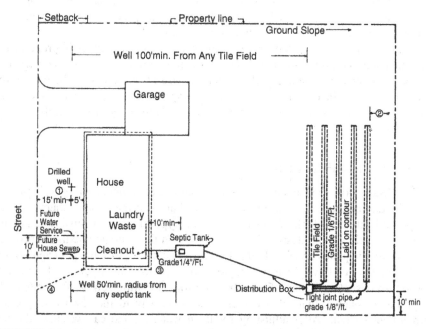

FIGURE 3.1 Typical private water supply and septic tank disposal systems. (Notes: (1) Watertight footing drain within 25 ft of well. (2) Tile field to be 50 ft or more from any lake, swamp, ditch, or watercourse and 10 ft or more from any waterline under pressure. (3) Cast-iron pipe, lead caulked joints within 50 ft of any well. (4) Discharge footing, roof, and cellar drainage away from sewerage system and well. (5) Grade lot to drain surface runoff away from the subsurface absorption system.)

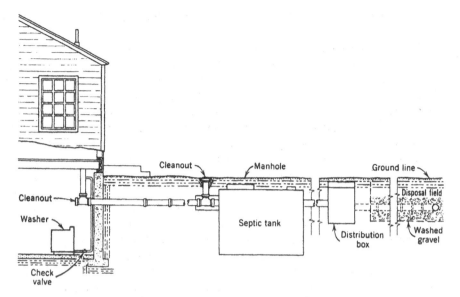

FIGURE 3.2 Cross-section of septic tank disposal system.

Septic Tank

A septic tank is a watertight tank designed to slow down the movement of raw sewage passing through it so that solids can settle out and be broken down by liquefaction and anaerobic bacterial action. Septic tanks do not purify the sewage, eliminate odors, or destroy all solid matter, but rather, simply condition sewage so that it can be disposed of using a subsurface absorption system. Suspended

TABLE 3.5 A Fixture Basis of Estimating Sewage Flow

Type of Fixture	Gallons per Day per Fixture, Country Clubs[a]	Gallons per Hour per Fixture, Public Parks[b]	Gallons per Hour per Fixture (Average), Restaurants[c]
Shower	500	150	17
Bathtub	300	—	17
Washbasin	100	—	8.5
Water closet	150	36	42 (flush valve)
			21 (flush tank)
Urinal	100	10	21
Faucet	—	15	8.5, 21 (hose bib)
Sink	50	—	17 (kitchen)

[a]John E. Kiker, Jr., "Subsurface Sewage Disposal," *Fla. Eng. Exp. Sta.*, Bull. No. 23 (December 1948).
[b]National Park Service.
[c]After M. C. Nottingham Companies, CA.

solids removal in such tanks is 50 to 70 percent; 5-day BOD removal is about 60 percent.

Recommended septic tank sizes based on estimated daily flows are given in Table 3.6. The septic tank should have a liquid volume of not less than 750 gallons. When a tank is constructed on the job, its liquid volume can be increased at a nominal extra cost, thereby providing capacity for possible future additional flow, garbage grinder, and sludge storage. A plastic sludge and gas deflector on the outlet as shown in Figure 3.3 is highly recommended.* The detention time for large septic tanks (see Table 3.6) should not be less than 24 to 72 hours. Schools, camps, theaters, factories, and fairgrounds are examples of places where the total or a very large proportion of the daily flow takes place within a few hours. For example, if the total daily flow takes place over a period of 6 hours, the septic tank should have a liquid volume equal to four times the 6-hour flow to provide a detention equivalent to 24 hours over the period of actual use. Septic tanks should be constructed of good-quality reinforced concrete, as shown in Figures 3.3 and 3.4. Precast-reinforced concrete and commercial tanks of metal, fiberglass, polyethylene, and other composition materials are also available. Because some metal tanks have a limited life, it is advisable that their purchase be predicated on their meeting certain minimum specifications (e.g., guaranteed 20-year minimum life expectancy, 12- or 14-gauge metal thickness, and acid-resistant coating).

If the septic tank is to receive ground garbage, its capacity should be increased by at least 50 percent. Some authorities recommend a 30 percent increase. Others recommend against garbage disposal to a septic tank.

Open-tee inlets and outlets as shown in Figure 3.3 are generally used in small tanks, and high quality reinforced concrete baffled inlets and outlets as shown in Figure 3.4 are recommended for the larger tanks.[16] Precast open concrete tees or baffles have, in some instances, disintegrated or fallen off; vitrified clay, cast-iron, PVC, ABS, or PE tees should be used. Cement mortar joints are unsatisfactory. Compartmented tanks are somewhat more efficient. The first compartment should have 60 to 75 percent of the total volume. A better distribution of flow and detention is obtained in the larger tank with a baffle arrangement of preferably rigid acid-resistant plastic. A minimum 16-inch manhole over the inlet of a small tank, and a 20- to 24-inch manhole over both the inlet and outlet of a larger tank are preferred.

An efficient septic tank design should provide for a detention period longer than 24 hours; an outlet configuration with a gas baffle to minimize suspended solids carryover (see Figure 3.3); maximized surface area to depth ratio for all chambers (ratio more than 2); and a multichamber tank with interconnections similar to the outlet design (open-tee inlet and outlet).[17]

---

*First suggested by Salvato in *Environmental Sanitation*, John Wiley & Sons, Inc., New York, 1958, p. 208.

TABLE 3.6 Suggested Septic Tank Dimensions

Population		Sewage Flow (gpd)[a]	Septic Tank[b,c]				Tile Field Laterals[c,d]	
Bedrooms	Persons		Length (ft)	Width (ft)	Depth (ft)	Volume (gal)	No. Length (ft)	Trench Width (in.)
2 or less	4	300	7.5	3.5	4	750		
3	6	450	9	4	4	1,000		
4	8	600	11	4	4	1,250		
5	10	750	10.5	5	4	1,500		
			12	5	4	1,750		
	(Tanks for		14	5	4	2,000		
	multiple		13	6	4	2,250		
	residences,		14.5	6	4	2,500	(DETERMINED BY SITE AND	
	schools,		16	6	4	2,750	SOIL PERCOLATION TEST)	
	camps,		17	6	4	3,000		
	fairgrounds,		16	6	5	3,500		
	factories)		16	7	5	4,000		
			19.5	7	5	5,000		
			20.5	8	5	6,000		
			24	8	5	7,000		
			23	8	6	8,000		
			28	8	6	10,000		

[a]The design basis is 75 gal per person and 150 gal per bedroom.
[b]Includes provision for home garbage grinder and laundry machine. Larger than minimum size septic tank is strongly recommended.
[c]See detail drawings for construction specifications.
[d]Based on the results of soil percolation tests. Discharge all kitchen, bath, and laundry wastes through the septic tank, but *exclude* roof and footing drainage, surface and groundwater, and softener wastes.

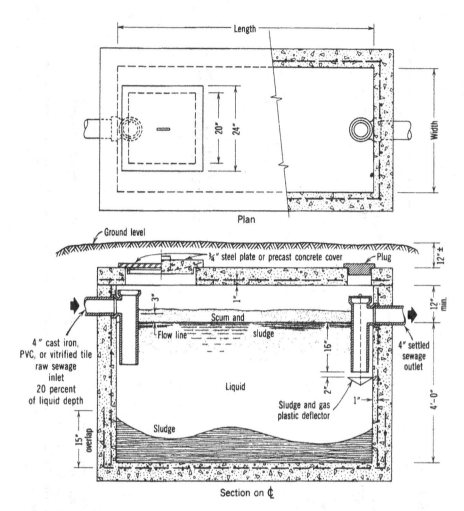

FIGURE 3.3 Details for small septic tanks. Recommended construction for small septic tanks: Top—Reinforced concrete poured 3- to 4-in. thick with two 3/8-in. steel rods per ft, or equivalent, and a 20- × 20-in. manhole over inlet, or precast reinforced concrete 1-ft slabs with sealed joints. Bottom—Reinforced concrete 4in. thick with reinforcing as in "top" or plain poured concrete 6in. thick. Walls—Reinforced concrete poured 4in. thick with 3/8-in. steel rods on 6-in. centers both ways, or equivalent; plain poured concrete 6in. thick; 8-in. brick masonry with 1-in. cement plaster inside finish and block cells filled with mortar. Concrete mix—One bag of cement to 2.25 in. of sand to 3 ft^3 of gravel with 5 gal of water for moist sand. Use 1:3 cement mortar for masonry and 1:2 mortar for plaster finish. Gravel or crushed stone and sand shall be clean, hard material. Gravel shall be 0.25 to 0.5 in. in size; sand from fine to 0.25 in. Bedding—At least 3 in of sand or pea gravel, leveled.

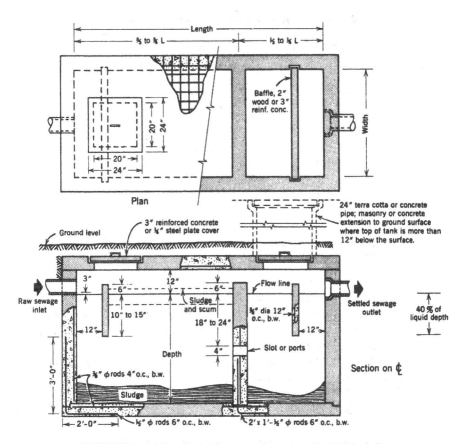

FIGURE 3.4 Details for large septic tanks. (See Table 3.6.)

Care of Septic Tank and Subsurface Absorption Systems

A septic tank for a private home will generally require cleaning every 3 to 5 years, depending on occupancy, but in any case, it should be inspected once a year. If a garbage disposal unit is used, more frequent cleaning is needed. Septic tanks serving commercial operations should be inspected at least every 6 months. When the depth of settled sludge or floating scum approaches the depth given in Table 3.7, the tank needs cleaning.[18] Sludge accumulation in a normal home septic tank has been estimated at 69 to 80 liters (18 to 21 gallons) per person per year.[19] The appearance of particles or scum in the effluent from a septic tank going through a distribution box is also an indication of the need for cleaning. Routine inspection and cleaning will prevent solids from being carried over and clogging leaching systems.

Septic tanks are generally cleaned by septic tank–cleaning firms. Sludge sticking to the inside of a tank that has just been cleaned would have a seeding effect and assist in renewing the bacterial activity in the septic tank. The septic tank

TABLE 3.7   Allowable Sludge Accumulation (in.)

Tank Capacity (gal)	Sludge Depth (in.) Tank Liquid Depth (in.)			
	30	36	48	60
250	4			
300	5	6		
400	7	9	10	
500	8	11	13	15
600	10	14	16	18
750	13	16	19	23
900	14	18	22	26
1,000	14	18	23	28
1,250		18	24	30

Source: Adapted from *Manual of Septic-Tank Practice*, U.S. PHS Pub. 526, DHEW, Cincinnati, OH, 1967.

should not be scrubbed clean. The use of septic tank cleaning solvents or additives containing halogenated hydrocarbon, aromatic hydrocarbon, or hazardous chemicals can cause carryover of solids, and clogging of absorption field as well as contamination of groundwater and should not be permitted.

An individual should never enter a septic tank that has been emptied, regardless of whether it is open or covered. Cases of asphyxiation and death have been reported due to the lack of adequate oxygen or presence of toxic gases in the emptied tank. If it should become necessary to inspect or make repairs, the tank should first be checked with a gas detector[20] for oxygen and toxic gases and thoroughly ventilated using a blower, which is kept operating.

Although soap, drain solvents, disinfectants, and similar materials used individually for household purposes are not harmful to septic tank operation unless used in large quantities, organic solvents and cleaners, pesticides, and compounds containing heavy metals could contaminate the groundwater and well-water supplies and should not be disposed of in a septic tank system. Also, sanitary napkins, absorbent pads, and tampons should not be disposed of in septic systems.

High weeds, brush, shrubbery, and trees should not be permitted to grow over an absorption system or sand filter system. It is better to crown the bed and seed the area with grass. If trees are near the sewage disposal system, difficulty with roots entering poorly joined sewer lines can be anticipated. About 2 to 3 pounds of copper sulfate crystals flushed down the toilet bowl once a year will destroy roots that the solution comes into contact with, but will not prevent new roots from entering. The application of the chemical should be done at a time, such as late in the evening, when the maximum contact time can be obtained before dilution. Copper sulfate will corrode chrome, iron, and brass, hence it should not be allowed to come into contact with these metals. Copper sulfate in the recommended dosage will not interfere with operation of the septic tank. A U.S. EPA registered herbicide, or a chemical foam, is also reported to be effective.[21]

Common causes of septic tank system failures are seasonal high groundwater; carryover of solids into the absorption field due to use of septic tank cleaning compounds, lack of routine cleaning of the septic tank, or outlet baffle disintegration or loss; excessive water use or hydraulic overloading; settlement of the septic tank, connecting pipe, or distribution box; and improper design and construction of the absorption system, including compaction and smearing of absorption trench bottom and sidewalls.

Corrective measures, once the cause is identified, include water conservation measures such as reduced water usage, low-flow toilets and showerheads, and reduced water pressure. Other measures to consider are cleaning of septic tank and flushing out distribution lines, and installation of additional leaching lines; installation of a separate absorption system and division box or gate for alternate use with the annual resting of existing system; lowering the water table with curtain drains; discontinuation of use of septic tank cleaning compounds; replacement of corroded or disintegrated baffles; replacement or releveling of distribution box; cleaning of septic tank at least every 3 years; and disconnection of roof, footing, and area drains.

Subsurface Soil Absorption Systems

The conventional system used for treating septic tank effluent is an absorption field or leaching pit. Where the soil is not suitable for subsurface disposal, a sand filter, evapotranspiration system, modified tile field system, aeration system, system in fill, mound system, stabilization pond, or some combination may be used (these systems are discussed in later sections). In all cases, it is important to avoid compaction of trench bottoms and soil of the absorption system or construction during wet weather.

Absorption Field System

Design standards and details for absorption fields are shown in Figures 2.1, 2.2, and 2.5 and in Table 3.8. Absorption field laterals should be laid in narrow trenches (18 to 24 inches), parallel to the contour and perpendicular to the groundwater flow, preferably not more than 24 inches below the ground surface, and spaced as shown in the Figure 3.5, to provide dispersion of the septic tank effluent over a larger area and promote aerobic conditions in the trenches. The highest seasonal groundwater level should be at least 2 feet, and preferably 4 feet, below the bottom of trenches. Where laterals must be laid at a greater depth, gravel fill around the laterals should extend at least to the topsoil (see Figure 3.5). After settlement and grading, the absorption field area should be seeded to grass.

When the total length of the laterals is between 500 and 1,000 linear feet, a siphon or pump should be installed between the septic tank and absorption system to distribute the sewage to all the laterals. If the lateral length is 1,000 to 3,000 linear feet, the system should be divided into two or four sections with alternating

TABLE 3.8 Suggested Minimum Standards—Subsurface Absorption Fields

Item	Material	Size	Grade	Minimum Governing Distances		
				To Building or Property Line	To Well or Suction Line	To Water Service Line
Sewer to septic tank	Cast iron for 10 ft from bldg. recommended.	4 in. min. diameter recommended	0.25 in. per ft max., 0.125 in. per ft min.	5 ft or more recommended	2 ft if cast-iron pipe, otherwise 50 ft	10 ft hor.[a]
Septic tank	Concrete or other app'd matl. Use a 1:2, 0.25:3 mix.	Min. 750 gal 4 ft liquid depth, with min. 16 in. M.H. over inlet.	Outlet 2 in. below inlet.	10 ft	50 ft	10 ft
Lines to distribution box and disposal system	Cast iron, vit. clay, concrete, or composition pipe.	Usually 4 in. diameter on small jobs.	$\frac{1}{8}$ in. per ft; but $\frac{1}{16}$ in. per ft with pump or siphon.	10 ft	50 ft	10 ft
Distribution box	Concrete, clay tile, masonry, coated metal, etc.	Min. 12 × 12 in. inside carried to the surface. Baffled.	Outlets at same level.	10 ft	100 ft	10 ft
Absorption field[b]	Clay tile, vit tile, concrete, composition pipe, laid in washed gravel or crushed stone, 0.75 in. to 2.5 in. size, min. 12 in. deep.	4 in. dia., laid with open joint or perforated pipe. Depth of trench 24 ft to 30 in.	$\frac{1}{16}$ in. per ft, but $\frac{1}{32}$ in. per ft with pump or siphon.	10 ft except when fill used, in which case 20 ft required	100 ft	10 ft (25 ft from any stream; 50 ft recommended)

Item	Construction/Material	Dimensions/Specification	Slope			Distance
Sand filter[b]	Clean sand, all passing 0.25 in. sieve with effective size of 0.30 to 0.60 mm and uniformity coefficient less than 3.5. Flood bed to settle sand.	Send 2-lb sample to health dept. for analysis 15 days before construction.	Laterals laid on slope $\frac{1}{16}$ in. per ft; but $\frac{1}{32}$ in. per ft with pump or siphon.	10 ft	50 ft	10 ft (25 ft from any stream; 50 ft recommended)
Leaching or seepage pit[b]	Concrete block, clay tile, brick, fieldstone, precast.	Round, square, or rectangle	Line to pit 0.125 in. per ft.	20 ft	150 ft plus in coarse gravel	20 ft (50 ft from any stream)
Chlorine contact-inspection tank	Concrete, concrete block, brick, precast.	2 × 4 in. and 2 in. liquid depth recommended	Outlet 2 in. below inlet	10 ft	50 ft	10 ft

[a] Water service and sewer lines may be in same trench, if cast-iron sewer with lead-caulked joints is laid at all points 12 in. below water service pipe; or sewer may be on dropped shelf at one side at least 12 in. below water service pipe, provided sound sewer pipe is laid below frost with tight and root-proof joints, which is not subject to settlement, superimposed loads, or vibration. Separate trenches are strongly recommended.

[b] *Manual of Septic-Tank Practice*, U.S. PHS Pub. 526 (1967), states that the leaching area should be increased by 20 percent where a garbage grinder is installed, and by 40 percent where a home laundry machine is also installed. It recommends that the gravel in the tile field extend at least 2 in. above pipe and 6 in. below the bottom of the pipe.

Note: A slope of $\frac{1}{16}$" per ft = 6.25' per 100' = 0.0052 ft per ft = 0.52 percent.

Note: All parts of disposal and treatment system shall be located above groundwater and *downgrade* from sources of water supply. The architect, builder, contractor, and subcontractor shall establish and verify all grades and check construction. Laundry and kitchen wastes shall discharge to the septic tank with other sewage. Increase the volume of the septic tank by 50 percent if it is proposed to also install a garbage grinder. No softening unit wastes, roof or footing drainage, surface water or groundwater shall enter the sewerage system. Where local regulations are more restrictive, they govern, if consistent with county and state regulations.

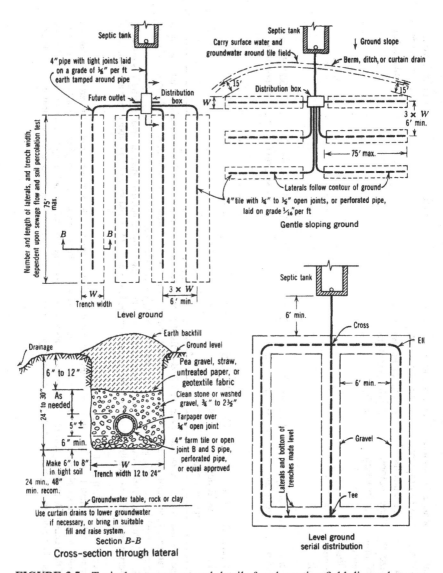

FIGURE 3.5   Typical arrangement and details for absorption field disposal system.

feed to each section. When lateral length exceeds 3,000 linear feet, it is advisable to investigate a secondary treatment process, although larger absorption systems can operate satisfactorily if the site and soil permeability are suitable to disperse effluent and prevent groundwater mounding.

The bottom of the absorption lateral trenches should be practically level to prevent sewage from running out the end of a trench or onto the ground surface. Laterals for absorption fields of less than 500 feet in total length, without siphon or pump, should be laid on a slope of $\frac{1}{16}$ inch/feet or 3 inches/50 feet. When siphons or pumps are used, the laterals should be laid on a slope of 3 inches/100 feet. Hydraulic loading rates should be kept between 0.25 and 0.5 gpd/ft^2.

Leaching Pit

Leaching pits, also referred to as seepage pits, are used for the disposal of settled sewage from septic tanks where the soil is suitable and a public water supply is used or where private well-water supplies are at least 150 to 200 feet away and at a higher elevation. Leaching pits work like a vertical absorption field, although they lack the areal extent of such fields. Pits are usually 10 to 20 feet deep and 6 to 12 feet in diameter. The bottom of the pit should be at least 2 feet, and preferably 4 feet, above the highest groundwater level. If this cannot be ensured, lateral absorption fields should be used. In special instances, where public water supply is available, suitable soil is found at greater depths, and groundwater can be protected, pits can be dug 20 to 25 feet deep or more, using precast perforated wall sections.

As part of the pit design, soil percolation tests should be conducted at mid-depth, at changes in the soil profile, and at the bottom of the proposed leaching pit. According to the 1980 *EPA Design Manual*, the weighted average of these tests should be used to obtain a design figure. Soil strata whose test results exceed 30 minutes per inch should not, however, be used in calculating the effective absorption area. Hydraulic loading rates for leaching pits should generally be kept between 0.4 and 0.8 gpd/ft^2, although the *EPA Manual* allows for up to 1.2 gpd/ ft^2 depending on the results of the percolation tests and the soil type present. The effective leaching area provided by a pit is equal to the vertical wall area of the pit below the inlet. Credit is not usually given for the pit bottom. A leaching pit is usually round to prevent cave-in. If precast perforated units are not used, the wall below the inlet is drywall construction—that is, laid with open joints, without mortar. Fieldstones, cinder or stone concrete blocks, precast perforated wall sections, or special cesspool blocks are used for the wall construction. Concrete blocks are usually placed with the cell holes horizontal. Crushed stone or coarse gravel should be filled in between the outside of the leaching pit wall and the earth hole. Table 3.9 gives the appropriate sizes for circular leaching pits and Figure 3.6 provides design details.

Since leaching pits concentrate pollution in a small area, their use should generally be avoided where the groundwater is a drinking water source. For this reason, use of such pits is frequently discouraged by regulatory agencies in favor of more diffuse systems, such as absorption fields.

Cesspool

Before septic tanks were widely used, sewage from individual dwellings was frequently discharged to cesspools. Cesspools are covered, open-joint or perforated walled pits that receive raw sewage. Their use is not recommended where the groundwater serves as a source of water supply. Many health departments

TABLE 3.9 Sidewall Areas of Circular Seepage Pits (ft²)[a]

Seepage[b] Pit Diameter (ft)	Thickness of Effective Layers Below Inlet (ft)									
	1	2	3	4	5	6	7	8	9	10
1	3.1	6	9	13	16	19	22	25	28	31
2	6.3	13	19	25	31	38	44	50	57	63
3	9.4	19	28	38	47	57	66	75	85	94
4	12.6	25	38	50	63	75	88	101	113	126
5	15.7	31	47	63	79	94	110	126	141	157
6	18.8	38	57	75	94	113	132	151	170	188
7	22.0	44	66	88	110	132	154	176	198	220
8	25.1	50	75	101	126	151	176	201	226	251
9	28.3	57	85	113	141	170	198	226	254	283
10	31.4	63	94	126	157	188	220	251	283	314
11	34.6	69	104	138	173	207	242	276	311	346
12	37.7	75	113	151	188	226	264	302	339	377

[a] Areas for greater depths can be found by adding columns. For example, the area of a 5-ft diameter pit, 15 ft deep is equal to 157 + 79, or 236 ft.
[b] Diameter of excavation.

Source: Design Manual, Onsite Wastewater Treatment and Disposal Systems, U.S. EPA, October 1980, p. 237.

prohibit the installation of cesspools because pollution could easily travel from cesspools to wells used for water supply. Where cesspools are allowed, they should be located downgradient and 200 to 500 feet away from sources of water. The bottom of the cesspool should be at least 4 feet above the highest groundwater level. Cesspool construction is the same as a leaching pit. The cesspool system can be made more efficient under such circumstances by providing a tee outlet, as shown in Figure 3.6, with the overflow discharging to an absorption field or leaching pit. A preferable alternative would be to replace the cesspool with a septic tank and absorption field.

Dry Well

A dry well, which is constructed in the same way as a leaching pit, is used where the subsoil is relatively porous for the underground disposal of clear rainwater, surface water, or groundwater from roofs and/or basement floor drains. Roof or basement drainage should never be discharged to a septic tank because its volume will seriously overload such systems. Dry wells should not be used for the disposal of toilet, bath, laundry, or kitchen wastes. These wastes should be discharged to a septic tank. In some cases, roof drainage may be discharged to a nearby watercourse if permitted by local regulations. Dry wells should be located at least 50 feet from any water well, 20 feet from any leaching portion of a sewage disposal system, and 10 feet or more from building foundations or footings.

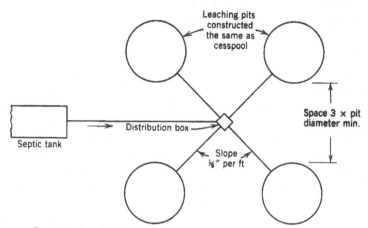

Septic tank – distribution box – leaching (seepage) pit plan

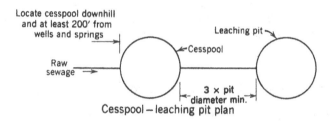

Cesspool – leaching pit plan

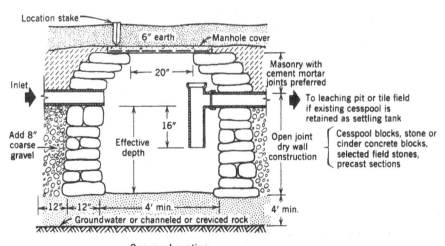

Cesspool section

FIGURE 3.6 Leaching pit details.

SMALL WASTEWATER DISPOSAL SYSTEMS
FOR UNSUITABLE SOILS OR SITES

General

Wastewater treatment systems in this category are usually more complex in design and more costly than the conventional septic tank systems previously described. These systems include the modified absorption system, the capillary seepage trench, the absorption-evapotranspiration system, the sand filter system, the aerobic treatment unit, the septic tank mound system, the raised bed system, the underdrained absorption system, and the evapotranspiration system. Effluent from the sand filter system requires must be disposed of to a subsurface or land disposal system, if discharge to surface waters is not permitted. Low-pressure, vacuum, and cluster systems may also be appropriate in certain situations. Alternative systems are considered when a conventional system cannot be expected to function satisfactorily because of high groundwater, a relatively impervious formation is close to the surface, space is limited, or where a highly porous formation exists and protection of nearby well-water or surface water supplies is a major concern. Alternative systems are usually quite expensive and plans and an engineer's report are normally required for review and approval *prior to construction.*

Modified Septic Tank Soil Absorption System

A conventional subsurface soil absorption system is usually designed on the basis of soil percolation rates not exceeding 1 inch in 60 minutes. However, less permeable soils can still be used as long as application rates are reduced. Ryon in his original notes recommended the following application rates for 60-minutes or poorer soils.[22]

Time to fall 1 in. (hours)	Safe application rate (gpd/ft^2)
1	0.4
1	0.3
2.5	0.2
3	0.14
5	0.07
10	0.03

Construction of the modified system is similar to a conventional system. Intermittent or alternating dosing (siphon, pump, tipping bucket) is usually required, particularly if the total length of distributors exceeds 500 feet. Design for a relatively tight soil still uses the conventional soil percolation test but carries it beyond the 1 inch/60 minute test to a point of constant rate. Moisture loss due to evaporation and transpiration is not credited but taken as a bonus.

The following example gives the modified system design for a "tight" soil site of fairly uniform composition.

Example

Design a subsurface leaching system for a daily flow of 300 gal. The soil test shows 0.25 in./hr and a permissible settled sewage application of 0.10 gpd/ft^2.

$$\text{Required leaching area} = \frac{300}{0.10} = 300\,\text{ft}^2$$

If trenches 3 feet wide with 18-inches of gravel underneath lateral distributors are provided, each linear foot of trench can be expected to provide 5 ft^2 of leaching area. The required trench = 3,000/5 = 600 linear feet, or 8 laterals, each 75-feet long, spaced 9 feet on center. Two gravel beds, 50 by 60 feet, can also provide the leaching area needed to compensate for the loss of the sidewall trench infiltration and dispersion area (see Figure 3.7). Use an alternating dosing device. This occupies the same land area as the absorption field. Evapotranspiration can be enhanced by incorporating sand trenches or funnels in the gravel between the distributors.

Capillary Seepage Trench

Another alternative to traditional absorption field design is the use of a capillary seepage trench. The capillary seepage trench is similar to a conventional seepage trench except that it has an impermeable liner at the bottom of the trench, which extends part way up the trench's sidewalls. As a result, sewage effluent collects

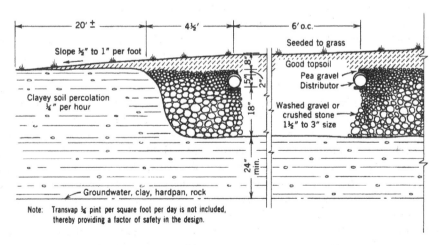

FIGURE 3.7 Absorption bed for a tight soil. Curtain drains may be needed to lower groundwater level and beds may need to be crowned to shed rainwater.

along the entire length of the trench and moves both upward and horizontally by capillary action before percolating downward. This modification in trench design results in a more even distribution and a longer time of contact. Fly ash is often used as trench fill material because it allows for more rapid capillary movement of the effluent and provides an increased surface area for microbial growth. Because a capillary seepage trench does not use the bottom of the trench as an absorption area, the trenches need to be longer than a conventional absorption trench.

Raised Bed Absorption-Evapotranspiration System

If clay, hardpan, groundwater, or rock is found within 4 feet of the ground surface, sewage disposal by a conventional absorption field is not recommended. Instead, a raised or build up area can be constructed using 12 to 18 inches of porous earth having a percolation rate of at least 1 inch in 120 minutes. A sandy, loamy, gravely soil is preferred and should be approximately 70 to 80 percent (by weight) medium to coarse sand (0.25 to 2.0 mm E.S.); 10 to 20 percent silt, fine sand and clay (0.25 or less mm E.S.); and not more than 10 percent gravel (2.0 mm to 7.5 cm E.S.). Preliminary percolation tests of the undisturbed soil can give an indication of its suitability. A long, narrow absorption field with fewer and longer laterals (75 to 100 feet), perpendicular to the groundwater flow, will provide greater area for underground wastewater dispersion, minimize possible groundwater mounding, and make seepage out of the toe of the feathered fill less likely.

The fill should be spread in 6-inch layers using a lightweight crawler tractor to achieve a uniform soil density without channels or holes. The fill soil should not be spread when it is wet or compacted. Also, sufficient fill should be provided so that the bottom of the absorption trenches are at least 2 feet above the highest ground water level, rock, clay, or hardpan. After soil stabilization (at least 6 months minimum), percolation tests should be run at four to six locations. The resultant percolation rates should be between 1 inch in 8 minutes and 1 inch in 31 minutes to prevent premature clogging, ensure effluent retention in the fill, and obtain maximum purification of sewage effluent before it reaches groundwater or ground surface.

An example of a raised bed system that uses both absorption and evapotranspiration is presented in Figure 3.8. In light of the uncertainties associated with these systems (e.g., uneven soil settlement and unreliability of percolation tests in fill), a conservative design is considered prudent. A fill percolation rate of 1 inch in 31 minutes, or 0.45 gpd/ft^2 (EPA, Table 3.4), which would correspond with a basal area application rate of 0.14 gpd/ft^2, is recommended. The basal area is the absorption field area extending 2.5 feet beyond the outer edges and ends of the distribution trenches. The absorption system should be dosed two to three times per day using a pump or siphon. The dose should be about 60 percent of the volume of the distribution lines. Intermittent operation will permit full dosage of the distribution laterals and enhance dispersion of the wastewater over the entire absorption field.

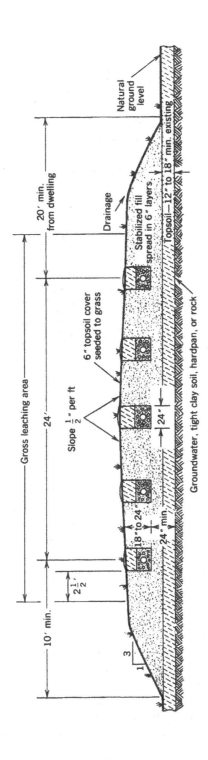

Section A–A

FIGURE 3.8 Raised bed sewage disposal system. Design basis: 300 gpd and a transvap-percolation rate of 0.45 gpd/ft² or trench bottom (total of 335 feet, 2 feet wide) or 0.14 gpd/ft² of system gross leaching area (72 × 31 feet). Pump or siphon distribution is usually required with dosing two or three times per day.

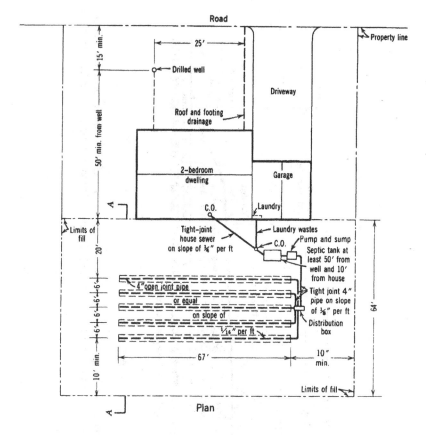

FIGURE 3.8 (*continued*)

The entire surface of the raised bed should be covered with at least 6 inches of topsoil, graded to enhance rainfall runoff, and seeded to grass. A diversion ditch or berm should be installed upgrade to divert surface runoff around the system. Also, a curtain drain may be needed in areas of high groundwater if the bottom of the trenches cannot be kept at least 2 feet above groundwater. If clay or hardpan is intercepted, the curtain drain trench and collection pipe should extend at least 6 inches into the impervious formation.

Septic Tank Sand Filter System

Sand filters can be used when conventional subsurface absorption systems are unlikely to function satisfactorily because of soil conditions or rock, or where space is very limited and discharge to a surface water or ditch is permissible. Settled sewage is typically distributed over the top of a sand filter bed by means of perforated, open-joint pipe as shown in Figure 3.9. The sewage is then filtered and oxidized through 24 to 30 inches of sand, on which a film of aerobic and nitrifying

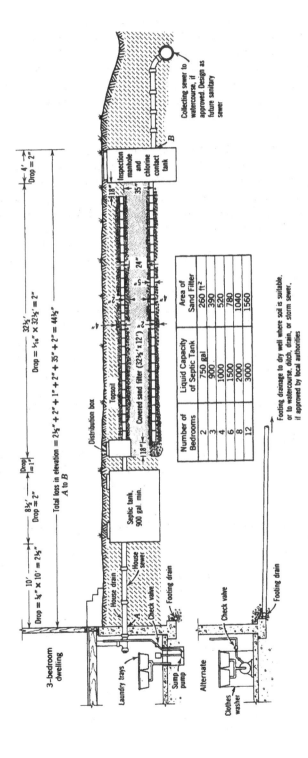

Number of Bedrooms	Liquid Capacity of Septic Tank	Area of Sand Filter
2	750 gal	260 ft²
3	900	390
4	1000	520
6	1500	780
8	2000	1040
12	3000	1560

FIGURE 3.9 Section through covered sand filter system. Design basis is 150 gal per bedroom and filter rate 1.15 gpd/ft². Larger capacity septic tank (50 to 100 percent larger than minimum) is strongly recommended.

organisms form. Sand filters can provide relatively high removal efficiencies for most constituents (see Table 3.10). Such effluents should not cause a nuisance in undeveloped areas, but should be chlorinated if discharged in locations accessible to children or pets because microorganisms associated with disease transmission, although greatly reduced, are still present. Because operation of a sand filter is dependent on the rate and strength of sewage application, it is essential that suspended solids carryover from the septic tank be kept at a minimum.

The recommended sizes of covered sand filters for private homes are shown in Figure 3.9 and are based on a flow of 150 gpd per bedroom and a settled sewage application rate of 1.15 gal/ft^2 of sand filter area per day. It is extremely important to *use a proper sand* (i.e., coarse sand passing 0.25 inch mesh screen with effective size of 0.3 to 0.6 mm and a uniformity coefficient not greater than 3.5). Distributor and collector lines for the filter should laid at exact grade and a topsoil cover, preferably not exceeding 8 to 12 inches, should be placed over the gravel-covered distributor lines. Geotextile fabric should be installed between the topsoil and gravel. Filter rates for covered sand filters should not exceed 50,000 gpd/acre for settled domestic sewage and 100,000 gpd/acre for temporary summer. Rates for open filters can range from 75,000 to 100,000 gpd/acre for settled sewage and 200,000 to 400,000 gpd/acre for secondary treated sewage. Loading should normally not exceed 2.5 lb. of either 5-day BOD or suspended solids per 1,000 ft^2/day. Recommended filter rates related to climate and sand size are also given in the *Manual of Septic-Tank Practice*.[23]

In freezing weather open filters will require greater operational control and maintenance. Scraping the sand before freezing weather into furrows about 8 inches deep with ridges 24 to 48 inches apart will help maintain continuous operation, as ice sheets will form between ridges and help insulate the relatively warm sewage in the furrows. Greenhouse covers are very desirable and help ensure continuous operation of the filters; however, such covers are expensive.

TABLE 3.10   Typical Efficiencies of Subsurface Sand Filters[a]

Determination	Percent Reduction
Bacterial per ml, Agar, 36°C, 24 hr	99.5
Coliform group, MPN per 100 ml	99.6
BOD, 5-day (mg/l)	97
Suspended solids (mg/l)	88
Oxygen consumed (mg/l)	75
Total nitrogen (mg/l)	42
Free ammonia	94
Organic	72

[a]Effluent will contain 5.2 mg/l dissolved oxygen and 17 mg/l nitrates.

Source: J. A. Salvato, Jr., "Experience with Subsurface Sand Filters," *Sewage Ind. Wastes*, 27, *8*, 909–916 (August 1955).

Wood-frame covers may also be used. Regular maintenance is essential, including raking and weeding.

Aerobic Sewage Treatment Unit

Another type of treatment unit that can be used when subsurface absorption systems are not practical is the self-contained, prefabricated aeration unit. Although effluent from such units is low in suspended solids and BOD, it will still need further treatment for other constituents, such as nitrate. Such treatment can include sand filtration and/or chlorination prior to discharge to a stream, if permit, or discharge to an oxidation pond or irrigation system. Routine maintenance and operation of the unit must be ensured by a maintenance contract or other means. Design details for extended aeration and activated sludge treatment plants are given later. Small rotating biological contractors with 2- to 4-foot diameter discs and for flows of 350 to 1,500 gpd or more are also available. Their application and limitations are similar to the above aeration units. In some locations, where tight soil exists and ample property is owned, the waste stabilization pond, irrigation, oxidation ditch, or overland flow system design principles may be adapted to small installations. Design information is given under "Sewage Works Design—Small Treatment Plants."

Septic Tank Mound System

The septic tank mound system was originally developed in North Dakota in late 1947. Numerous refinements to this system have been made over the years, including those noted by Salvato,[24] Bouma, et al.,[25] and Converse, et al.[26] In the mound system, the absorption area is raised above the natural soil to keep the bottom of the trenches at least 2 feet above groundwater, bedrock, or relatively impermeable soil. In this respect, the system serves the same purpose as the built-up soil absorption system previously described. Where it differs is in the type and size of fill material and method used to apply septic tank effluent to the mound system (see Figure 3.10).

The texture and structure of the fill soil used (see Table 3.3) will affect its tendency to clog and its purification capacity; appropriate settled sewage application rates for the range of soil types used in mounds are listed in Table 3.11. In the mound system developed by Converse et al., a 2-foot bed of clean coarse sand was used (i.e., 0.5 to 1.0 mm effective size with less than 5 to 6 percent silt and clay and less than 15 to 16 percent fine and very fine sand).[27] To provide sufficient vertical and lateral spreading of the percolating wastewater, 2 feet of sand and 1 foot of natural topsoil is typically used.[28] However, experience indicates that mound systems can also be constructed (1) wholly in the natural soil, (2) partly in the natural soil, or (3) completely above the natural ground surface. Because design of a mound system is complex (i.e., involving hydraulic conductivity determinations, hydraulic analyses, pump/siphon selection and sand analyses), it should only be done by a qualified professional, preferably one with experience with mounds.

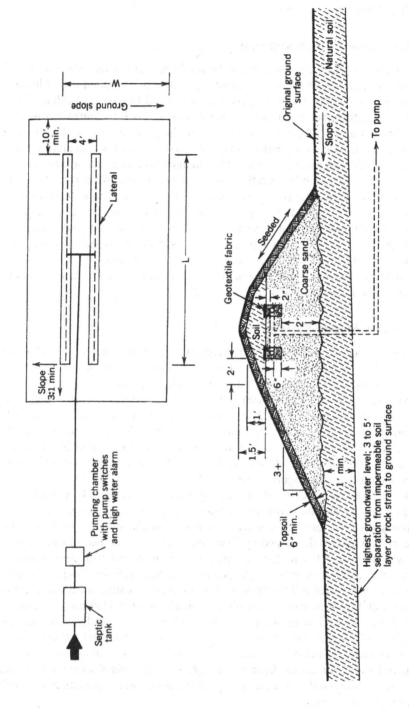

FIGURE 3.10 Details of a mound system using trenches. Basal area is L × W. Sloping ground. Laterals may be placed in a gravel bed.

TABLE 3.11 Suggested Settled Sewage Application Rates

Percolation Rate (min/inch)	Soil Type	Application Rate[a] (gpd/ft^2)	Hydraulic Conductivity[b] (gpd/ft^2)
Less than 1	Gravel, coarse sand	Unsatisfactory	—
1–5	Coarse[d] to medium sand	1.2[c]	9,600[d]–1,200
6–15	Fine sand,[d] loamy sand	0.8	540[d]–100
16–30	Sandy loam, loam	0.6	100–20
31–60	Loam, porous silt loam	0.45	20–5
61–120	Silty clay loam, clay loam	0.2	1–0.1
121 and greater	Clay loam, clay	See Ref. 14.	—

[a] Approximate vertical hydraulic conductivity.
[b] Very approximate horizontal hydraulic conductivity. Make field or laboratory determination.
[c] Reduce to 0.8 gpd/ft^2 if groundwater supplies may be affected.
[d] Fletcher G. Driscoll, *Groundwater and Wells*, Johnson Division, St. Paul, MN, 1986, p. 78. Coarse sand 0.84 to 1.17 mm size; fine sand 0.2 to 0.3 mm size.

Example 1

Design of a residential mound system on level ground for a flow of 300 gpd. Natural soil has a percolation of 1 inch in 120 minutes, or 0.2 gpd/ft^2.

1. Absorption area $= \dfrac{\text{Daily flow}}{\text{Sand infiltration rate}} = \dfrac{300}{1.2} = 250 \, \text{ft}^2$
   This area can be provided by two 2-foot wide trenches 62.5 feet long.

2. Trench and lateral spacing = space laterals 4 feet on center.* Gravel trenches may be combined into one gravel bed $(4 + 1 + 1) = 6$ feet wide.

3. Mound height (at center) = Sand depth + Gravel depth + Soil cap and topsoil (See Figure 3.10.) $= 2 + 0.75 + 1.5 = 4.25$ feet

4. Mound length = Lateral length + End barriers (mound height $\times$ 3 on 1 slope $\times$ 2 $= 62.5 + (4.25 \times 3) \times 2 = 88$ feet

5. Mound width (including topsoil) = 0.5 $\times$ Trench width $\times$ 2 + Trench spacing on center + (Mound height at edge of trench + 3 on 1 slope) $\times$ 2 $= (0.5 \times 2 \times 2) + 4 + (3.75 \times 3) \times 2 = 2 + 4 + 22.5 = 28.5$ feet

6. Basal area: Required $300/0.2 = 1{,}500 \, \text{ft}^2$.
   Provided $62.5 \times 28.5 = 1{,}781 \, \text{ft}^2$, excluding end areas

7. Distribution system: See Table 3.11 for lateral length and diameter and corresponding hole diameter and spacing. Make manifold 2 inches in diameter for pumping.

*A 1-in. diameter pipe holds .041 gal; a 1.25-in. pipe .064 gal; a 1.5-in. pipe .092 gal; and a 2-in. pipe .164 gal

8. Pressure distribution: For pumping chamber volume, pump size, and dosing volume, see Converse[29], and for siphon discharge. Include valve on pump discharge line for fine adjustment of pump head and discharge.

9. Pumping chamber = 500 gal capacity for 1-, 2-, 3-, or 4-bedroom dwelling is recommended.

10. Dose volume = 0.25 daily flow and at least 10 times lateral volume when pump is used.

11. Pump size: The pumping head is the difference in elevation between pump and lateral invert, plus friction loss in the pump discharge line, manifold, fittings, valve, laterals, orifices, plus head at end of lateral (2 feet). Pump capacity is 20 gpm for 1-bedroom dwelling (150 gpd) and 0.25-inch diameter orifice spaced 30 inches on center; 36 gpm for 2-bedroom, 54 gpm for 3-bedroom, and 70 gpm for 4-bedroom. For 7/32-inch diameter orifice, use a 15-gpm pump for 1-bedroom dwelling, 28-gpm for 2-bedroom, 41-gpm for 3-bedroom, and 54-gpm for 4-bedroom.

Based on historic experiences with sand filters, sands with effective size less than 0.2 to 0.35 mm can be expected to clog with a dosage of 1 to 1.5 gpd/ft^2. Also, during construction, compaction of the sand fill and the natural soil under and around the dispersal area should be kept to a minimum.

Electric Osmosis System

In this process, septic tank effluent discharged to a conventional subsurface absorption system in a soil having a percolation slower than 1 inch in 60 minutes is disposed of by evapotranspiration. Mineral rock-filled anodes adjacent to the trench and coke-filled cathodes with graphite cores a short distance away generate a 0.7 to 1.3 V potential, causing soil water, claimed to be removed by evapotranspiration, to move to the cathodes. These systems have been used successfully in several states.

Septic Tank Evapotranspiration System

An evapotranspiration system can be used, when the available soil has no absorptive capacity or where little or no topsoil exists over clay, hardpan, or bedrock, provided that a water balance study shows the evapotranspiration plus runoff exceed precipitation infiltration plus inflow. It can also be used when the groundwater level is high, provided the system is provided with a watertight liner on the bottom and sides to exclude the groundwater from the transvap bed. If an impermeable liner is not provided, elevation of the bed or curtain drains may be necessary if seasonal high water is a problem. The design of a transvap system is based on maintenance of a favorable input-output water balance.

Evaporation from water surfaces varies from about 20 inches per year in the northeastern United States to 100 to 120 inches in some southwestern areas, and that evaporation from land areas will be approximately one-third to one-half these values. Brandes found that over a 15-month period, 58 percent of the total precipitation on a sand filter in Ontario, Canada, left the filter through evapotranspiration.[30] This value is significantly higher than the average values for hydrologic water recycling given historically by McGauhey:[31]

Evaporation	30 percent
Evapotranspiration	40 percent (from soil mantle)
Surface runoff	20 percent
Groundwater storage	10 percent

Successful operation of a transvap system is largely dependent on runoff, surface vegetation, soil cover, capillarity, and evapotranspiration, in addition to controlled wastewater flow *to maintain a favorable water balance*. Plant roots can reach a depth of about 24 inches in well-developed absorption beds and take up wastewater. Maintenance of a permeable soil structure and microbial and macroscopic organisms is essential to minimize system clogging and failure, as previously explained.

Figure 3.11 provides design and construction details of a transvap disposal system, which uses sand and gravel beds to provide storage during the periods when transpiration and evaporation is low or zero. The sand ridges and sand bed are essential to provide capillarity. Soil evaporation can average one-third to one-half of lake evaporation for 6 months of a year in which average lake

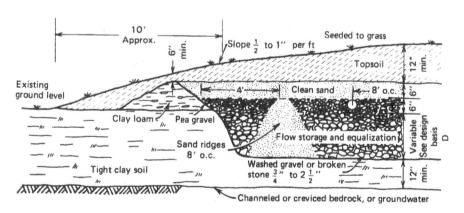

FIGURE 3.11 Transvap sewage disposal system in tight soil. (Raise bed as necessary if groundwater or bedrock is a problem.) Clean washed sand, 0.1 mm effective size for up to 12 to 16 in. gravel and sand depth, and 0.05 mm sand for up to 24 in. gravel and sand depth. Sand ridges are necessary to obtain capillarity and promote evapotranspiration. Permeable geotextile fabric is recommended over the sand ridges and in place of the 6 in. of sand over the gravel. Add 6- or 8-in.diameter perforated risers in and to bottom of gravel bed for inspection and emergency pump-out. Pressure distribution is usually required. Silt in sand will increase capillary rise.

evaporation is 30 inches/year. Sublimation during the snow-covered nongrowing season, although small, can contribute to moisture removal from the system.

Example 2

Rational design of a transvap sewage disposal system for year-round occupancy.[32]

Design basis

1. Sewage flow: 200 gpd = 6,083 gal/month = 73,000 gal/year
2. Bed surface cover: sandy, silty, clayey loam topsoil, and lawn grass, crowned 1 in./ft.
3. Use gravel bed (40 percent void space) with sand ridges, or with gravel ridges. See Figure 3.11.

Required area of evapotranspiration bed

$$\text{Outflow} = \text{Inflow}$$

$$(ET \times A) + (E \times A) = Q + (I \times A),$$

where

ET = evaporation from bed during the growing season, gal/ft^2 (as noted above)

A = area of bed, ft^2

E = land evaporation from bed during the nongrowing period, except when the bed is frozen or snow covered, gal/ft^2 (as noted above)

Q = septic tank inflow, gal/year

I = infiltration, precipitation inflow, gal/ft^2 (as noted above)

The above equation may be rewritten:

$$A = \frac{Q}{ET + E - 1};$$

Using the design basis and the parameters listed in Table 3.12:

$$A = \frac{200 \times 365}{3.12 \times 5 + .934 \times 4 - 12.076} = \frac{73,000}{7.26} = 10,055 \text{ ft}^2$$

Storage (Y) required during 7-month nongrowing period (J,F,M,S,O,N,D) (Or make monthly water balance study):

$$Y = Q_1 + I_1 E_1 = \text{sewage flow} + \text{infiltration} - \text{soil evaporation}$$

$$= 6083 \times 7 + (.206 + .196 + .810 + .966 + .810 + .872 + .271)$$
$$\times 10,055 \ (.934 \times 4) \times 10,055$$

$$= 42,583 + 41,537 - 37,565$$

$$= 46,555 \text{ gal}$$

TABLE 3.12 Precipitation, Infiltration, and Evaporation Rates

	Jan	Feb	Mar	Apr	May	Jun
Precipitation (ppt.), in.	2.2	2.1	2.6	2.7	3.3	3.0
Percent ppt. infilt. in bed	15	15	50	85	85	85
Infiltration ppt., gal/ft^{2a}	0.206	0.196	0.810	1.430	1.748	1.589
Land evaporation, gal/ft^{2b}	0	0	0.934	in ET	in ET	in ET
Evapotranspiration, gal/ft^{2c}	0	0	0	3.12	3.12	3.12
	Jul.	Aug	Sep	Oct.	Nov	Dec.
Precipitation (ppt.), in.	3.1	2.9	3.1	2.6	2.8	2.9
Percent ppt. infilt. in bed	85	85	50	50	50	15
Infiltration ppt., gal/ft^{2a}	1.642	1.536	0.966	0.810	0.872	0.271
Land evaporation, gal/ft^{2b}	in ET	in ET	0.934	0.934	0.934	0
Evapotranspiration, gal/ft^{2c}	3.12	3.12	0	0	0	0

[a]Infiltration = percent infiltration × precipitation/month × 0.623, in gal/ft^2/month.
[b]Land evaporation = 0.6 lake evaporation of 30 in./year = 0.6 × 30 ÷ 12 × 0.623 = 0.934 gal/ft^2/month.
[c]Evapotranspiration = 25 in. in growing season × 0.623 ÷ 5 = 3.12 gal/ft^2/month.
Note: 0.623 = gal/ft^2 per in. precipitation.

Bed depth (D) to provide required storage

$$D = \frac{Y}{A \times 7.5\,\text{gal/ft}^2 \times \text{ void surface}} = \frac{46{,}555}{10{,}055 \times 7.5 \times .4}$$

$$= 1.54\,\text{ft (This is within the fine sand capillarity range.)}$$

The storage required can be determined by means of a water balance study and by the graphical mass diagram or Rippl method. The weekly or monthly inflow (consisting of the precipitation minus runoff, or infiltration, plus wastewater input flow) and the outflow (evaporation and transpiration) is tabulated or plotted against time in weeks or months. The difference between cumulative inflow and cumulative outflow is the storage required at any point in time. Beck[33] recommended an evapotranspiration rate of 0.482 gpd/ft^2 for raised sand beds, while Lomax found that evapotranspiration systems composed of 1.65 feet depth of sand could dispose of 0.08 gpd/ft^2 satisfactorily in area, which had an annual precipitation of 55 inches.[34]

Water Conservation

Although not technically a disposal system, water conservation can provide a simple and economic way of reducing hydraulic loads on an existing septic system. Water conservation measures include the installation of low-flow fixtures, such as toilets and faucet aerators, which in and of themselves can reduce wastewater flows by up to 50 percent, maintenance of proper water pressures, elimination of leaks and drips, and discontinuation of the use of garbage disposals.

SEWAGE WORKS DESIGN – SMALL TREATMENT PLANTS

Small treatment plants typically discharge treated effluents to a surface water body as contrasted to septic systems, which discharge to groundwater systems. Surface discharges require a permit from a regulatory agency and allowable effluents are based on the minimum average 7-day flow expected to recur once in 10 years (MA7CD10), upstream and downstream discharges, and downstream uses. Some of the more common flow diagrams for small sewage treatment plants are illustrated in Figure 3.12 and predesigned and prefabricated units are available.

As noted in Figure 3.12, bar screens or comminutors and grit chambers are provided ahead of pumping equipment and settling tanks to remove larger solids. If secondary treatment is needed, primary treatment units should be designed to have water level of sufficient height to permit gravity flow to the both the secondary units and to the receiving stream without additional pumping.

Location of small treatment plants should take into consideration the type of plant desired, the availability of supervision, the location of the nearest residence, the receiving watercourse, the likelihood of flooding, prevailing winds and natural barriers, and the cost of land. A distance of 400 feet from the nearest residence is frequently recommended, although distances of 250 to 300 feet should prove adequate with good plant supervision. Oxidation ponds and lagoons should be located at least 0.25 to 0.5 mile from the nearest human habitation.

Disinfection

The need for disinfection of sewage effluent depends on the probability of disease transmission by the ingestion of contaminated water or food including shellfish, by contact, and by aerosols.

This probability should be balanced against the effects that chlorination can have on aquatic life.[35] Normal chlorination does not destroy all pathogenic viruses, fungi, bacteria, protozoa, and helminths. Although chlorine as a disinfectant is discussed here, it does not preclude dechlorination or the use of other disinfectants. Also, wastewater must be adequately treated in the first place for the disinfectant (usually chlorine) to be effective.

Chlorination treatment of raw sewage is not reliable for the destruction of pathogenic organisms since solid penetration is limited. The dosage of chlorine required to produce a 0.5 mg/l residual after 15-minute contact has been estimated for different kinds of sewage (see Table 3.13). An early study found that less than 250 coliform organisms per 100 ml remained in treated sewage if a chlorine residual of 2.0 to 4 mg/l is maintained in the effluent after 10-minute contact.[36] Other tests found that with no mixing, at least twice the chlorine residual had to be maintained in the treated sewage for 10 minutes to get results approximately equal to those obtained with mixing.

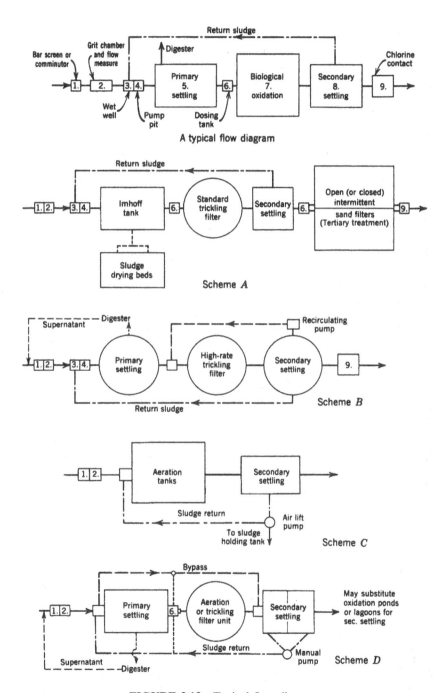

FIGURE 3.12   Typical flow diagrams.

TABLE 3.13 Probable Chlorine Dosages to Give a Residual of at Least 0.5 mg/l after 15-Min Retention in Average Sanitary Sewage or Sewage Effluent[a]

Type of Sewage Effluent	N.Y. State[d]	Dunham[e]	White[f]	Suggested Chlorine Dosages, (mg/l)[b] Griffin[g,h]	Imhoff and Fair[i]	Metcalf & Eddy[l]	GLUMRB[j]	EPA[k]
Raw sewage			8-15 15-30	6-12 fresh to stale 12-25 septic	6-25 fresh to stale	6-25		8-15 fresh
Imhoff tank or settled sewage	20-25	5-20	8-15	5-10 fresh to stale 12-40 septic	5-20	5-20	20	
Imhoff tank or settled sewage	20-25	5-20	8-15	5-10 fresh to stale 12-40 septic	5-20	5-20	20	
Trickling filter	15	3-15	3-10	3-5 normal 5-10 poor	3-20	3-15	15	
Activated sludge	8	2-8	2-8	2-4 normal 5-8 poor	2-20	2-8	8	10-15 pack plant
Intermittent sand	6	2	1-5	1-3 normal 3-5 poor	1-10		6[c]	1-5
Chemical precipitation				3-6	3-20	2-6		

[a]WHO suggests 0.5 mg/l free residual chlorine after 1 hour to inactivate viruses (after secondary treatment) with turbidity < 1.0 JTU. Combined chlorine, mostly monochloramine, is normally produced which is a slow-acting disinfectant. Eight to 10 mg/l of chlorine is needed to neutralize each mg/l of ammonia before free chlorine is produced. Most secondary effluents contain more than 1.0 mg/l ammonia.

[b]12 mg/l = 1 lb per 10,000 gal. Each mg/l chlorine in sewage effluent reduces the BOD about 2 mg/l. No appreciable industrial wastes.

[c]Tertiary filtration effluent and for nitrified effluent.

[d]Manual of Instruction for Wastewater Treatment Operators, Vol. 1, New York State Dept. of Environmental Conservation, Albany, May 1979, pp. 6–9.

[e]Military Preventive Medicine, Military Publishing Co., Harrisburg, Pa., 1940.

[f]Handbook of Chlorination and Alternative Disinfectants, Van Nostrand Reinhold, New York, 1992.

[g]Public Works Magazine, Ridgewood, N.J. (October 1949), p. 35.

[h]Operation of Wastewater Treatment Plants, Water Pollution. Control Fed., Washington, DC, 1970, p. 144.

[i]Sewage Treatment, John Wiley & Sons, Inc., New York, 1956.

[j]Recommended Standards for Sewage Works, 1997 ed., Great Lakes-Upper Mississippi River Board of State and Provincial Public Health and Environmental Managers.

[k]Design Manual, Onsite Wastewater Treatment and Disposal Systems, U.S. EPA, October 1980, p. 165.

[l]Wastewater Engineering Treatment Disposal Reuse, revised by George Tchobanoglous, McGraw-Hill, New York, 1979, p. 376.

Although chlorine is available as a pressurized liquid, liquid chlorine is not ordinarily required or economical to use at small sewage treatment plants. Instead, either calcium hypochlorite, which is a powder containing 70 percent or less available chlorine, or sodium hypochlorite, which is a solution containing 15 percent or less available chlorine, is generally used. Both the powder and solution are mixed and diluted with water to make a 0.5 to 5.0 percent solution, which is then added to the sewage by means of a feeder known as a hypochlorinator. Positive-feed hypochlorinators are generally preferred because of their greater dependability, but gravity flow stack or tablet erosion-type chlorinators are also available.

Combined chlorine, mostly monochloramine, is produced in the conventional chlorination of sewage effluent. This result is to be expected, since most secondary effluents contain substantially more than 1 mg/l of ammonia, which alone requires 8 to 10 mg/l of chlorine for neutralization, before any free residual chlorine is produced. Although slow-acting, combined chlorine is effective in reducing fecal coliforms to 200 mg/l or less, with sufficient contact time. Conventional chlorination of municipal wastewater to the combined chlorine residual level also yields relatively small amounts of chlorinated organic compounds, which are suspected of being carcinogenic. This outcome is in contrast to the chlorination of surface drinking water supplies to the free chlorine residual level in which the formation of relatively high concentrations of chlorinated organic compounds (200 to 500 ppb) such as trihalomethanes occur. Thus, controlled chlorination of sewage to below the free residual chlorine level would seem to have public health and economic merit, although free chlorine is recognized as the more rapid, effective disinfectant.

Chlorine, chloramines, and other chlorine products formed during chlorination are toxic to aquatic organisms at very small concentrations. For that reason, U.S. EPA has recommended a maximum total residual chlorine limit of 0.002 mg/l in salmonoid fish areas and 0.01 mg/l for marine and other freshwater organisms. Dechlorination with sulfur dioxide, sodium thiosulfate, and sodium biosulfite, can remove all residual toxicity to aquatic life from chlorination. Although dechlorination is believed to be beneficial, the toxicity of compounds formed by chlorination, of dechlorinating agents, and of compounds formed by dechlorination, has been a source of concern. Because of this, pilot plant studies and the possible use of alternative disinfectants are advised.[37]

Alternative disinfectants include ultraviolet (UV) radiation, ozone, ozone plus UV, and chlorine dioxide. Some viruses, bacterial spores, and protozoan cysts survive normal UV doses. Because the UV rays must penetrate the microorganism to damage or kill it, microorganisms may be protected within particulates, making prior filtration of effluent necessary. Studies have shown that municipal wastewater that has received tertiary treatment can be disinfected using UV in a cost-effective manner.[38] Ozone is a good virucide and nontoxic to aquatic organisms, and adds dissolved oxygen to treated effluents. Because it has been found to be both reliable and effective, ozone has been received greater attention in recent years as a wastewater disinfectant, even though its cost is still higher than chlorine disinfection.[39] For example, an ozone dosage of 1.5 mg/l can meet fecal coliform permit requirements at an activated sludge plant including nitrification. Although chlorine dioxide does not result in the formation of appreciable concentrations of trihalomethanes, it can cause a drop in pH and dissolved oxygen, requiring treatment adjustment.

Physical-Chemical Treatment

Physical-chemical treatment includes a range of treatment processes, including chemical coagulation, flocculation, sedimentation, and filtration. All of these unit processes remove suspended matter. One of the better texts on the application of physical-chemical methods to wastewater treatment was written by Weber and is still extensively used.[40] Another useful text was published by Sincero in 2003.[41]

Sedimentation

Sedimentation, which is one of the most widely used processes in wastewater treatment, involves the gravitational settling of suspended particles. Sedimentation typically takes place in a settling tank, or clarifier. The three main clarifier types are horizontal flow, solids contact, and inclined surface. Horizontal flow clarifiers can be rectangular, square or circular in shape. In rectangular basins, flow is parallel to the long axis of the basin. In circular basins, it is from the center outwards. In either case, the basins must be designed to keep the influent velocity as uniform as possible to prevent currents, which would prevent settling, from forming. Flow velocities within the clarifier must be reduced to 1 to 2 feet/minute to promote settling. Also, clarifier bottoms are sloped to facilitate collection and removal of the sludge that settles out. Solids contact clarifiers make use of a suspended layer of sludge to enmesh and capture incoming solids. Inclined surface clarifiers, which are also known as high-rate settling basins, use inclined trays to divide the basin into shallower sections. These trays give the basins a larger surface area and reduce particle-settling times.

Settling basins are used in a variety of sewage treatment processes, including primary settling (particulate removal), activated sludge settling (biological floc removal), and chemical coagulation (chemical floc removal). Also, septic tanks (see Figures 3.4 and 3.5) rely primarily on sedimentation to treat sewage wastes. To be effective, settling basins must be designed to produce both a clarified effluent and a concentrated sludge. Mechanical scrappers along the bottoms of these basins are used to push the settled sludge into a hopper, where it can be pumped to a sludge-processing unit. Any floating materials (e.g., oil or scum) are skimmed from the surface, and the clarified effluent is discharged over weirs to a collection trough.

Coagulation/Flocculation

Chemical coagulation of sewage prior to sedimentation promotes the flocculation (i.e., aggregation) of finely divided solids into more easily settable flocs, which,

TABLE 3.14 **Removal Efficiency of Sedimentation Compared to Coagulation plus Sedimentation**

Constituent	Sedimentation Only	Coagulation followed by Sedimentation
Total suspended solids	40%–90%	60%–90%
Five-day BOD	25%–40%	40%–70%
COD		30%–60%
Phosphorus	5%–10%	70%–90%
Bacteria levels	50%–60%	80%–90%

Source: *Design of Municipal Wastewater Treatment Plants*, MOP 8, Water Environment Federation and American Society of Civil Engineers, 1998.

in turn, enhances the efficiency of sedimentation processes as noted in Table 3.14. Other advantages of coagulation include the ability to use higher flow rates and achieve more consistent performance. However, on the negative side, coagulation produces larger quantities of sludge, which must be thickened and dewatered and has higher operational costs and more attention on the part of plant operators.

Chemicals used in coagulation to remove suspended matter include polyelectrolytes, ferric chloride or ferric sulfate, aluminum sulfate (alum) and sodium aluminate, and lime. Amounts required range from 45 to 90 mg/l ferric chloride for 85 to 90 percent phosphorus reduction to 75 to 250 mg/l alum for 55 to 90 percent phosphorus reduction. Because orthophosphate is converted to an insoluble form at a pH of 9.5, lime in doses ranging from 200 to 400 mg/l can raise pH above 9.5 and precipitate out phosphorus. Polymers are costly when used alone, but are attractive as settling and filtration aids when used in conjunction with the above coagulants. At average flows, rapid mix should be achieved in 2 minutes, flocculation in 15 minutes, and sedimentation achieved as long as application rates average 900 gpd/ft^2 (1,400 gpd/ft^2 at peak hourly flow). When alum or lime is used, pH control is also necessary before filtration.

As regulations regarding the formation of disinfection byproducts have become more stringent, considerable research has been conducted into to ways that coagulation and flocculation can be used to reduce byproduct concentrations. For example, one study found that removal efficiencies for organic matter, including disinfection byproducts such as trihalomethane, could be improved if polyaluminum chloride was used in conjunction with alum and ferric chloride.[42] Another found that use of double-step coagulation (i.e., addition of coagulants in a two-step sequence) reduced the need for high coagulant doses to remove organic matter and turbidity.[43]

Filtration

Although filtration of wastewater treatment effluents is a relatively new practice, it has gained widespread acceptance as a method for removing suspended solids from chemical and biological treatment effluents. Application rates of 5 gpm/ft^2 are recommended for mixed media filtration units, but rates of up to 10 gpm/ft^2 can be used. A flow-equalization tank is often installed ahead of the filter to ensure filtration at a relatively constant rate. Typical physical-chemical treatment for the removal of heavy metals includes flash mix using calcium hydroxide or sodium hydroxide, flocculation, clarification, and sand filtration. Lime coagulation, mixed media filtration, and use of activated carbon filtration can greatly reduce U.S. EPA priority pollutants.

The removal efficiencies of various filter media have also received attention in recent years. In one study, metal removal rates for conventional and sorptive filter media, including plain sand, granular activated carbon, cementitious media, and oxide coated/admixture media, were analyzed. Of the four media, the oxide coated/admixture media provided the highest removal efficiency for lead, copper and zinc.[44]

Activated Carbon Adsorption

Carbon adsorption can be used to remove soluble organics that remain after wastewater treatment by other processes. The two most common types of activated carbon adsorption are granular activated carbon, which has a diameter of greater than 0.1 millimeter, and powdered activated carbon, whose diameter is less than 200 mesh. Granular carbon is commonly used in separate carbon adsorption units, while powdered carbon is added directly to the biological or chemical treatment unit, where it is allowed to settle out and is then removed.

Carbon adsorption units come in several types, including fixed-bed, expanded-bed and moving-bed contactors. In fixed-bed units, the wastewater is applied to the top of a column of activated carbon and withdrawn at the bottom. In expanded-bed units, the wastewater is introduced at the bottom of the column and allowed to expand. In moving-bed units, spent carbon is continuously replaced with fresh carbon. Granular activated carbon units are generally designed to provide about 30-minutes wastewater contact time. Use of countercurrent flow patterns is believed to provide more efficient utilization of the carbon, which will need to be regularly backwashed and regenerated when its adsorption capacity is exhausted. Nitrogen can be removed by adding ammonia stripping, ion exchange, and breakpoint chlorination to the treatment process.

Biological Treatment

Trickling Filters Trickling filters are the most commonly used biological treatment process for removal of organic matter from wastewater. These filters are composed of a bed of highly permeable media, such as rock or plastic packing material, to which biologic organisms are attached, forming a biological slime layer, and through which the sewage is percolated. Typically, trickling filters are used following a primary settling tank to provide secondary treatment of the sewage. Seeding of the filter stone and development of a gelatinous film of aerobic microorganisms is necessary before good results (i.e., adsorption of organic matter by the slime layer) can be produced.

While noticeable BOD reduction can be obtained within 7 days of start-up, as long as 3 months may be required to obtain equilibrium, including high nitrification. Nitrification is the aerobic process in which the ammonia from sewage is acted upon by the oxygen in the air to form nitrate and carbon dioxide. Continuous operation, particularly during cold months of the year, is necessary to maintain nitrification efficiency. High nitrification is important in reducing the nitrogenous oxygen demand on the receiving water body to which the treated sewage effluent is discharged. Analysis of historical data from trickling filter systems has found that the degree of nitrogen removal and biological denitrification is strongly influenced by the BOD load, hydraulic loading and the media size.[45] Better nitrification can be obtained if the organic loading rate is sufficiently low enough so that all of the biodegradable organic matter is removed and filter space is available for the growth of nitrifying bacteria. Organic loading rates for trickling filters and other packed bed reactors that are supportive of combined organic oxidation and nitrification are provided in Table 3.15.

Small standard-rate trickling filters are usually 6 feet deep and designed to handle application rates of 200,000 to 300,000 gpd/acre-foot. Filter loading is also expressed in terms of 5-day BOD in the sewage being applied to the filter. Typically, 35 percent of the BOD in a raw sewage is removed by the primary settling unit. Application rates for standard-rate trickling filters range from 200 to 600 pounds of BOD/acre-ft/day with an average loading are 400 pounds in northern states and 600 pounds in southern states. Since dosage must be controlled, dosing siphons may be used for very small filters and dosing tanks with siphons or pumps containing revolving distributors or stationary spray nozzles on standard filters. Also, periodic dosing with interim resting usually produces a better effluent than continuous dosing.

A trickling filter should be followed by a secondary settling tank to remove the biological growths sloughed off the filter stone, from which sludge should be removed at least twice a day. The resultant sludge is removed by pumping or by gravity flow if possible to the sludge digester or Imhoff tank.

Because odors and filter flies can be expected with a standard rate filter, filters should be at least 400 feet from any private residence. Filter flies can be

TABLE 3.15 Organic Loading Rates for Combined Carbon Oxidation and Nitrification

Reactor Type	Lb $BOD_5/1,000\,ft^3$-day	Kg BOD_5/m^3-day
Rock media trickling filter	5–10	0.1–0.2
High rate media trickling filter	10–25	0.2–0.4
Submerged packed bed reactor	120–160	2–3

Source: Glen Diagger, "Nutrient Removal in Fixed-Film Processes: Current Design Practices," *Advances in Water and Wastewater Treatment*, American Society of Civil Engineers, 2004.

controlled by weekly chlorination (1 mg/l in effluent for 4 to 8 hours), flooding (24 hours), increased hydraulic loading, and insecticide treatment. For odor control or disinfection of the sewage effluent for bacterial reduction, chlorination of the final effluent is often required. Trickling-filter treatment can be supplemented by sand filtration, oxidation pond, solids contact basin, flocculator-clarifiers, or chemical coagulation and settling, where a higher-quality effluent is required. Variations of the standard-rate trickling filter, include high-rate filters with recirculation; biological towers (20 to 30 feet), which use a plastic media; biological aerated filters, which use a submerged media and forced air; and rotating biological contractors. Flow diagrams, which include trickling filters, are shown in Figure 3.12.

Rotating Biological Contactors   A rotating biological contactor is another type of attached-growth biological process in which large closely spaced circular disks, which are mounted on horizontal shafts, rotate slowly through wastewater. The plastic disks, which are typically on 25-foot-long horizontal shafts, rotate at two to five revolutions/minute, while partially submerged (40 percent) in wastewater that has already received primary treatment. At least four sets of contactors are typically needed to achieve secondary treatment standards and, in most instances, prior trash and grit removal is considered necessary in addition to primary settling tanks.[46] The biological growth (biomass) that forms on the wetted area of the disks through contact with organic material in the wastewater is maintained by contact with air during the rotation, which makes oxygen transfer to the wastewater possible as it trickles down the disks. Some of the growth is stripped or sloughed off from the disk as it passes through the moving wastewater and is removed in the secondary settling tank.

Rotating biological contactor design is based on hydraulic and organic loading data from pilot plant and full-scale studies. These studies have shown that hydraulic loading rates need to generally range from 2 to 4 gpd/ft^2 of contactor surface, while organic loading per stage should range from 1 to 4 lb BOD/day/1,000 ft^2. A loading range of 2.5 to 3.0 lb soluble BOD is also recommended. Lower BOD loading rates (1 gpd/ft^2) are needed to produce a high-quality (10 mg/l BOD and suspended solids) effluent. Better effluent BOD quality and nitrification are also possible by controlling pH (8.4), dissolved oxygen, and raw wastewater alkalinity levels.[47]

Removal efficiencies of 85 percent BOD removal or higher can be expected if the contactors are not overloaded, but efficiencies are reduced below 55°F (13°C). Removal efficiencies can also be improved by operating the rotating biological contactors in a step-feed mode as compared to single-feed mode and by recirculating system effluent to the inlet stage.[48] Contactors should be covered to protect them from low temperatures, as well as from rainfall and heavy winds, which flush off growths, and sunlight, which embrittle the plastic disks. However, complete enclosure is not desirable because it promotes accumulation of hydrogen sulfide and high humidities, which can corrode metal parts. Although contactors are reliable and can withstand shock loading, when peak flow exceeds 2.5 times the average daily flow, or when a large organic loading occurs, appropriate flow equalization or more disks may be added.

Rotating biological contactors can also be used for carbonaceous removal (i.e., BOD, COD, and TOC reductions) as well for as nitrification, and for sulfide and methane removal. Smaller-diameter disks, 2 to 4 feet, can achieve greater BOD removals than larger diameter disks. Operation and maintenance costs typically average less than activated sludge are higher than trickling filters.[49] Contactors are usually followed by final settling tanks, which should provide at least 1.5 hours detention, a maximum surface settling rate of 600 gpd/ft^2, and an overflow rate not greater than 5,000 gpd/linear ft.[50]

Aerobic Digestion   Aerobic digestion systems are frequently used in small treatment plants with activated sludge units being one of most commonly used. Activated sludge involves the use of a mass of activated microorganisms in an aeration basin, which aerobically degrade organic matter from wastewater. An aerobic environment is maintained by means of diffused or mechanical aeration, which also serves to keep the contents of the basin completely mixed. After a specified retention time, the content of the basin (the mixed liquor) is passed into a secondary clarifier where the sludge is allowed to settle. A portion of the settled sludge is then recycled back to the aeration basin to maintain the required activated sludge level. Design data for activated sludge units and other aerobic digestion processes are presented in Table 3.16.

Contact stabilization is a modification of the conventional activated sludge process in which two aeration tanks are used. In the first tank, the return sludge is re-aerated for at least 4 hours before it is permitted to flow into the second tank to be mixed with the wastewater is to be treated. An oxidation ditch is a ditch in which a revolving drumlike aerator supplies air to the circulating wastewater and by doing so reduces the organic matter in the wastewater by aerobic action. Design requirements for extended aeration units typically include the following parameters:

Average sewage flow. 400 gal/dwelling or 100 gpd/capita.

Aeration tanks. At least two tanks to treat flows greater than 40,000 gpd with 24- to 36-hour detention period for average daily flow, not including recirculation, and 1000 feet^3 per 7.5 to 15 lb of BOD, whichever is greater. If raw sewage goes directly to aeration tank, primary tank is omitted.

Air requirements. 3 cfm/foot of aeration tank length, or 2,000 to 4,000 ft^3/lb of BOD entering the tank daily, whichever is larger. Additional air required if air is needed for airlift pumping of return sludge from settling tank.

Settling tanks. At least two tanks to treat flows greater than 40,000 gpd with a 4-hour detention period based on average daily sewage flow, not including recirculation. For tanks with hopper bottoms, upper third of depth of hopper may be considered as effective settling capacity.

Rate of recirculation. At least 1:1 return activated sludge based on average daily flow.

TABLE 3.16 Aeration Digestion Tank Loading Rates and Removal Efficiencies

Process	Organic Organic Loading (lb BOD$_5$/day per 1000 ft^3)	F/Mª Ratio (lb BOD$_5$/day per lb MLVSS^b)	MLVSS^c (mg/l)	Detention Time (hr)	Overall BOD Removal Efficiency
Activated sludge					
Plug flow	20–40	0.2–0.5	1,000–3,000	4–8	85–95%
Completely mixed	50–120	0.2–0.5	3,000–6,000	3–6	85–95%
Step aeration				3–6	85–95%
Contact stabilization	60–75^d	0.2–0.6	1,000–3,000	0.2–1.5	80–90%
Extended aeration	10–25	0.05–0.2	3,000–6,000	18–36	75–90%

^{a}Food to microorganism ratio (F/M)
^{b}Mixed liquor volatile suspended solids (MLVSS)
^{c}MLVSS values are dependent upon the surface area provided for sedimentation and the rate of sludge return as well as the aeration process.
^{d}Total aeration capacity, includes both contact and reaeration capacities. Normally the contact zone equals 30 to 35 percent of the total aeration capacity.
Source: *Technical Criteria Guide for Water Pollution Prevention, Control and Abatement Programs*, U.S. Department of Army, Washington, DC. April 1987, p. 6-2.

> *Sludge holding tanks.* Provide 8 feet^3/capita. A minimum of 1,000 gallons capacity per 15,000 gallons design flow and 20- to 40-day retention. Tanks should be aerated.

Extended aeration systems require daily operational control because air blowers must be operated continuously and sludge must be returned. Aeration tubes or orifices require periodic cleaning; and dissolved oxygen and mixed liquor suspended solids concentration must be watched. Clogging of the airlift for return sludge is also a common cause of difficulty. However, with proper controls, a 90 to 97 percent BOD and suspended solids removal, as well as good nitrification of ammonia nitrogen, can be expected. Typically, a 3-month adjustment period is needed to produce an acceptable effluent, which is why these systems are not recommended for seasonal operations, such as camps and schools.

Stabilization Ponds   In areas where ample space is available, stabilization ponds can be a relatively inexpensive way to treat sewage. Small stabilization ponds have even been used at resorts or motels with a septic tank ahead of the pond and not produced any odor problems.[51] Stabilization ponds, also known as oxidation ponds, are operated as high-rate aerobic ponds, aerobic-anaerobic (facultative) ponds (the most common), aerated ponds or anaerobic ponds. Table 3.17

TABLE 3.17 Types of Lagoons

Type	Detention[d] (days)	Depth (ft)	Loading, (lb/5-day/ BOD/acre/day)	BOD Removal or Conversion (percent)
High-rate aerobic pond	2–6	1–1.5	60–200	80–95
Facultative pond	7–50	3–8 (2–5)[c]	15–80	70–95
Anaerobic pond	5–50	Variable[a]	200–1000	50–80
Maturation pond[b]		3–8	<15	Variable
Aerated lagoon	2–10	Variable[a]	Up-400	70–95

[a]Usually 10- to 15-ft deep

[b]Generally used for polishing effluents from conventional secondary treatment plants.

[c]These depths are more common.

[d]W. Wesley Eckenfelder, Jr., *Water Quality Engineering for Practicing Engineers*, New York, 1970, p. 210.

Source: *Upgrading Lagoons*, U.S. EPA, EPA-625/4-73-00 lb (August 1973): 1.

summarizes the detention times, depths, loading rates, and efficiencies for each type of pond.

Facultative ponds with a minimum of three cells in series and a 20-day actual detention time, and aerated ponds with a separate settling pond prior to discharge, provide more than adequate helminth (ascaris, trichuris, hookworm, tapeworm) and protozoa (giardia, amoeba) removal. The physical, chemical, and biological activities in the ponds, as well as competing organisms, all serve to reduce the numbers of surviving enteric bacteria and viruses.[52] Using a water balance analysis, ponds can be designed for zero discharge (i.e., pond surface area is sufficient to provide an annual net evaporation after precipitation that is greater than the wastewater inflow).

BOD removal of 85 to 90 percent is not unusual for stabilization ponds, and removal of viruses, bacteria, protozoa, and helminths is also reported to be very high. Ponds in open areas and in series (a minimum of three) give best results due to their additional detention time. Pond performance is affected by temperature, solar radiation, wind speed, loading, actual detention time, and other factors.[53] Primary treatment of sewage with grit chamber, comminutor and rack prior to discharge to ponds is also recommended. Design criteria for facultative-type stabilization ponds include:

Detention time. 90 to 180 days, depending on climatic conditions; 180 days for controlled discharge pond; 45 days minimum for small systems.

Liquid depth. 5 feet plus 2 feet freeboard, with minimum liquid depth of 2 feet.

Embankment. Top width of 6 to 8 feet; inside and outside slope 3 horizontal to 1 vertical. Use dense impervious material; liner of clay soil, asphaltic coating, bentonite, plastic or rubber membrane, or other material required, if seepage expected.

Pond bottom. Level, impervious, no vegetation. Soil percolation should be less than 0.25 in./hr after saturation.

Inlet. 4-inch diameter minimum at center of square or circular pond; at one-third point if rectangular with length not more than twice width. Submerged inlet 1 foot off bottom on a concrete pad or at least 1.5 feet above highest water level.

Outlet. 4-inch minimum diameter; controlled liquid depth discharge using baffles, elbow, or tee fittings; drawoff 6 to 12 inches below water surface to control, avoid short-circuiting and minimize algae removal; discharge to concrete or paved gutter.

Normally, stabilization ponds are aerobic to some depth, depending on surface aeration, microbial activity, wastewater clarity, sunlight penetration, and mixing. In deeper ponds, wastewater at lower layers becomes facultative and then anaerobic. Anaerobic and aerated ponds are usually followed by aerobic ponds to reduce suspended solids and BOD to acceptable levels for discharge. In general, increased detention time will increase BOD removal, and decreased BOD areal loading will increase BOD removal. Thus, required BOD and suspended solids removal and effluent quality will determine the detention time and areal loading. Pond efficiency can be improved by recirculation, inlet and outlet arrangements, supplemental aeration and mixing, and algae removal by coagulation-clarification, filtration, and land treatment of the effluent.[54]

Algae formed in ponds, particularly from seasonal blooms, are the main cause of solids carryover and increased oxygen demand in receiving streams from pond discharges. Thus, further treatment of pond effluent using coagulation, settling, filtration, centrifugation or microstraining may be required to remove the algae before discharge. It is also possible to prevent algal growths by copper sulfate treatment in the final cell, effluent withdrawal from below the surface, or effluent disposal to a wetland or a wastewater reuse facility.

The practicability of using waste stabilization ponds, lagoons, or land treatment will depend on local conditions. For example, the risk of odor, nuisance, or health hazards should be evaluated before a selection is made. However, these ponds should not be dismissed too quickly either, as they can provide an acceptable answer when no other treatment is practical at a reasonable cost. Control of aquatic vegetation, embankment weeds, and floating mats is often necessary to minimize mosquito and other insect breeding.

Wastewater Reuse

Treated wastewater can be reused as long as it does pose a health hazard due to six factors:

1. Possible inhalation of aerosols containing pathogenic microorganisms
2. Consumption of raw or inadequately cooked vegetables from crops irrigated with wastewater or possible ingestion of heavy metals or other toxic materials taken up by crops during growth

3. Contamination of groundwater through infiltration of wastewater chemicals into a groundwater aquifer serving as a source of drinking water

4. Runoff, from land areas receiving wastewater effluent, to surface waters used as sources of drinking water, shellfish, bathing water, or other recreational purposes

5. Possible cross-connection between potable and nonpotable water systems

6. Buildup of detrimental chemicals in the soil

Although most pathogenic microorganisms can be removed from wastewater as it infiltrates through an adequate distance of unsaturated fine sand, loamy, or sandy soil, these microorganisms have been observed to travel several hundred feet in saturated soil and up to 2,000 feet in coarse gravel and creviced rock.

In addition, wastewater can contain a variety of chemicals such as heavy metals (cadmium, copper, nitrates, lead, mercury, zinc, nickel, and chromium), pesticides (insecticides, fungicides, and herbicides), and commercial and industrial wastes, such as trichloroethylene and polychlorinated biphenyls. Nitrification of organic material in sewage can add nitrate-nitrogen to the groundwater if not immediately used by plants and endanger sources of drinking water used by infants (methemoglobinemia).

In light of the many pathogens normally found in wastewater, irrigation, and spraying of crops with wastewater should be restricted to those foods that are not eaten raw. Many of the metals normally found in wastewater do not appear to be a problem when applied to crops because they are not significantly absorbed by plant roots.[55] However, some heavy metals (cadmium, copper, molybdenum, nickel, and zinc) can accumulate in soil and become toxic unless good management practices are followed. For that reason, crop tissue and grain analyses may be necessary to monitor crop uptake. In a study of the effects of 20 years of irrigation with secondary *domestic sewage* effluent, which contained no major industrial wastes, soil and crop (alfalfa) analyses found no accumulation of lead, copper, zinc, nickel, chromium, or cadmium.[56] However, late-crop irrigation with wastewater high in nitrate can lead to high-nitrate concentrations in both soils and crops. Excessive nitrate levels are known to be injurious when fed to animals, resulting in cyanosis. Also, boron, a constituent not normally removed by conventional treatment, is toxic to citrus crops.[57]

Wastewater Aerosol Hazard

The potential hazard from aerosols is related primarily to wastewater treatment by the activated sludge, trickling-filter, and spray irrigation processes. The presence of microbiological pathogens in sewage aeration products downwind has been well documented (particularly for *E. coli*). Although pathogens can be recovered from such aerosols, the risk human disease from the aerosols has not been clearly demonstrated in the United States.[58] Such risk has been demonstrated, however, among workers exposed to nondisinfected spray irrigation in India.[59] Also, a study of medical records at a kibbutzim showed apparent seasonal increase in enteric disease in the 0- to 4-year-olds who lived within 1,000 meters of fields sprinkler-irrigated with stabilization pond effluent.[60] In view of the potential risk, it is advisable to chlorinate wastewater and provide adequate buffer zones (1,000 feet or more) as a precautionary measure.

Regulation Standards for controlling the reuse of wastewater have been recommended by a number of agencies, including the World Health Organization; their standards for wastewater use in agriculture are presented in Table 3.18. However, the standards established by the California Department of Public Health, are among the most explicit[61] and cover "wastewater constituents, which will assure that the practice of directly using reclaimed wastewater* for the specified purposes does not impose undue risks to public health." The California fecal coliform standards for reclaimed water are complete wastewater treatment for direct discharge, including disinfection and tertiary treatment to less than 2.2 MPN per 100 ml when public access is possible; less than 23 MPN per 100 ml for secondary effluent used for golf courses, cemeteries, landscaping; and less than 2.2 per 100 ml when water is used at parks, schoolyards, playgrounds, or near residential areas for irrigation.

Wastewater Disposal by Land Treatment

Land treatment and disposal of wastewater on natural biological, physical, and chemical processes in the soil to treat wastewater. Methods used include spray or sprinkler irrigation; ridge-and-furrow and border strip irrigation; overland flow; subsurface disposal in a soil absorption system; and wetland treatment. Table 3.19 compares the design features for land treatment processes and Table 3.20 shows expected quality of wastewater treated by processes.

Irrigation Spray irrigation is the most common method of applying wastewater to land. Wastewater application is generally limited to 8 hours, followed by a 40-hour rest period to permit drainage of the soil, reaeration, plant nutrient uptake, and microbial readjustment. Physical, biological, and chemical treatment takes place during percolation, particularly in the upper soil, including BOD and COD removal. Dissolved solids and chlorides may cause a soil problem where the wastewater is high in these constituents. Phosphorus and cadmium are accumulated by plants.

Because cadmium may be hazardous in edible crops, its level should be kept below 2.5 mg/kg in the soil.

Slow-rate spray irrigation rates are generally between 0.5 to 4 inches per week, depending on soil permeability, climate, and wastewater strength. Ground slope

---

*"Reclaimed wastewaters" are waters, originating from sewage or other waste, which have been treated or otherwise purified so as to enable direct beneficial reuse or to allow reuse that would not otherwise occur. "Disinfected wastewater" means wastewater in which the pathogenic organisms have been destroyed by chemical, physical, or biological methods.

TABLE 3.18 Recommended Microbiological Quality Guidelines for Wastewater Use in Agriculture[a]

Category	Reuse Conditions	Exposed Group	Intestinal Nematodes[b] (arithmetic mean no. of eggs per litre[c])	Fecal Coliforms (geometric mean no. per 100 ml[c])	Wastewater Treatment Expected to Achieve the Required Microbiological Quality
A	Irrigation of crops likely to be eaten uncooked, sports fields, public parks[d]	Workers, consumers, public	≤1	≤1, 000[d]	A series of stabilization ponds designed to achieve the microbiological quality indicated, or equivalent treatment
B	Irrigation of cereal crops, industrial crops, fodder crops, pasture and trees[e]	Workers	≤1	No standard recommended	Retention in stabilization ponds for 8–10 days or equivalent helminth and fecal coliform removal
C	Localized irrigation of crops in category B if exposure of workers and the public does not occur	None	Not applicable	Not applicable	Pretreatment as required by the irrigation technology, but not less than primary sedimentation

[a]In specific cases, local epidemiological, sociocultural, and environmental factors should be taken into account, and the guidelines modified accordingly.
[b]Ascaris and Trichuris species and hookworms.
[c]During the irrigation period.
[d]A more stringent guideline (≤200 fecal coliforms per 100ml) is appropriate for public lawns, such as hotel lawns, with which the public may come into direct contact.
[e]In the case of fruit trees, irrigation should cease 2 weeks before fruit is picked, and no fruit should be picked off the ground. Sprinkler irrigation should not be used.

Reproduced, with permission, from *Health Guidelines for the Use of Wastewater in Agriculture and Aquaculture*, World Health Organization, 1989 (WHO Technical Report Series No. 778).

TABLE 3.19 Comparison of Design Features and Site Characteristics for Land Treatment Processes

Feature	Principal Processes			Other processes	
	Slow rate	Rapid infiltration	Overland flow	Wetlands	Subsurface
Application techniques	Sprinkler or surface[a]	Usually surface	Sprinkler or surface	Sprinkler or surface	Subsurface piping
Annual application rate (ft)	0.5 to 6	6 to 125	3 to 20	1.2 to 30	2.4 to 27
Field area required (acres)[b]	23 to 280	3 to 23	6.5 to 44	4.5 to 113	5.3 to 57
Typical weekly application rate (in.)	1.3 to 10	10 to 240	6 to 15[c]	2.5 to 64	5.1 to 51
Land slope	0 to 20% Cultivated 35% Uncultivated	Not critical	15 to 40[d] 2 to 8% for final slopes		
Soil permeability	Moderate to slow	Rapid	Slow to none		
Groundwater depth (ft)	2 to 10	3 during application 5 to 10 during drying	Not critical		
Minimum pretreatment required	Primary sedimentation[e]	Primary sedimentation	Grit removal and screening	Primary sedimentation	Primary sedimentation

	Evapotranspiration and percolation	Mainly percolation	Surface runoff and evapotranspiration with some percolation	Evapotranspiration, percolation, and runoff	Percolation with some evapotranspiration
Disposition of applied wastewater					
Need for vegetation	Required	Optional	Required	Required	Optional

[a]Includes ridge-and-furrow and border strip.
[b]Field area in acres not including buffer areas, roads, or ditches for 1 million gpd (43.8 litre/s) flow.
[c]Range for application of screened wastewater.
[d]Range for application of lagoon and secondary effluent.
[e]Depends on the use of the effluent and the type of crop.
1 cm = 0.394 in
1 m = 3.28 ft
1 hectare = 2.47 acre

Source: Process Design Manual for Land Treatment of Municipal Wastewater, U.S. EPA, Washington, D.C., EPA/625/R-06/016, September 2006, p. 1-1 and *Process Design Manual for Land Treatment of Municipal Wastes*, U.S. EPA, Cincinnati, OH, October 1981, p. 1-3.

TABLE 3.20 Expected Effluent Water Quality for Land Treatment Processes (mg/l unless otherwise noted)

Constituent	Slow Rate[a]	Rapid Infiltration[b]	Overland Flow[c]
BOD	<2	5	10
Suspended solids	<1	2	10
Ammonia nitrogen as N	<0.5	0.5	<4
Total nitrogen as N	3	10	5
Total phosphorus as P	<0.1	1	4
Fecal coliform (#/100 ml)	<1	10	200+

[a]Percolation of primary or secondary effluent through 5 ft (1.5 m) of soil.
[b]Percolation of primary or secondary effluent through 15 ft (4.5 m) of soil.
[c]Runoff of comminuted municipal wastewater over about 150 ft (45 m) of slope.
Source: Process Design Manual for Land Treatment of Municipal Wastewater, U.S. EPA, EPA/625/R-06/016, September 2006, p. 1-2.

should be less than 20 percent on cultivated land and 40 percent on noncultivated land. Soil permeability should be moderately slow to moderately rapid* and the depth to groundwater a minimum of 2 to 3 feet, although 5 feet is preferred. High-rate spray irrigation rates are 4 to 40 and up to 120 inches per week. Soil permeability for high-rate systems should be rapid with a permeable soil depth of 15 feet or more. Nitrogen and phosphorus removal is usually not complete in high-rates systems.

Ridge-and-furrow ditch systems are typically 100 to 1,500 feet in length, with depth and spacing varying depending on the type of crop and the soil's ability to transmit water laterally. Border strips are 30 to 60 feet long.[62] Application rates for ridge-and-furrow (gpm/100 ft) and border strip irrigation are similar to those for spray irrigation and will vary with soil permeability, spacing, and slope of the furrow.

Overland Flow Overland flow is a treatment process in which wastewater is treated as it flows down a series of vegetated terraces. Application to the top of a grassed, slightly permeable slope (2 to 8 percent) as sheet flow allows for both physical (grass filtration and sedimentation) and chemical-biological (oxidation) treatment. Treated runoff is collected in ditches and discharged to a watercourse. Surface runoff may be 50 percent or more. Grasses, which have high nitrogen uptake capacity,[†] are usually chosen for cover vegetation. Viruses and bacteria are not removed. Overland flow treatment is more effective during warm weather.

---

*0.2 to 6.0 or more inches per hour permeability corresponding roughly to a soil percolation rate of 1 inch in 45 minutes to less than 10 minutes.
[†]Bent grass, Bermuda grass, Reed Canary grass, Sorghum-Sudan, Vetch; also Alfalfa, Clover, Orchard grass, Broome grass, and Timothy.

Natural or Constructed Wetlands Secondary wastewater effluent can be applied to either existing or artificial wetlands. Wetlands include inundated areas having water depths of less that 2 feet, which support emergent plants such as bulrush, reeds, hyacinths, or sedges. These plants provide surfaces to which bacterial films can attach, aid in the filtration and adsorption of wastewater constituents, and add oxygen to the water column. Natural wetlands include marshes, bogs, peat lands, and swamps. Constructed wetlands include both free water surface systems and subsurface flow systems.

Free water surface systems are generally composed of a series of parallel shallow basins from 0.3 to 2 feet in depth, which have relatively impermeable bottom soil and emergent vegetation. Wastewater is treated in these systems as it flows through the stems and roots of the emergent vegetation. Subsurface flow systems are composed of beds or channels filled with gravel, sand, or similar permeable material in which emergent plants have been planted. Wastewater is treated in these systems as it flows horizontally through the media-plant filter.

Advanced Wastewater Treatment

Advanced (tertiary) wastewater treatment may be needed in some instances to protect the water quality of the receiving surface and groundwaters from undesirable nutrients, toxic chemicals, or pathogenic organisms, which are not removed by conventional secondary treatment. For example, nitrogen and phosphorus in plant effluent may promote the growth of plankton; toxic organic and inorganic chemicals may endanger fish and other aquatic life and endanger sources of water supply, recreation, and shellfish growing; and pathogens, such as the infectious hepatitis virus and giardia, that are not removed by conventional sewage treatment increase the probability of waterborne disease outbreaks. Figure 3.13 shows wastewater treatment unit process including advanced or tertiary treatment.

Advanced wastewater treatment may include combinations of the following unit processes depending on the water quality objectives to be met. This list is meant to be illustrative and should not be considered all-inclusive.

For Nitrogen Removal

Breakpoint chlorination—to reduce ammonia nitrogen levels (nitrate and organic are not affected).

Ion exchange, after filtration pretreatment—to reduce nitrate nitrogen and ammonium levels using selective resin for each; phosphate also reduced.

Nitrification followed by dentrification, ammonia (if present) removed or converted to nitrate and then to nitrogen gas—ammonia stripping* (degasifying) to remove ammonia nitrogen, or biological oxidation of ammonia

---

*Wastewater pH is raised to 10.0 to 10.5 or above, usually by the addition of lime or sodium hydroxide, at which pH the nitrogen is mostly in the form of ammonia, which can be readily removed by adequate aeration, but pH adjustment of the effluent will be needed to meet stream standards. Organic or nitrate nitrogen are not removed. Ammonia stripping equipment includes tray towers, cascade aerators, step aerators, and packed columns.

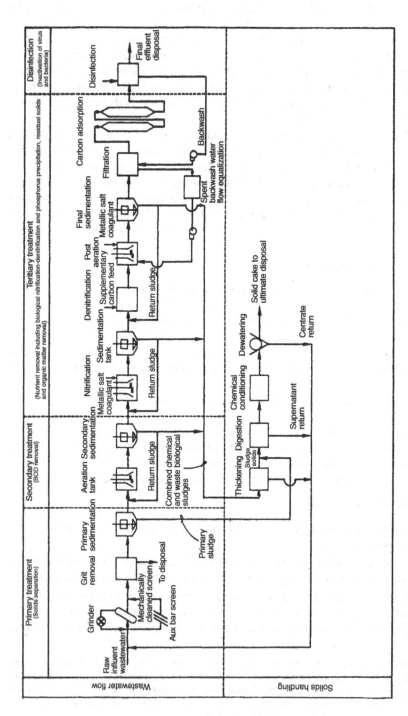

FIGURE 3.13 Typical schematic flow and process diagram. (*Source: Operation of Wastewater Treatment Plants*, MOP 11, Water Pollution Control Fed., Washington, DC, 1996.® 1996 WPCF, used with permission.)

in the activated sludge process to nitrate; denitrification (organic nitrogen) achieved by filtration through sand or GAC, or by biological denitrification, usually under anaerobic conditions, following activated sludge treatment (nitrification and denitrification).

Methanol—to reduce nitrate levels.

Reverse osmosis, following treatment to prevent fouling of membranes—to reduce total nitrogen levels; also dissolved solids.

Electrodialysis, following pretreatment—to reduce ammonia, organic, and nitrate nitrogen levels; also dissolved solids.

Oxidation pond—to reduce total nitrogen levels.

Land treatment, low-rate irrigation to overland flow—to reduce total nitrogen and phosphorous levels. Rapid infiltration is also effective.

For Phosphorus Removal

Coagulation (lime, alum, or ferric chloride, and polyelectrolyte) and sedimentation—to reduce phosphate levels, TDS increased, additional nitrogen removal, also some heavy metals.

Coagulation, sedimentation, and filtration (mixed media)—to further reduce phosphate levels; also suspended solids, TDS increased, additional nitrogen also removed.

Lime treatment, after biological treatment, followed by filtration—to reduce phosphorus (pH above 11) also suspended solids.

Ion exchange, with selected specific resins—to reduce phosphate, dissolved solids, and nitrogen.

For Dissolved Organic Removal

Activated carbon (granular or powdered) absorption, following filtration—to reduce COD including dissolved organics; also chlorine.

Reverse osmosis, following pretreatment—to reduce dissolved solids.

Electrodialysis following pretreatment—to reduce dissolved solids.

Distillation, following pretreatment—to reduce dissolved solids.

Biological wastewater treatment—to reduce dissolved organics.

Aeration—to remove volatile organics.

For Heavy Metals Removal

Lime treatment—to reduce heavy metals levels.

Coagulation and sedimentation—to reduce heavy metals levels.

For Dissolved Inorganic Solids Removal (Demineralization)

Ion exchange, using anionic and cationic resins, following pretreatment—to reduce total dissolved solids.

Coagulation and sedimentation—to reduce heavy metals.

Reverse osmosis—to reduce total dissolved solids.

Electrodialysis—to reduce total dissolved solids.

For Suspended Solids Removal

Filtration (sand, lime, or ferric chloride and possible polyelectrolytes), sedimentation, filtration—to reduce suspended solids, also ammonia nitrogen, and phosphate if high alum or lime dosage used; adding ammonia stripping will reduce total nitrogen further; adsorption using activated carbon will reduce dissolved organics and total nitrogen.

For Recarbonation

Carbon dioxide addition—to reduce pH where wastewater pH has been raised to 10 to 11; this is necessary to reduce deposition of calcium carbonate in pipelines, equipment, or the receiving watercourses.

TYPICAL DESIGNS FOR SMALL TREATMENT PLANTS

The following design analyses are for a treatment plant, which is meant to serve 150 persons at 100 gal per capita per day (gpcd) = 15,000 gpd.

Standard-Rate Trickling Filter Plant with Imhoff Tank

See Figure 3.14 for design details for an Imhoff tank.

1. Flowing through channel provides 2.5-hr detention.

$$\frac{15,000}{24} \times 2 = 1250 \text{ gal} = 167 \text{ ft}^3$$

2. Sludge storage at 5 ft^3 per capita = $5 \times 150 = 750$ ft^3.
3. Sludge drying beds at 1.25 ft^2 per capita = $150 \times 1.25 = 188$ ft^2.
4. Trickling filter loading at 4000 lb of BOD/acre-ft = 0.25 lb/yd^3. Loading based on 0.17 lb of BOD/capita with 35 percent removal in primary settling = $150 \times 0.17 \times 0.65 = 16.6$ lb/day.

$$\text{Filter volume required} = \frac{400}{43,560} = \frac{16.6}{x}; \quad x = 1800 \text{ ft}^3$$

Hence, the required filter diameter,

$$\text{assuming a 6-ft depth} = D = \sqrt{\frac{1800 \times 4}{\pi \times 6}} = 19.5 \text{ ft}$$

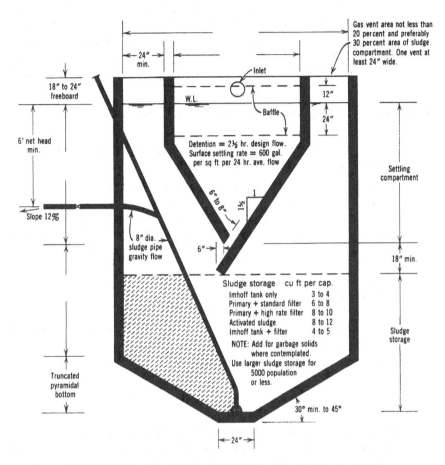

FIGURE 3.14 Section through Imhoff tank, with design details.

The volumetric loading

$$\frac{15,000}{\frac{\pi \times 20 \times 20}{4}} = \frac{x}{43,500}; \ x = 2,080,000 \text{ gpd / acre on a}$$

6-ft deep filter.

5. Final settling provides 2-hour detention. $\frac{15,000}{24} \times 2 = 1250 \text{ gal} = 167 \text{ ft}^3$
With a surface

$$\text{settling rate} = 500 \text{ gpd / ft}^2 = \frac{180 \times \text{tank depth}}{2\text{-hour detention}}; \ \text{tank depth} = 5.6 \text{ ft.}$$

6. If the BOD in the raw sewage is 200 mg/l, and the Imhoff tank removes 35 percent, the applied BOD $= 0.65 \times 200 = 130$ mg/l. According to the National Research Council Sanitary Engineering Committee formulas,* a filter loaded at 400 lb of BOD per acre-ft will produce an average settled effluent containing 14 percent of that applied, or $0.14 \times 130 = 18$ mg/l.

High-Rate Trickling Filter Plant with Imhoff Tank

1. Flowing through channel same as with standard rate filter $= 209$ ft^3.
2. Sludge storage at 8 ft^3/capita $= 8 \times 150 = 1,200$ ft^3.
3. Sludge drying beds at 1.50 ft^2/capita $= 150 \times 1.50 = 225$ ft^2.
4. Trickling filter loading at 3000 lb of BOD/acre ft $= 1.86$ lb/yd^3. Loading based on 0.17 lb of BOD per capita with 35 percent removal in primary settling $= 150 \times 0.17 \times 0.65 = 16.6$ lb/day. The BOD in the raw sewage is $150 \times 0.17 = 25.5$ lb. Filter volume required $= \dfrac{3000}{43,560} = \dfrac{25.5}{x}$ $x = 370$ ft^3. Hence, the required filter diameter, assuming 3.25 ft depth $\times D = \sqrt{\dfrac{370 \times 4}{\pi \times 3.25}} = 12.0$ ft. The volumetric surface loading on a 12-ft diameter filter with influent + recirculation $[(I + R = 1 + 1 = 2)$ or $2(15,000)] = 30,000$ gal/day is $x = 11,500,000$ gpd/acre on a 3.25/ft deep filter.
5. Final settling provides 2-hour detention at flow $I + R$.

$$\dfrac{30,000}{24} \times 2 = 2500 \text{ gal} = 334 \text{ ft}^3$$

6. Without recirculation, an applied BOD of 130 mg/1 (0.65×200), at a rate of 3000 lb/acre-ft will be reduced to $0.32 \times 130 = 42$ mg/l in the settled effluent. With recirculation of $R/I = 1$, the efficiency of the high-rate filter and clarifier can be determined from the following formulas:[63]

$$F = \dfrac{1 + \dfrac{R}{I}}{\left(1 + 0.1\dfrac{R}{I}\right)^2}$$

where

    F = recirculation factor
    R = volume of sewage recirculated $= 1$
    I = volume of raw sewage $= 1$
    $F = (1 + 1)/(1 + 0.1^2) = 1.65$

---

*$E = \dfrac{100}{1 + 0.0085\sqrt{u}}$; E = percent BOD removed, standard filter and final clarifier. u = filter loading, lb of BOD per acre-ft.

and from

$$u = \frac{w}{VF}$$

where

u = unit loading on high-rate filter, lb of BOD/acre ft
w = total BOD to filter, lb/day = 16.6
V = filter volume, acre-ft based on raw sewage strength = 0.0084
F = recirculation factor = 1.65

$$u = \frac{16.6}{0.0084 \times 1.65} = 1.98 \text{lb /acre-ft}$$

and from

$$E = \frac{100}{1 + 0.0085\sqrt{u}}$$

where

E = percent BOD removed by a high-rate filter and clarifier

$$E = \frac{100}{1 + 0.0085\sqrt{1198}} = 77 \text{ percent}$$

Hence, the BOD will be reduced to $(1-0.77)130 = 30 \text{ mg/l}$.

Intermittent Sand Filter Plant with Imhoff Tank or Septic Tank

1. Flowing through channel of Imhoff tank provides 2.5-hour detention = 209 ft^3.
2. Sludge storage at 4 ft^3/capita = $4 \times 150 = 600$ ft^3.
3. Sludge drying bed provides 188 ft^2. OR: Septic tank provides 24-hour detention = 15,000 gal = 2,000 ft^3.
4. Sand filter, covered, designed for loading of 50,000 gpd/acre. Filter area is

$$\frac{50,000}{43,560} = \frac{15,000}{x}; = x = 13,000 \text{ ft}^2$$

If filter is open, the required area = 6,550 ft^2. Make filters in two sections. Provide dosing tank to dose each covered filter section at volume equal to 75 percent of the capacity of the distributor laterals or to dose each open filter section to depth of 2 to 4 in.

If the efficiency of BOD removal of a sand filter is 90 percent, the BOD of the effluent would be $130 \times 0.10 = 13 \text{ mg/l}$.

DESIGN OF LARGE TREATMENT PLANTS

Although the design details for large sewage treatment plants are beyond the scope of this text, some of the major design elements are presented in the following sections for general information. Also, state and federal regulatory agencies have recommended standards and guidelines and should be consulted.[64] Treatment processes and bases of design are summarized in Figure 3.15 and Table 3.21, while process efficiencies are given in Table 3.22. Typical flow diagrams are shown in Figures 3.12 and 3.13.

Larger sewage treatment plants should be designed for a population at least 10 years in the future, although 15 to 25 years is preferred, and a per capita flow of not less than 100 gpd. Where available, actual flow studies, population trends, zoning, and growth potential should be used. Plants should be accessible from highways but as far as practical from habitation and sources of water supply, and protected from a 100-year flood. The required degree of treatment should be based on the water quality standards and objectives established for the receiving waters and other factors, as noted previously.

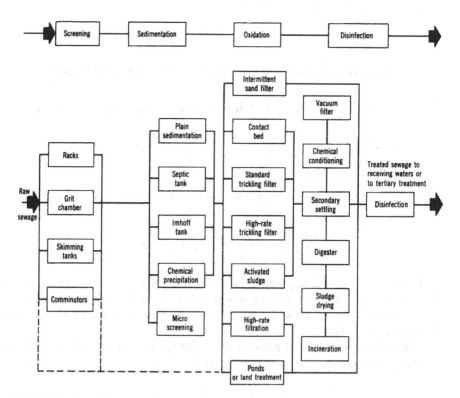

FIGURE 3.15 Conventional sewage treatment unit processes. Tertiary treatment may include denitrification, phosphorus removal; coagulation, sedimentation, and/or filtration; adsorption, ion exchange, electrodialysis, reverse osmosis, or any combination of processes depending on the end use of the renovated wastewater.

TABLE 3.21 Conventional Sewage Treatment Plant Design Factors

Preliminary Treatment	Coagulation and Sedimentation Treatment
Racks Area: 200% plus sanitary sewer; 300% plus combined sewer. Bar space: 1–1.75 in., dual channels. *Screens* Net submerged area: 2 ft^2/mgd for sanitary sewer; 3 ft^2/mgd for combined sewer. Slot opening 0.125 in. min. Dual units, preceded by racks. *Grit Chamber* Sewage velocity: 1 fps mean, 0.5 fps, minimum. Detention: 45–60 sec, floor 1 ft below outlet. Minimum of 2 channels. *Skimming Tank* Air or mechanical agitation with or without chemicals. Detention: 20 min for grease removal, 5–15 min for aeration, 30 min for flocculation. *Comminutors* Duplicate or bypass, downstream from grit chamber.	*Sedimentation* Surface settling rates at peak flows: primary and intermediate set—tanks 1,500 gpd/ft^2; final set—tanks 1,200 gpd/ft^2 after trickling filters or rotating biological contactors and for activated sludge for conventional, step aeration, contact stabilization, and carbonaceous staged of separate-stage nitrification; following extended aeration 1,000 gpd/ft^2; for physicochemical treatment using lime: 1,400 gpd/ft^2. Weir rates: 10,000 gpd per linear foot for average flows to 1.0 mgd and up to 15,000 for larger flows Sludge hopper: 1 horizontal to 1.7 vertical. Sludge pipe: 6 in./min. *Chemical Precipitation* Rapid mix, coagulation, sedimentation. Ferric chloride, ferrous sulfate, ferric sulfate, alum, lime, or a polymer. *Imhoff Tank* Detention period: 2–2.5 hours. Gas vent: 20% total area of tank minimum. Bottom slope: 1.5 vertical to 1 horizontal. Sludge compartment: 3–4 ft^3 per capita 18 in. below slot; 6–10 ft^3 per capita secondary treatment. Bottom slope: 1 to 1 or 2. Slot and overlap: 8 in. Sludge pipe: 8 in. minimum under 6 ft head. Velocity: 1 fpm. Surface settling rate: 600 gpd/ft^2

(*continues*)

TABLE 3.21 (continued)

Preliminary Treatment	Coagulation and Sedimentation Treatment

Flow Basis
100 gal per capita plus industrial wastes. Usual to assume total flow reaches small plants in 16 hr.

Flow Equalization
Based on 24-hr plot to smooth out hydraulic and organic loading.

Chemical Treatment
For odor control, oxidation, corrosion control, neutralization.

Biological Treatment

Intermittent Sand
Filter rate: 50,000–100,000 (gpad)[b] with plain settling and 400,000 gpad with trickling filter or activated sludge. Sand: 24 in. all passing 0.25-in. sieve, Effective size 0.35–0.6 mm. Uniform coefficient <3.5.

Contact Bed
Filter rate: 75,000–100,000 gpad/ft.

Trickling Filter
Standard rate: 400–600 lb BOD/acre-ft/day; or 2–4 mgad[c], 6 ft deep. High rate: 3000+ lb BOD/acre-ft/day, or 30 mgad for 6 ft deep. Minimum filter depth 5 ft, maximum 10 ft. 1-1/34 ft[2] with activated sludge and 2 ft[2] with chemical coagulation. Glass covered; reduce area by 25%,

Tube and Inclined Plate Settlers
PVC or metal tubes, at 45° 60°. from horizontal, 2 in. × two 6-in, 4 ft long. May be installed in existing basin.

Sludge Treatment

Digester[a]
Capacity: with plain sedimentation 2–3 ft[3] per capita heated or 4–6 ft[3] unheated. With standard trickling filter 3–4 ft[3] heated and 6–8 ft[3] unheated; 4–5 ft[3] heated and 8–10 ft[3] unheated with a high-rate filter. With activated sludge 4–6 ft[3] per capita heated and 8–12 ft[3] unheated. Bottom slope: 1 on 4, gravity.

Sludge Drying Bed
Open: 1 ft[2] per capita with plain sedimentation, 1.5 ft[2] with trickling filter.

Vacuum Filtration
Pounds per square foot per hour dry solids. Primary 6 to 10, trickling filter 1.5–2.0, activated sludge 1–2.

Activated Sludge
Normally 2 hr retention in primary and final sedimentation and 6–8 hr aeration.

Rapid Filtration—Tertiary Treatment
2–5 gpm/ft², 1–4 mm sand, 48 in., backwash 15–25 gpm/ft².[d]

Land Treatment
See text.

Stabilization Pond—Facultative
15–35 lb BOD/acre-ft/day, 3–5 ft liquid depth, center inlet; variable withdrawal depth, 3-ft freeboard, detention 90–180 days; multiple units; winter flow retention. Use up to 50 lb BOD loading in mild climate and 15–20 lb in cold areas.

Rotating Biological Contactors
See text.

Disinfection
Chlorine, ozone: see text.

Centrifuging
Flow rate based on gallons per minute per horsepower.

Wet Combustion
Sludge thickener: loading of 10 lb/day/ft².

Land Disposal
Stabilized sludge only. See text.

Incineration
Tons per hr depending on moisture and solids content. Temperature 1,250–1,400°F. Pyrolysis temperature higher.

Gas Production
A properly operated heated digester should produce about 1 ft³ of gas per capita per day from a secondary treatment plant and about 0.8 ft³ from a primary plant. The fuel value of the gas (methane) is about 640 Btu/ft³.

[a] Anaerobic sludge digestion will require approximately 65 days at 55°F, 56 days at 60°F, 42 days at 71°F, 27 days at 86°F, 24 days at 95°F, 20 days at 113°. The optimum temperature is 86°–95°F. Mixing of sludge can reduce digestion time up to 50%. In large plants, sludge is usually digested in two stages. Temperature of 140°F causes caking on pipes.
[b] Gallons per acre per day = gpad.
[c] Million gallons per acre per day = mgad.
[d] For multimedia, see state standards.

Note: Surface setting rate = gpd/ft² = $\frac{180 \times \text{tank depth in ft}}{\text{detention, hr}}$.

351

TABLE 3.22 Sewage Treatment Plant Unit Combinations and Efficiencies: Approximate Total Percent Reduction

Treatment Plant	Suspended Solids	Biochemical Oxygen Demand
Sedimentation plus sand filter	90–98	85–95
Sedimentation plus standard trickling filter, 600 lb BOD/acre-ft maximum loading	75–90	80–95
Sedimentation plus single-stage high-rate trickling filter	50–80	35–65[a]
Sedimentation plus two-stage high-rate trickling filter	70–90	80–95[a]
Activated sludge	85–95	85–95
Chemical treatment	65–90	45–80
Preaeration (1 hr) plus sedimentation	60–80	40–60
Plain sedimentation	40–70	25–40
Fine screening	2–20	5–10
Stabilization (aerobic) pond	—	70–90
Anaerobic lagoon	70	40–70

[a]No recirculation. Efficiencies can be increased within limits by controlling organic loading, efficiencies of settling tanks, volume of recirculation, and number of stages; however, effluent will be less nitrified than from standard rate filter but will usually contain dissolved oxygen. Filter flies and odors are reduced. Study first cost plus operation and maintenance.

Biosolids Treatment and Disposal

The U.S. EPA established standards for the use and disposal of biosolids (i.e., sewage sludge and septage) in 1993.[65] These standards, which are referred to as the Part 503 Rule, set pollutant limits and management practices for biosolids that are applied to land, placed in a surface disposal site, or fired in an incinerator.

Pollutant limits for pathogens and metals are specifically mandated under the Part 503 Rule. For biosolids applied to the land, limits on pathogen and metal levels must be met with the requirements for pathogens vary depending on the type of land application. Class A pathogen requirements must be met when biosolids are applied to lawns or home gardens. For such use, pathogens such as *Salmonella* sp. bacteria, enteric viruses, and viable helminth ova must be below detectable levels. Class B pathogen requirements, which require that pathogens be reduced to levels that are not likely to pose a threat to public health or the environment, include various site restrictions. For example, lands receiving Class B biosolids must prohibit human access to or animal grazing of the site for a specified period of time. Similarly, crops from farmlands receiving Class B biosolids must not be harvested until a set time has elapsed after sludge application has stopped. The Class B requirements and site restrictions have to be met when biosolids are applied to farmlands, forests, public contact sites (i.e., parks or sports fields), or reclamation sites. U.S. EPA pathogen standards for Class A and B sludges are presented in Table 3.23, while numerical limits for heavy metals in biosolids are listed in Table 3.24.

TABLE 3.23 Microbiological Standards for Class A and B Biosolids

Standard	Class A	Class B
Fecal coliforms per grams dry solids	<1,000	<2,000,000[a]
Salmonellae per 4 grams dry solids	<3	
Enteroviruses pfu per 4 grams dry solids	<1[b]	
Parasite ova per 4 grams dry solids	<1[b]	

[a]Geometric mean of seven samples.
[b]For processes unable to satisfy specific operational requirements.

The Part 503 Rule also stipulates that various management practices must be met when biosolids are applied to the land. For bagged biosolids, certain labeling requirements must be met. For bulk biosolids, prohibitions against its application on frozen or flooded ground or within 10 meters of surface waters are required.

U.S. EPA requirements for biosolids placed in a surface disposal site vary depending upon whether the biosolids are placed in a landfill that only accepts biosolids (i.e., a monofill) or are codisposed with municipal solid waste and on whether the site is lined and has a leachate collection system. In most instances, liners and leachate collection systems are only used in association with codisposal sites. For sites not having liners and leachate collection systems, biosolids must meet specified limits for arsenic, chromium, and nickel. These limits vary, depending on the distance between the boundary of the active biosolids disposal area and the actual property line of the disposal site. For example, the limits for arsenic range from 30 mg/kg at distances less than 25 meters up to 73 mg/kg for

TABLE 3.24 Metal Concentration Limits for Biosolids Applied to Land

Parameter	Ceiling Concentration (mg/kg)	Monthly Average Concentration (mg/kg)	Annual Loading Rate (kg/ha-yr)	Cumulative Loading Rate (kg/ha)
Arsenic	75	41	2.0	41
Cadmium	85	39	1.9	39
Chromium	N/A	N/A	N/A	N/A
Copper	4,300	1,500	75	1,500
Lead	840	300	15	300
Mercury	57	17	0.85	17
Molybdenum	75	N/A	N/A	N/A
Nickel	420	420	21	420
Selenium	100	100	5.0	100
Zinc	7,500	2,800	140	2,800

Note: Concentrations and loading rates are on dry-weight basis. A February 25, 1984, *Federal Register* notice deleted chromium, deleted the molybdenum values for all but the ceiling concentration, and increased the selenium limit for monthly average concentration from 36 to 100.

distances greater than 150 meters. The limits for chromium and nickel also vary in a similar fashion (i.e., 200 to 600 mg/kg for chromium and 210 to 420 mg/kg for nickel). However, for sites with relatively impermeable liners (i.e., hydraulic conductivity values of 10^{-7} cm/sec or less) and leachate collection systems, the above limits do not apply.

Biosolid disposal sites must also comply with certain siting criteria and management practices. For example, biosolid landfills cannot impede the flow of a 100-year flood, be located in geologically unstable areas, or in a wetland unless a special permit is obtained. The landfills used for biosolid disposal must also be able to divert runoff from a 25-year, 24-hour storm event.

U.S. EPA requirements for biosolids that are incinerated include limits on metal concentrations and total hydrocarbons. Levels of beryllium and mercury emitted from biosolid incinerators must meet the National Emission Standards for Hazardous Air Pollutants. Arsenic, cadmium, chromium, and nickel must meet risk-specific concentrations, which range from 0.023 to 2.0 ug/m^3 and are based on a combination of biosolid feed rates, dispersion factors, and incinerator control efficiencies.

Biosolid treatment and disposal can be time-consuming and costly. Biosolid handling prior to final disposal may involve collection, thickening, stabilization, conditioning, dewatering, heat drying, air-drying, lagooning, composting, and final disposal of the sludge.[66] Figure 3.13 shows some sludge treatment processes and Table 3.20 gives some treatment design parameters. Sludges, including septic tank sludge, can be expected to contain numerous organic and inorganic chemicals and pathogens that can pose a hazard to agricultural produce, grazing animals, surface water, groundwater, and human health if not properly handled. Anaerobically digested sludge has been found to contain ascaris, toxocara, and trichuris ova, which remained viable in storage lagoons for up to 5 years.[67]

Biosolid thickening processes include gravity settling, flotation and centrifugation. Biosolid stabilization is usually achieved by aerobic or anaerobic digestion, lime treatment, or composting. Digestion reduces the organic solids and pathogens in sludge. Anaerobic two-phase (first digester acid, second digester methane producer) digestion of municipal sludge at 127.4°F (53°C) for 10 days "reduces to essentially undetectable levels indicator bacteria (fecal coliforms, *Escherichia coli*, fecal streptococci), enterovisuses, and viable *Ascaris* eggs."[68] Lime treatment and composting can also reduce pathogen levels. Also, sludge can be heated and mixed to accelerate the rate of digestion with sludge usually added at a rate of about 200 lb volatile solids per 1,000 ft^3 per day.

Biosolids can also be conditioned, prior to thickening or dewatering, by the addition of chemicals. Heat treatment by means of a furnace or dryer reduces sludge moisture content. Dewatering is accomplished by means of drying beds, centrifuges, vacuum filters, continuous belt presses, plate and frame presses, or evaporation lagoons.

Final disposal of biosolids can be by composting, incineration in multiple-hearth or fluidized bed (Figures 3.16 and 3.17), pyrolysis, sanitary

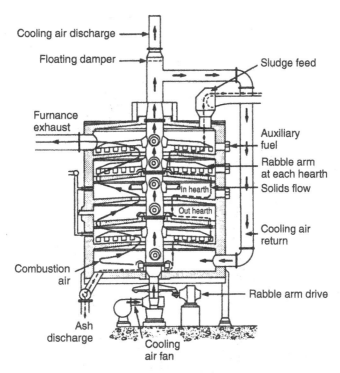

FIGURE 3.16 Cross-section of a multiple-hearth sludge incineration furnace. Temperature 1,400 to 1,500°F (769° to 816°C) in middle hearths. (*Source*: Environmental Regulations and Technology, Use and Disposal of Municipal Wastewater Sludge, U.S. EPA, Washington, DC, September 1984, p. 49.)

landfill,* land application or reclamation, or sod growing. Composting may be by the (1) window method including 5 turnings over 15 days and mixture temperatures of not less than 131°F (55°C) 6 to 8 inches below the surface, (2) static pile method in which the pile is kept at a temperature of not less than 130°F (55°C) for at least three consecutive days, or 3) enclosed vessel method in which the mixture is maintained at a temperature not less than 130°F (55°C) for at least three consecutive days.[69] Sawdust is often mixed with the finished compost.

Incineration can be combined with other industrial processes, such as cement manufacturing, to reduce the cost of disposal. This approach was used by the Los Angeles County Sanitation district to dispose of a portion of the 1,250 tons of biosolids they produce each day. Since 1996, the county has agreed to provide a local cement manufacturer with between 240 and 480 tons of biosolids per day. The biosolids are injected into the cement plant's hot exhaust gases where ammonia in the biosolids reduces plant nitrogen oxide emissions by up to 45 percent.

---

*Sludge dewatered to at least 20 percent solids.

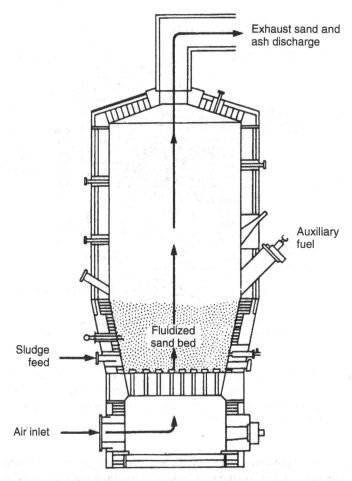

FIGURE 3.17   Cross-section of a fluidized-bed sludge incineration furnace. Tempera-
ture of bed 1,400 to 1,500°F (769° to 816°C). (*Source*: Environmental Regulations and
Technology, Use and Disposal of Municipal Wastewater Sludge, U.S. EPA, Washington,
D. C., September 1984, p. 49.)

While land disposal of stabilized sludge can promote the growth of vegetation
and control erosion, certain precautions must be taken to ensure that sludge
contaminants do not endanger the public health. For example, cadmium and zinc
are known to accumulate in food crops grown on sludge disposal sites. For that
reason, U.S. EPA has set limits under the Part 503 Rule for both maximum and
average monthly metal levels that are not to be exceeded (see Table 3.21) for
biosolids that are applied to the land.

Authority for implementation of the Part 503 Rule biosolid disposal requirements has been delegated to the states. As a result, a number of states had
imposed more restrictive limits for the specified pollutants (13 states) and required

testing for additional pollutants (22 states).[70] In some cases, communities (e.g., Kern County in California) have responded imposition of additional restrictions that have essentially prohibited the application of Class B biosolids to land. In response, many municipalities have converted to Class A treatment of biosolids. The treatment options usually selected for upgrading to Class A standards have been heat drying, composting, lime pasteurization, the N-Viro process (an alternative type of lime pasteurization), and thermophilic aerobic digestion.[71]

Cost of Sewage Treatment

The cost of sewage treatment systems can vary widely based on location, system size, and degree of treatment required. In general, costs can be divided into two categories: capital, and operation and maintenance. Cost estimates can be adjusted to present-day costs using the *Engineering News-Record* or other appropriate construction cost index (see Table 3.23).

The cost of individual septic tank systems will vary based on dwelling size, site conditions, and type of system. Typical costs (2006) for various on-site sewage treatment systems are given in Table 3.24.

For sewage treatment plants, typical costs can be estimated by adding 10 to 15 percent to the construction cost for contingencies. To these costs an additional 15 to 20 percent for engineering and 2 to 3 percent for legal/administrative costs needs to be added. The resulting total would be considered the total project construction cost. To this total cost an additional 3 to 6 percent needs to be added for financing costs. Taken together, these additions increase the total project cost by between 36 to 48 percent from the construction cost.

Cost comparisons should also consider the total annual costs—that is, the initial cost of construction and the annual cost of operation and maintenance (*O&M*). Typical average capital, O&M, and unit costs for selected sewage treatment processes are presented in Table 3.25.

Sometimes advanced wastewater treatment (Figure 3.13) is desired without fully realizing the large additional cost to obtain a small incremental increase in plant efficiency. Advanced wastewater treatment to remove an additional 3.8 to 10 percent BOD, 5.2 to 13 percent suspended solids, and approximately 61 to 68 percent phosphorus and ammonia-nitrogen has been found to increase capital costs by 42 to 99 percent and O&M costs by 37 to 55 percent.[72] This finding suggests that the other more cost-effective alternatives should be explored before making a decision to add advanced wastewater treatment.

Because the cost of advanced treatment can be high, some operators have opted to use natural or constructed wetlands for effluent polishing. The cost of constructing and, in particular, operating such systems can be significantly lower than those for advanced treatment processes. Typical construction and O&M costs for the natural treatment systems are given in Table 3.26 (See also Tables 3.27 and 3.28).

Many treatment plant operators have installed computer-based monitoring and control systems in an effort to reduce their operational costs. Also, the Internet revolution has offered the means for not only accessing real-time operational data,

TABLE 3.25 Cost Indices (Average per Year)

Year	Marshall & Stevens Installed Equipment Indices: 1926, 100 (All Industry)	Engineering News-Record Construction Index: 1913, 100 (Annual Average)	Handy-Whitman Index for Water Treatment Plants[a]: 1936, 100		Engineering News-Record Building Cost Index: 1913, 100 (Annual Average)	Chemical Engineering Plant Construction Index: 1957–1959, 100	U.S. EPA Sewage Treatment Plant Construction Index: 1957-1959, 100
			(Large Plant)	(Small Plant)			
1956	209	692	275	276	491	94	
1957	225	724	288	289	509	99	
1958	229	759	296	296	525	100	102
1959	235	797	311	309	548	102	104
1960	238	824	317	317	559	102	105
1961	237	847	315	315	568	101	106
1962	239	872	324	322	580	102	107
1963	239	901	330	327	594	102	109
1964	242	936	340	336	612	103	110
1965	245	971	350	346	627	104	112
1966	252	1,019	368	362	650	107	116
1967	263	1,074	380[b]	374[b]	676	110	119
1968	273	1,155	398	389	721	114	123
1969	285	1,269	441	424	790	119	132
1970	303	1,381	480	462	836	126	143
1971	321	1,581			948	132	160
1972	332	1,753			1,048	137	172
1973	344	1,895			1,138	144	182
1974	398	2,020			1,205	165	217
1975	444	2,212			1,306	182	250
1976	472	2,401			1,425	192	262
1977	491[c]	2,576			1,545	199[d]	271[d]
1978		2,776			1,654		
1979		3,003			1,919		
1980		3,237			1,941		
1981		3,535			2,097		
1982		3,825			2,234		
1983		4,066			2,384		
1984		4,146			2,417		
1985		,195			2,428		
1986		4,295			2,483		
1987		4,406			2,541		
1988		4,519			2,598		
1989		4,615			2,634		

TABLE 3.25   (*continued*)

Year	Marshall & Stevens Installed Equipment Indices: 1926, 100 (All Industry)	Engineering News-Record Construction Index: 1913, 100 (Annual Average)	Handy-Whitman Index for Water Treatment Plants[a]: 1936, 100		Engineering News-Record Building Cost Index: 1913, 100 (Annual Average)	Chemical Engineering Plant Construction Index: 1957-1959, 100	U.S. EPA Sewage Treatment Plant Construction Index: 1957-1959, 100
			(Large Plant)	(Small Plant)			
1990		4,732			2,702		
1991		4,835			2,751		
1992		4,985			2,834		
1993		5,210			2,996		
1994		5,408			3,111		
1995		5,471			3,112		
1996		5,620			3,203		
1997		5,826			3,364		
1998		5,920			3,391		
1999		6,059			3,456		
2000		6,221			3,539		
2001		6,334			3,574		
2002		6,538			3,623		
2003		6,695			3,693		
2004		7,115			3,984		
2005		7,446			4,205		
2006		7,888			4,441		

[a] Based on July of year.
[b] Based on January of year.
[c] Based on first quarter of year.
[d] Based on March of year. Example: 2006 index ÷ 1995 index = multiplier to obtain 2006 cost for a 1995 project cost.

Source: *Engineering News-Record* and *Process Design, Wastewater Treatment Facilities for Sewered Small Communities*, U.S. EPA, Environmental Research Information Center Technology Transaction, EPA-625/1-77-009, Cincinnati, OH, October 1977.

but also managing customer relations and billings more efficiently. Because of the risk-adverse nature of treatment plant operators—and regulators, implementation of such management innovations has been slow. As noted in a 1998 Environmental Law Institute report, treatment plant operators are generally slow to install innovative technologies whether they be tools like the Internet or technological innovations, such as ozonation, UV disinfection, enzyme treatment, or biological nutrient removal.

TABLE 3.26   Estimated O&M and Capital Costs for Individual Septic Systems

Type of system	Annual O&M cost	Initial Capital cost (2006)
Septic tank—absorption system	$55/year	$3,700–4,300[a]
Septic tank—built-up absorption system (excluding pumping station)	55/year	19,000–29,000[b]
Septic tank—subsurface sand filter, including chlorine and contact tank	75/year	13,000–18,000[c]
Aerobic system—excluding absorption field: including service contact[d]	850/year	10,000–24,000
Chemical recirculation toilet[e]	—	8,000–12,000

[a]3-bedroom home for which septic tanks are pumped out every 3 years.
[b]9,000 to 10,000 ft^2
[c]390 to 520 ft^2
[d]*Cleaning Up the Water, Private Sewage Disposal in Maine*, Maine Dept. of Environmental Conservation, Augusta, Me., July 1974, pp. 16–17. Estimated updated costs.
[e]Peter T. Silbermann, "Alternatives to Sewers," *Wastewater Treatment Systems for Private Homes and Small Communities*, Paul S. Babar, Robert D. Hennigan, and Kevin J. Pilon, eds., Central New York Regional Planning and Development Board, Syracuse, 1978, pp. 127–188. Updated cost.

INDUSTRIAL WASTES

Hazardous and Toxic Liquid Wastes

Hazardous wastes are usually byproducts of the chemical industry, which, if not recovered, require proper treatment and disposal. Toxic wastes are chemical substances that present an unreasonable risk of injury to public health or the environment. A toxic substance is one that kills or injures an organism through chemical, physical, or biological action, having an adverse physiological effect on humans. Examples include cyanides, pesticides and heavy metals. The terms toxic and hazardous are sometimes used interchangeably.

Treatment requirements for industrial wastes are typically based on effluent standards, receiving water quality standards or, if discharged to a municipal system, the pretreatment requirements of the publicly owned wastewater treatment plant. The U.S. EPA has published national recommended water quality criteria for 157 toxic pollutants pursuant to Section 304a of the Clean Water Act. These criteria are used in a number of state and federal environmental programs, such as the National Pollutant Discharge Elimination System (NPDES) permits, for setting discharge limits.

The Clean Water Act also requires designated states and authorities operating pretreatment programs to notify industrial/commercial users of hazardous wastes of their responsibilities under Resource Conservation and Recovery Act (RCRA) and Comprehensive Environmental Response, Compensation and Liability Act

TABLE 3.27 Estimated Total Annual and Unit Costs for Selected Sewage Treatment Processes (Design Flow: 1.0 MGD)

Process	Initial Capital Cost (2006 Dollars)[a,b]	Annual Cost [b] (2006 Dollars)			Unit Cost (Dollars per 1,000 gal)[b]
		Captial[c]	O&M[d]	Total	
Imhoff tank	1,300,000	143,000	53,000	196,000	0.54
Rotating biological disks	2,700,000	297,000	196,000	493,000	1.35
Trickling-filter processes	3,100,000	341,000	199,000	540,000	1.48
Activated sludge processes					
With external digestion	3,400,000	374,000	253,000	627,000	1.72
With internal digestion[e]	1,700,000	187,000	166,000	353,000	0.97
Stabilization pond processes[f]	850,000	93,500	81,000	174,500	0.48
Land disposal processes[g]					
Basic system	1,200,000	132,000	141,000	273,000	0.75
With primary treatment	3,200,000	279,965	221,330	501,295	1.72
With secondary treatment	3,370,000	352,000	277,000	629,000	2.35

[a]Estimated average cost.
[b]Original 1975 costs adjusted to expected 2006 costs using *Engineering News-Record* Building Cost Index.
[c]Capital recovery factor = 0.11 (15 years at 7 percent).
[d]Original 1975 process O&M costs adjusted to expected 2006 O&M costs using ENR index.
[e]Extended aeration, aerated lagoon, oxidation ditches.
[f]High-rate aerobic, facultative, and anaerobic.
[g]Irrigation and overland flow.

Source: George Tchobanoglous, "Wastewater Treatment for Small Communities," *Water Pollution Control in Low Density Areas*, University Press of New England, Hanover, N.H., 1975, p. 424.

(CERCLA). However, hazardous wastes when mixed with sewage are excluded from RCRA requirements and instead regulated under Clean Water Act pretreatment programs. Since the federal and state requirements are quite complex, affected persons should consult the numerous regulations that U.S. EPA has published regarding industrial waste pretreatment. As a general rule, waste disposal should not transfer a hazardous waste from one environmental medium (i.e., land, air, water) to another.

TABLE 3.28 Typical Capital and O&M Costs for Constructed Wetlands

Type of Constructed Wetland	Average Capital Cost (2006)[a]	Yearly O&MCost[a]
Free Water Surface Wetland		
<1 MGD	$62,320acre	$1760–$5900/acre
>1 MGD	$23,310/acre	$1070–$2330/acre
Subsurface Flow Wetland (<1 MGD)	$318,910/acre	Minimal

[a]Original 1998 costs adjusted to expected 2006 costs using *Engineering News-Record* Building Cost Index.

Source: Ronald Crites et al, *Natural Wastewater Treatment Systems*, CRC/Taylor & Francis, Boca Raton, FL, 2006, pp. 325-327 and 373.

Pretreatment

Industrial/commercial wastes that are hazardous or that cannot be treated by the municipal treatment plant must be excluded unless given adequate pretreatment. Examples include synthetic organic wastes and inorganic wastes that interfere with plant operation or treatment; are toxic; ignitable; emit hazardous fumes; damage wastewater treatment plant, pumping stations, or sewer system; endanger personnel; are explosive; will pass through the treatment process; or contaminate sewage sludge. Toxics of concern include mercury, cadmium, lead, chromium, copper, zinc, nickel, cyanide, phenol, and PCBs.[73] In addition, other metals and numerous organics may be prohibited or regulated. In some instances, the joint treatment of industrial wastes and municipal wastewater may be mutually advantageous and should be considered on an individual basis.[74]

One method of simplifying a waste problem is simply to spread its treatment and disposal over 24 hours rather than over a 4- or 6-hour period by means of a holding tank to equalize flows and strength of waste, accompanied by a constant discharge over 24 hours. Other approaches for dealing with liquid industrial wastes include raw material and process changes, waste volume and toxicity reduction, waste recovery and reuse.

Possible treatment of these wastes may vary from recovery, solids removal and disposal, to involved physical, chemical, and biological processes. Possibilities for recovery of waste oils include separators, air flotation, and ultrafiltration. Methods for the recovery of metals include evaporation, reverse osmosis, ultrafiltration, and ion exchange. Treatment of organic waste might consist of biological or chemical processes, activated carbon filtration, or air stripping. Possible solids removal processes include sedimentation, chemical treatment, and filtration. The sludge collection will require special handling, possibly further treatment such as dewatering, disposal to an approved facility, or incineration. More detailed information concerning the treatment of specific wastes can be obtained from standard texts, periodicals, and other publications devoted to this subject.[75]

The actual volume of a liquid waste to be discharged should be determined because many municipalities levy a charge for the treatment of industrial wastes to help pay the cost of operating their treatment works. The cost levied typically is based on the volume and strength of the waste based on COD, BOD, chlorine demand, certain organic and inorganic compounds, and/or other parameters. Pretreatment is often required if waste characteristics exceed certain predetermined values; if the waste as released may damage the facility, upset the treatment process, or is not amenable to treatment in a municipal treatment plant; or if the waste would cause a hazardous condition in the sewers or create a water quality problem in the receiving water.

Manuals prepared by the Water Environment Federation are an excellent source of information on standards and recommendations for treatment of industrial wastes and plant operation.[76] These manuals also provide regulations to exclude unacceptable or hazardous materials, protect sewers, and control the discharge of wastes that may upset municipal treatment plant operation. Unacceptable wastes include large volumes of uncontaminated wastes that may cause hydraulic overloading, storm waters, acid and alkaline wastewaters, explosive and flammable substances, toxic substances, large volumes of organic wastes unless adequately pretreated, and oil and grease.

REFERENCES

1. F. L. Bryan, "Diseases Transmitted by Foods Contaminated by Wastewater", *J. Food Protection* (January 1977): 45–56, a historical review of medical and engineering literature covering 180 selected references; *Process Design Manual: Land Application of Sewage Sludge and Domestic Septage*, U.S. EPA, National Risk Management Research Laboratory, Cincinnati, OH, EPA 625/K-95/001, September 1995; R. G. Feachem et al., *Sanitation and Disease: Health Aspects of Excreta and Wastewater Management*, The International Bank for Reconstruction and Development, The World Bank, Washington, DC, 1983.

2. Feachem et al.

3. Process Design Manual.

4. S. A. Esrey, R. G. Feachem, and J. M. Hughes, "Interventions for the control of diarrheal diseases among young children: Improving water supplies and excreta disposal facilities", *WHO Bulletin*, **63** 4 (1985): 637–640. In *Community Water Supply*, The World Bank, Washington, DC, 1987, p. 24.

5. J. A. Salvato, *Guide to Sanitation in Tourist Establishments*, WHO, 1976, pp. 36–37.

6. D. D. Mara, *Bacteriology for Sanitary Engineers*, Churchill Livingstone, London, 1974, p. 102.

7. *Glossary Water and Wastewater Control Engineering*, APHA, ASCE, AWWA, WPCF, 1981, pp. 247, 276, 302.

8. Ibid.

9. *Design Manual, Onsite Wastewater Treatment and Disposal Systems*, U.S. EPA, October 1980, pp. 367–381.

10. Thomas C. Peterson and Robert C. Ward, "Bacterial retention in soils", *J. Environ. Health* (March/ April 1989): 196–200.

11. William A. Drewry and Rolf Eliassen, "Virus Movement in Groundwater", *J. Water Pollution. Control Fed.*, **40**, *8*, R271, (August 1968); Gordon Culp, *Summary of A "State-of-the-Art" Review of Health Aspects of Waste Water Reclamation for Ground Water Recharge*, Supplement to *Report of the Consulting Panel on Health Aspects of Wastewater Reclamation for Ground Water Recharge*, CA, PB-268 540, U.S. Dept. of Commerce, National Technical Information Service, Springfield, VA, June 1976, p. 29.

12. F. M. Wellings, A. L. Lewis, and C. W. Mountain, *"Virus Survival Following Wastewater Spray Irrigation of Sandy Soils"*, in *Virus Survival in Water and Wastewater Systems*, J. F. Malina, Jr., and B. P. Sagik, eds., Austin, TX, Center for Research in Water Resources, 1974, pp. 253–260.

13. Paper given at Environmental Engineering Division Conference, ASCE, Seattle, WA, July 1976.

14. Henry Ryon, *Notes on the Design of Sewage Disposal Works With Special Reference to Small Installations*, 1924, pp. 11–14; also *Notes on Sanitary Engineering*, Albany, NY, 1924, pp. 31–33.

15. *Manual of Septic-Tank Practice*, U.S.PHS Pub. 526, Washington, DC, 1967, pp. 4–8, 75–76; John T. Winneberger, "Correlation of Three Techniques for Determining Soil Permeability", *J. Environ. Health* (September/October 1974): 108–118. Also Rein Laak, *Wastewater Engineering Design for Unsewered Areas*, Ann Arbor Science Publishers, Inc., Ann Arbor, MI, pp. 12–15; Michael E. Peterson, "Soil Percolation Tests", *J. Environ. Health* (January/February 1980): 182–186.

16. *Manual of Septic-Tank Practice*, U.S.PHS Pub. No. 526 (1967).

17. Rein Laak, "Multichamber Septic Tanks", *J. Environ. Eng. Div.*, ASCE (June 1980): 539–546.

18. Ibid.

19. M. Brandes, *Accumulation Rate and Characteristics of Septic Tank Sludge and Septage*, Ontario Ministry of the Environment, Research Report W63, 1977, Toronto, Canada. Also *J. Water Pollution Control Fed.* (May 1978): 936–943.

20. Ted Pettit and Herb Linn, *A Guide to Safety in Confined Spaces*, DHHS (NIOSH) Pub. No. 87–113, July 1987.

21. Tim Casey, "Removing Roots from a Sewer System", *Public Works* (September 1989): 134–135.

22. Henry Ryon, *op. cit.*, p. 33.

23. *Manual of Septic-Tank Practice, op. cit.*, p. 66.

24. Joseph A. Salvato Jr., *Environmental Sanitation*, John Wiley & Sons, New York, 1958, pp. 221–225.

25. J. Bouma, J. C. Converse, R. J. Otis, W. G. Walker, and W. A. Ziebell, "A Mound System for On-Site Disposal of Septic-Tank Effluent in Slowly Permeable Soils with Seasonally Perched Water Tables", *J. Environ. Quality* (July-September 1975).

26. J. C. Converse, R. J. Otis, and J. Bouma, *Design and Construction Procedures for Fill Systems in Permeable Soils With High Water Tables*, Small Scale Waste Management Project, University of Wisconsin, Madison, April 1975.

27. James C. Converse and E. Jerry Tyler, *Use of Wisconsin Mounds to Overcome Limiting Soil Conditions*, pp. 337–354. In 3rd Annual National Environmental Health Association Mid-year Conference, February 7–9, 1988, Mobile, AL, National Environmental Health Association, Denver, CO.

28. *Alternatives for Small Wastewater Treatment Systems*, U.S. EPA Technology Transfer, Washington, DC, October 1977, pp. 8, 47–48.

29. James C. Converse, *Design and Construction Manual for Wisconsin Mounds*, Small Scale Waste Management Project, University of Wisconsin, Madison, September 1978.

30. Marek Brandes, "Effect of precipitation and evapotranspiration on a septic tank-sand filter disposal system", *J. Water Pollution Control Fed.* (January 1980): 74.

31. P. H. McGauhey, *Engineering Management of Water Quality*, McGraw-Hill, New York, 1968.

32. Joseph A. Salvato, "Rational Design of Evapotranspiration Bed", *J. Environ. Eng.* (June 1983): 653–657.

33. Arthur F. Beck, "Evapotranspiration Bed Design", *J. Environ. Eng. Div.*, ASCE (April 1979): 411–415.

34. Kenneth M. Lomax, "Evapotranspiration Method Works for Wastewater Disposal along Chesapeake Bay", *J. Environ. Health* (May/June 1979): 324–328.

35. "Assessing the Need for Wastewater Disinfection", WPCF Disinfection Committee, *J. Water Pollution Control Fed.* (October 1987): 856–864.

36. H. Heukelekian and R. V. Day, "Disinfection of Sewage with Chlorine, III", *Sewage Ind. Wastes* (February 1951): 155–163.

37. Paula A. Riley, "Wastewater Dechlorination—A Survey of Alternatives", *Public Works* (June 1989): 62–63.

38. David Kirkwold, "Disinfecting with Ultraviolet Radiation", *Civil Engineering*, ASCE, December 1984.

39. Kerwin L. Rakness, et al., "Design, Start-up and Operation of an Ozone Disinfection Unit", *J. Water Pollution Control Fed.* (November 1984): 1152–1159.

40. Walter J. Weber, Jr., *Physicochemical Processes for Water Quality Control*, Wiley-Interscience, New York, 1972.

41. Arcadio Sincero and Gregoria Sincero, *Physical-Chemical Treatment of Water and Wastewater*, CRC Press, Boca Raton, FL, 2003.

42. L. Rizzo et al., "Organic THMs Precursors Removal from Surface Water with Low TOC and High Alkalinity by Enhanced Coagulation", *Water Science Technology: Water Supply* (May 2004): 103.

43. D. Codiasse et al., "Improving Water Quality by Reducing Coagulant Dose?" *Water Science Technology: Water Supply* (May 2004): 271.

44. D. Liu et al., "Comparison of Sorptive Filter Media for Treatment of Metals in Runoff", *J. Environ. Engr. ASCE* (August 2005): 1178.

45. P. Pearce, "Trickling Filters for Upgrading Low Technology Wastewater Plants for Nitrogen Removal", *Water Science and Technology* (November/December 2004): 47.

46. B. C. G. Steiner, "Take a New Look at the RBS Process", *Water & Wastes Engr.* (May 1979): 41.

47. *Operation of Municipal Wastewater Treatment Plants*, MOP 11, Water Environment Federation, 1996, pp. 691–752.

48. G. Ayoub and Pl Saikaly, "The Combined Effect of Step-Feed and Recycling on RBC Performance", *Water Research* (2004): 3009.

49. "Review of Current RBC Performance and Design Procedures," U.S. EPA, EPA/600/52–85/033, June 1985.

50. *Design Standards for Wastewater Treatment Works 1988*, New York State Dept. of Environmental Conservation, Albany, p. 438.

51. Harvey F. Ludwig, "Industry's Idea Clinic", *J. Water Pollution Control Fed.* (August 1964): 937.

52. *"Wastewater Stabilization Ponds, An Update on Pathogen Removal,"* U.S. EPA (August 1985).

53. *Design Manual, Municipal Wastewater Stabilization Ponds*, U.S. EPA, EPA 625/1-83-015, Center for Environmental Research Information, Cincinnati, OH, October 1983.

54. D. H. Caldwell, et al., *Upgrading Lagoons*, U.S. EPA, EPA-625/4-73-001b, August 1973.

55. "Science", *Environ. Science Technology* (March 1986): 214.

56. James H. Reynolds, et al., "Long-term effects of irrigation with wastewater", *J. Water Pollution Control Fed.* (April 1980): 672–687.

57. *Health guidelines for the use of wastewater in agriculture and aquaculture*, Report of a WHO Scientific Group, Technical Report Series 778, World Health Organization, Geneva, 1989, p. 18.

58. *Process Design Manual for Land Treatment of Municipal Wastewater*, U.S. EPA, Cincinnati, OH, 1981.

59. D. C. J. Zoeteman, "Reuse of Treated Sewage for Aquifer Recharge and Irrigation", *Municipal Wastewater Reuse News*, AWWA Research Foundation, Denver, CO, November 1980, pp. 8–11.

60. Badri Fattal, et al., "Health Risks Associated with Wastewater Irrigation: An Epidemiological Study", *Am. J. Public Health* (August 1986): 977–979.

61. *California Administrative Code*, Title 22, Division 4, Environmental Health, Wastewater Reclamation Criteria, State of California, Department of Health Services, Berkeley, CA.

62. "Design of Effluent Irrigation Systems", *Municipal. Wastewater Reuse News*, AWWA Research Foundation, August 1980, pp. 16–22.

63. W. A. Hardenbergh, *Sewerage and Sewage Treatment*, 3rd ed., International Textbook Co., Scranton, Pa., 1950, p. 328.

64. *Recommended Standards for Wastewater Facilities*, Great Lakes-Upper Mississippi River Board of State and Provincial Public Health and Environmental Managers, Health Research, Inc., Health Education Services Division, Albany, NY, 1997.

65. *Standards for Use or Disposal of Sewage Sludge*, Federal Register 9248, February 19, 1993.

66. Charles R. Scroggin, Don A. Lewis, and Paul B. Danheiser, "Developing a Cost-Effective Sludge Management Approach", *Public Works* (June 1980): 87–91.

67. *Process Design Manual Land Application of Municipal Sludge*, U.S. EPA, EPA-625/1–83–016, Center for Environmental Research Information, Cincinnati, OH, October 1983; R. G. Arthur, et al., "Parasite Ova in Anaerobically Digested Sludge", *J. Water Pollution Control Fed.* (August 1981): 1334–1338.

68. Kun M. Lee, et al., "Destruction of enteric bacteria and viruses during two-phase digestion". *J. Water Pollution. Control Fed.* (August 1989): 1421–1429.

69. *Solid Waste Management Facilities*, 6NYCRR Part 360, New York State Dept. of Environmental Conservation, Albany, December 31, 1988, p. 5–5.

70. Goldstein, N., "National Overview of Biosolids Management", *Bio-Cycle* (December 1998): 69.

71. Goodfree, A. and J. Farrell, "Processes for Managing Pathogens", *Journal of Environmental Quality* (2005): 105–113.

72. "Advanced waste treatment: Has the wave crested?" *J. Water Pollution Control Fed.* (July 1978): 1706–1709.

73. G. W. Foess and W. A. Ericson, "Toxic Control—the Trend of the Future", *Waterwastes Eng.* (February 1980): 21–27.

74. *Pretreatment of Industrial Wastes*, MOP FD-3, 1994;   *Industrial Wastewater Control Program for Municipal Agencies*, MOP OM-4, 1982; and   *Preliminary Treatment for Wastewater Facilities*, MOP OM-2, 1980;   Water Pollution Control Fed., Alexandria, Va.

75. W. Wesley Eckenfelder Jr., *Industrial Water Pollution Control*, 2nd ed., McGraw-Hill, New York, 1989.

76. See Manuals of Practice listed in Bibliography.

BIBLIOGRAPHY

Alternatives for Small Wastewater Treatment Systems, U.S. Environmental Protection Agency, EPA-625/4–77–011, October 1977, 90 pp.

Alternative Systems for Small Communities and Rural Areas, U.S. Environmental Protection Agency, FRD-10, January 1980.

Anderson, Damon L., Robert L. Siegrist, and Richard J. Otis, *Technology Assessment of Intermittent Sand Filters*, Municipal Environmental Research Laboratory, U.S. Environmental Protection Agency, Cincinnati, Ohio, April 1985.

Barnes, D., and F. Wilson, *The Design and Operation of Small Sewage Works*, E. & F. N. Spon, Ltd., London, Halsted Press, John Wiley & Sons, New York, 1976, 180 pp.

Bernhart, Alfred P., *Treatment and Disposal of Waste Water From Homes by Soil Infiltration and Evapotranspiration*, University of Toronto Press, Canada, 1973, 173 pp.

Crites, R. and G. Tchobanoglous, *Small and Decentralized Wastewater Management Systems*, WCB/McGraw-Hill, New York, 1998.

Crites, R., J. Middlebrooks, and S. Reed, *Natural Wastewater Treatment Systems*, CRC/Taylor & Francis, Boca Raton, FL, 2006, 552 pp.

Design Manual Onsite Wastewater Treatment and Disposal Systems, U.S. Environmental Protection Agency, Office of Water Program Operations, Washington, DC, October 1980, 391 pp.

Eckenfelder, W. Wesley, Jr., *Water Quality Engineering for Practicing Engineers*, Barnes & Noble, Inc., New York, 1970, 328 pp.

Fair, Gordon M., and D. Barnes, *Water and Wastewater Engineering Systems*, Pitman Pub., Marshfield, MA, 1979, 513 pp.

Fair, Gordon M., and John C. Geyer, *Elements of Water Supply and Wastewater Disposal*, John Wiley & Sons, New York, 1971, 752 pp.

Feachem, R.G., et al., *Health Aspects of Excreta and Wastewater Management*, The World Bank, Washington, DC, 1983, 501 pp.

Flack, Ernest J., *Design of Water and Wastewater Systems*, Environmental Resource Center, Colorado State University, Fort Collins, 1976, 149 pp.

Goldstein, Steven N., and Walter J. Moberg, Jr., *Wastewater Treatment Systems for Rural Communities*, Commission on Rural Water, Washington, DC, 1973, 340 pp.

Grady, C. P. Leslie et al., *Biological Wastewater Treatment*, Marcel Dekker, New York, 1999, 1076 pp.

Guide to Septage Treatment and Disposal, U.S. Environmental Protection Agency, EPA/625/R-94/002, Cincinnati, OH, 1994.

Hammer, M., and M. Hammer, Jr., *Water and Wastewater Technology*, 3rd Edition, Prentice Hall, Englewood Cliff, NJ, 1996.

Handbook for Management of Onsite and Clustered (Decentralized) Wastewater Treatment Systems, U.S. Environmental Protection Agency, Cincinnati, OH, EPA 832-D-03-001, February 2003, 196 pp.

Hendricks, David, *Water Treatment Unit Processes: Physical and Chemical*, CRC/Taylor & Francis, Boca Raton, FL, 2006, 1266 pp.

Imhoff, Karl, and Gordon Maskew Fair, *Sewage Treatment*, John Wiley & Sons, New York, 1956, 338 pp.

Individual Onsite Wastewater Systems, Vols. 1–6, Nina I. McClelland, ed., Ann Arbor Science Publishers, Inc., Ann Arbor, MI, 1974–1980.

Individual Sewage Systems, Great Lakes-Upper Mississippi River Board of State Sanitary Engineers, Health Research, Inc., Health Education Services Division, Albany, NY, 1979, 51 pp.

Joint Task Force of WEF and ASCE, *Design of Wastewater Treatment Plants*, Vol. 1, MOP8, Water Environment Federation, 1998.

Kalbermatten, John M., et al., *Appropriate Sanitation Alternatives, A Planning and Design Manual*, World Bank Studies in Water Supply and Sanitation 2, The John Hopkins University Press, Baltimore, 1982.

Kreissl, James F., *Management of Small Waste Flows*, U.S. Environmental Protection Agency, Cincinnati, OH, September 1978, 800 pp.

Land Treatment of Municipal Wastewater, U.S. Environmental Protection Agency, U.S. Army Corp of Engineers, U.S. Department of Agriculture, EPA/625/1–77–008, October 1977, 505 pp.

Manual of Instruction for Wastewater Treatment Plant Operators, Vols. 1–2, New York State Department of Environmental Conservation, Albany, 1979.

Manuals of Practice—Water Environment Federation: *Alternative Sewer Systems* MOP FD-12, 1986. *Design of Municipal Wastewater Treatment Plants* MOP-8, 1998. *Gravity Sanitary Sewer Design and Construction* MOP FD-5, 1982. *Industrial Wastewater Control Program for Municipal Agencies* MOP OM-4, 1982. *Municipal Strategies for the Regulation of Sewer Use* MOP SM-7, 1988. *Odor Control for Wastewater Treatment Plants* MOP 22, 1995. *Wastewater Collection Systems Management* MOP 7, 1999. *Operation of Municipal Wastewater Treatment Plants* MOP 11, 1996. *Pretreatment of Industrial Wastes*, MOP FD-3, 1994. *Safety and Health in Wastewater Systems* MOP-1, 1983. *Water Reuse* MOP SM-3, 1989.

Manual of Septic-Tank Practices, U.S. Public Health Service Publication 526, DHEW, Washington, DC, 1967, 92 pp.

Mara, Duncan, *Sewage Treatment in Hot Climates*, John Wiley & Sons, New York, 1976, 168 pp.

National Recommended Water Quality Criteria-Correction, U.S. Environmental Protection Agency, EPA/822-Z-99-001, Washington, DC, 1999.

National Specialty Conference on Disinfection, American Society of Civil Engineers, 345 East 47th Street, New York, 1970, 705 pp.

Onsite Wastewater Management and Groundwater Protection, 3rd Annual National Environmental Health Association Mid-year conference, February 7–9, 1988, Mobile, AL, National Environmental Health Association, Denver, CO, 1988, 412 pp.

Onsite Wastewater Treatment, Proceedings of the Fourth National Symposium on Individual and Small Community Sewage Systems, December 10–11, 1984, New Orleans, LA, American Society of Agricultural Engineers, St. Joseph, MI, 1985, 381 pp.

Onsite Wastewater Treatment, Proceedings of the Ninth National Symposium on Individual and Small Community Sewage Systems, March 11-14, 2001, Fort Worth, TX, American Society of Agricultural Engineers, St. Joseph, MI, 2001, 691 pp.

Perkins, Richard J., *Onsite Wastewater Disposal*, Lewis Publishers, Chelsea, MI, 1989, 251 pp.

Process Design Manual-Land Application of Sewage Sludge and Domestic Septage, U.S. Environmental Protection Agency, EPA/625/R-95/001, Cincinnati, OH, 1995.

Process Design Manual for Land Treatment of Municipal Wastewater, U.S. Environmental Protection Agency, Washington, DC, EPA/625/R-06/016, September 2006, 194 pp.

Public Works Manual, Public Works Magazine, Ridgewood, NJ.

Qasim, S., "Small Flow Wastewater Treatment Technology for Domestic and Special Applications", *The Encyclopedia of Environmental Science and Engineering*, Gordon & Breach, Science Publishers, New York, 1998, pp. 1150–1162.

Qasim, S., *Wastewater Treatment Plants: Planning, Design and Operation*, 2nd Edition, Technomic Publishing, Lancaster, PA, 1999.

Quality Criteria for Water, U.S. Environmental Protection Agency, Washington, DC, EPA 440/5-86-001, May 1, 1986, 477 pp.

Recommended Standards for Sewage Works, Great Lakes-Upper Mississippi River Board of State Sanitary Engineers, Health Research, Inc., Health Education Services Division, P.O. Box 7126, Albany, N.Y. 12224, 1978 ed., 104 pp.

Recommended Standards for Wastewater Facilities, Great Lakes-Upper Mississippi River Board of State and Provincial Public Health and Environmental Managers, Health Research, Inc., Albany, NY, 1997.

Rural Wastewater Management, The California State Water Resources Control Board, Sacramento, 1979, 52 pp.

Standard Methods for the Examination of Water and Wastewater, 21st ed., L. S. Clesceri, A. E. Greenberg, and A. D. Eaton, eds., American Public Health Association, 1015 Fifteenth St., N.W., Washington, D.C. 20005, 2005.

Standards for the Use or Disposal of Sewage Sludge, Subpart B: Land Application, U.S. Environmental Protection Agency, 40 CFR Part 503, July 1, 1995.

Surampalli, R., and K. Tyagi, eds., *Advances in Water and Wastewater Treatment*, American Society of Civil Engineers, Reston, VA, 2005, 585 pp.

Tchobanoglous, George, and Franklin L. Burton, *Wastewater Engineering: Treatment, Disposal, Reuse*, Metcalf & Eddy, Inc., McGraw-Hill, New York, 1991.

Tchobanoglous, G., H. Theisen and S. Vigil, *Integrated Solid Waste Management: Engineering Principles and Management Issues*, 2nd Ed., McGraw-Hill, New York, 1993.

Technical Criteria Guide for Water Pollution Prevention, Control and Abatement Programs, U.S. Department of Army, Washington, DC, April 1987, 150 pp.

Tillman, G., *Wastewater Treatment Troubleshooting and Problem Solving*, Ann Arbor Press, Chelsea, MI, 1996.

Use and Disposal of Municipal Wastewater Sludge, Environmental Regulations and Technology, U.S. Environmental Protection Agency, Washington, DC, September 1984, 76 pp.

Velz, Clarence J., *Applied Stream Sanitation*, John Wiley & Sons, Inc., New York, 1984, 800 pp.

Vesilind, P. Aarne, and J. Jeffrey Peirce, *Environmental Engineering*, Ann Arbor Science, Ann Arbor, MI, 1982, 602 pp.

Wastewater Treatment: Alternatives to Septic Systems-Guidance Manual, U. S. Environmental Protection Agency, EPA/909/K-96/001, June 1996.

Wastewater Treatment/Disposal for Small Communities, U.S. Environmental Protection Agency, EPA/625/R-92/005, Washington, DC, September 1992.

Wastewater Treatment Units and Related Products and Components, Standards Nos. 40 and 41, National Sanitation Foundation, P.O. Box 1468, Ann Arbor, MI, June 1, 1990.

Weber, Walter J., Jr., *Physicochemical Processes for Water Quality Control*, Wiley-Interscience, New York, 1972, 640 pp.

Winneberger, John H. Timothy, *Manual of Gray Water Treatment Practice*, Ann Arbor Science, Ann Arbor, MI, 1974, 102 pp.

INDEX

371